Acoustic and MIDI Orchestration for the Contemporary Composer

Acoustic and MIDI Orchestration for the Contemporary Composer, Second Edition provides effective explanations and illustrations to teach you how to integrate traditional approaches to orchestration with the use of the modern sequencing techniques and tools available to today's composer. By covering both approaches, Pejrolo and DeRosa offer a comprehensive and multifaceted learning experience that will develop your orchestration and sequencing skills and enhance your final productions.

A leading manual on its subject, the second edition allows experienced composers and producers to be exposed to sequencing techniques applied to traditional writing and arranging styles. It also helps the young musician to learn about traditional orchestration in a very practical way. The book continues to provide a comprehensive and solid learning experience and has been fully revised to include the latest tools and techniques.

The new edition has been updated to include:

* A new chapter on writing and sequencing for vocal ensembles
* Coverage of writing for different ensemble sizes
* A new final chapter on writing and production techniques for mixed contemporary ensembles.
* All new techniques, tools, and sound libraries available to today's composer.

A companion website (www.routledge.com/cw/pejrolo) includes a wide selection of audio examples, templates, sounds, and videos showcasing operational processes, allowing you the opportunity to listen to the techniques discussed within the book.

Andrea Pejrolo is the assistant chair of the Contemporary Writing and Production Department at Berklee College of Music in Boston, as well as a composer, producer, music technology expert, audio engineer, and bassist. He is the author of *Creative Sequencing Techniques for Music Production* and *Acoustic and MIDI Orchestration for the Contemporary Composer*, and has written several articles for music magazines, including *Sound On Sound Magazine*, *MacWorld*, *Computer Studio Magazine*, and *Bass World Magazine*. Andrea has extensive professional experience as a sound designer, audio engineer/producer, MIDI programmer, and composer for film, TV, theater, and multimedia. He is also an active presenter and lecturer for prestigious conferences such as AES and MacWorld.

Richard DeRosa is a Grammy-nominated composer and a professor and the director of jazz composition and arranging at the University of North Texas. His work in jazz includes writing and conducting for the WDR Big Band in Cologne, Germany, and for Wynton Marsalis and the Lincoln Center Jazz Orchestra. His work in theater includes arrangements for *A Bed and a Chair*, featuring the music of Stephen Sondheim, the Broadway show *After Midnight*, orchestrations for *Frankenstein—the Musical*, and multiple original scores for the national touring company ArtsPower. Compositions for television include background music cues for *Another World*, *As the World Turns*, and *The Guiding Light*. For film, his work includes various documentaries broadcast on PBS and orchestrations for independent films *Gray Matters*, *Falling For Grace*, and *Standard Time*.

Acoustic and MIDI Orchestration for the Contemporary Composer

A Practical Guide to Writing and Sequencing for the Studio Orchestra

SECOND EDITION

Andrea Pejrolo and Richard DeRosa

Routledge
Taylor & Francis Group

NEW YORK AND LONDON

Second edition published 2017
by Routledge
711 Third Avenue, New York, NY 10017

and by Routledge
2 Park Square, Milton Park, Abingdon, Oxon OX14 4RN

Routledge is an imprint of the Taylor & Francis Group, an informa business

First published 2007 by Focal Press

Library of Congress Cataloging in Publication Data
Names: Pejrolo, Andrea, author. | DeRosa, Rich, author.
Title: Acoustic and MIDI orchestration for the contemporary composer :
a practical guide to writing and sequencing for the studio orchestra /
Andrea Pejrolo and Richard DeRosa.
Description: Second edition. | New York ; London : Routledge, 2016. | "2016 |
Includes index.
Identifiers: LCCN 2016004762 | ISBN 9781138801509 (paperback)
Subjects: LCSH: Instrumentation and orchestration. | Composition (Music) |
MIDI (Standard) | Software sequencers. | Digital audio editors.
Classification: LCC MT70 .P45 2016 | DDC 781.3/74—dc23
LC record available at http://lccn.loc.gov/2016004762

ISBN: 978-1-138-69275-6 (hbk)
ISBN: 978-1-138-80150-9 (pbk)
ISBN: 978-1-315-75493-2 (ebk)

Typeset in Univers and Antique Olive
by Florence Production Ltd, Stoodleigh, Devon, UK

Contents

Contents

Illustrations

Figures

Tables

Foreword

This book is written to address concerns encountered by the twenty-first-century music composer, arranger, and producer. Developments in electronic music over the last 25 years have created a whole new world of possibilities for the consumer of commercial music and, as a result, a multitude of considerations for the modern musician and composer.

Since the history in this area is still relatively young, the music industry consists of several types of music creators. There are traditionalists (usually those who are more seasoned), who had to learn the conventional methods of orchestration and how to function within the conventional means of recording. There are technology experts (most commonly the youngest members of our musical world), who most likely never had the chance to experience what it is like to arrange and hear a piece for live instruments. Some may have bypassed any formal musical education, utilizing instead their natural talents with computers and musical sound to create "unwritten" works. Finally, there are the ones who have blended both approaches successfully, creating musical art through whatever means necessary with the highest respect for the art's esthetic as well as the client's budget.

As musicians who utilize both approaches in their creation of music, it is the objective of the authors to provide a concise and thoughtful method in each area in order to provide the reader with the knowledge necessary to function as a music creator in the twenty-first century.

Since 1983, the world of MIDI has continued to develop and redefine the possibilities within the process of creating music. The advent of digital audio sequencers and recorders/editors in the late 1990s has increased even further the possibilities that are available to the modern composer. These technological developments have created the need to adapt in accordance with the expectations and convenience of the consumer.

Today's commercial composer will most likely need to incorporate the use of electronic instruments in the form of samplers and/or synthesizers to create commercial soundtracks, for several reasons:

- The consumer has come to expect a "finished" demo that gives the truest representation of the final product. In earlier days—prior to the 1980s—the composer might have played at the piano a minimal representation of the larger-scale work. It took great imagination and trust on the part of the client and great inspiration on the part of the composer to convince the consumer that the endeavor was worthy of the money about to be invested.
- There is no question that the sound and impact of a recorded orchestra of musicians is greater than the sound generated from computers and, in a perfect world, most would choose that approach, but many times the budget or time constraints will not allow us to pursue that desire.
- A MIDI demo can be quite effective when working with vocalists. Each vocalist has a unique instrument that must be presented within its proper range. As a result, it is the composer's (or arranger's) job to find the right key for the singer to deliver the melody most comfortably. Of course, this can be done by playing at a piano, but the intensity and scope of the sound may not match the reality of the final orchestration. An electronic rendition will provide a truer sense of the energy required by the vocalist and will ultimately be a better indicator regarding the appropriateness of the key. It would be most unfortunate to have completed an orchestration (let alone the recording of it) only to find that the vocalist now has reservations about the key.
- Many electronic sounds are unique in timbre and cannot be created or simulated as effectively through conventional acoustic instruments.

Now that the scope of this book has been explained, it must be mentioned that, because of the breadth of these topics, the information within each area is designed to be as informative as possible within the space allowed. Specifically, regarding orchestration, the instruments discussed will be ones that are most commonly found within today's commercial and contemporary scoring situations. For this reason, we suggest that readers who want to study traditional orchestration further refer also to texts that are dedicated solely to that endeavor.

So, if you're ready, turn the page to begin your transformation from traditionalist or technology expert to a fully functional twenty-first-century music composer!

Notes for the "Acoustic and MIDI Orchestration Website"

Welcome to the exciting world of "Acoustic and MIDI Orchestration for the Contemporary Composer."

With this book you will find additional music examples, images, and original scores on the companion website. We recommend keeping the supporting files handy when reading and studying the concepts of the book since each acoustic and MIDI orchestration technique is explained and demonstrated through a series of scores and music examples specially written and recorded for this manual.

On the companion website you can find the following files:

- Images and Scores: here you will find a digital version in JPG format of all the Figures and Scores that are used in the book. For teaching purposes, I find that having a digital version of the Figures is of great help since you can easily project them from a computer on the screen in the classroom.
- Music Examples: here you have access to all the music examples recorded.

For information on the book and the DVD, you can contact Andrea Pejrolo at acoustic midiorchestration@apejrolo.com.

Enjoy!

Andrea Pejrolo
Richard DeRosa

Acknowledgments

The topic of this book has been always something very close to me and to my work as a composer, producer, performer, and educator. Its realization is truly a dream come true for me. The time, effort, energy, and inspiration that were required during the writing process could not have been available without the support of some truly amazing people who are part of my everyday life. First of all, my dear wife, Irache; she is an incredible source of inspiration and support. She is my trustworthy compass whom I could not do without. I gratefully acknowledge the educational background that my family and my dear friend Nella provided in my early years. All their work paid off today when organizational skills, logic, hard work, and striving for precision became crucial in completing this task. A special thank-you to my parents, who have taught me through the years how passion and commitment for what I do is the only way of living a full and meaningful life.

I am also extremely grateful to Richard DeRosa for such a great experience and collaboration in cowriting this book. His expertise and knowledge were always inspiring.

I am extremely grateful to all the staff at Focal Press for their great work and guidance. And, finally, a big thanks to my colleagues at Berklee College of Music.

Andrea Pejrolo

I would like to thank Andrea Pejrolo for inviting me to join him as a coauthor of this text; his evolution as a musician and teacher is continually impressive. Thanks also to William Paterson University, where I was a faculty member during the creation of this book's first edition, for granting me a one-semester sabbatical; the free time from my normal teaching duties enabled me to write this book while sustaining my professional career. And, finally, thanks to my lovely wife, Kimberly, whose support, wisdom, and companionship are invaluable.

Richard DeRosa

1 Basic Concepts for the MIDI Composer, Arranger, and Orchestrator

1.1 Introduction to MIDI and Audio Sequencing

If you are reading these pages you probably already have some basic experience of either composing or sequencing (or maybe both). The purpose of this chapter is to ensure that you are up to speed with some of the key concepts and techniques that are needed in order to learn advanced orchestration and MIDI production procedures. In this chapter we will learn the principles of the MIDI standard, and a detailed description of MIDI messages, digital audio, studio setup procedures, and more. After covering the technical part of the production process we will focus on the main principles on which orchestration, arranging, and composition are based. You will become familiar with such concepts as range, register, overtone series, transposition, balance, and intensity, and many others. These are all crucial and indispensable concepts that you will use to achieve coherent and credible hybrid MIDI productions.

As you will notice, in the majority of the chapters of this book we follow a structure in which the principles of MIDI sequencing and the traditional rules of orchestration alternate and collaborate, in order to give you a solid background on which to build your MIDI sequencing and production techniques. It is much easier to try to re-create a convincing string section if you first wrote the parts as if they were supposed to be played by a real set of strings. This is a basic concept that you always should keep in mind. No matter how sophisticated (and expensive) your setup and your sound libraries are, the final result of your production will always sound unconvincing and disappointing if you don't compose and orchestrate with the acoustic instrumentation and real players in mind.

Many composers believe that writing and orchestrating for a MIDI ensemble is easier than working with a real orchestra, because you don't have to deal with the stressful environment of live musicians. In fact, the opposite is true. Trying to re-create a live ensemble (or even an electronic one) with the use of a MIDI, a digital audio workstation (DAW), and a series of synthesizers and sound libraries is an incredibly challenging task, mainly because in most situations you will be the composer, the arranger, the orchestrator, the producer, the performer, the audio engineer, and the mastering engineer, all at the same time! While this might sound a bit overwhelming, this is what makes this profession so exciting and, in the end, extremely rewarding. There is nothing as rewarding as when you finish your production and you are completely satisfied with the final result.

Before we introduce more advanced orchestration techniques, let's review some of the basic concepts on which MIDI production and orchestration are based. While some of these concepts (such as the MIDI standard and common MIDI setup) will be reviewed only briefly, others (such as control changes, MIDI devices, and MIDI messages) will be analyzed in detail, as they constitute the core of more-advanced MIDI orchestration and rendition techniques. Keep in mind that to fit a comprehensive description of the MIDI standard and all its nuances into half a chapter is very hard. The following sections represent an overall review of the MIDI messages with an in-depth analysis of the control change messages, since we will frequently use this type of message to improve the rendition of our scores. For a more detailed look at how to set up your MIDI studio and at the basic of the MIDI standard, I recommend reading my book *Creative Sequencing Techniques for Music Production*, published by Focal Press, ISBN 0240522168.

1.2   Review of the MIDI Standard

MIDI (Musical Instrument Digital Interface) was established in 1983 as a protocol to allow different devices to exchange data. In particular, the major manufacturers of electronic musical instruments were interested in adopting a standard that would allow keyboards and synthesizers from different companies to interact with each other. The answer was the MIDI standard. With the MIDI protocol, the general concept of "interfacing" (i.e. establishing a connection between two or more components of a system) is applied to electronic musical instruments. As long as two components (synthesizers, sound modules, computers, etc.) have a MIDI interface, they are able to exchange data. In early synthesizers, the "data" were mainly notes played on keyboards that could be sent to another synthesizer. This allowed keyboard players to layer two sounds without having to play the same part simultaneously with both hands on two different synthesizers. Nowadays, the specifications of MIDI data have been extended considerably, ranging from notes to control changes, from system exclusive messages to synchronization messages (i.e. MTC, MIDI clock, etc.).

The MIDI standard is based on 16 independent channels on which MIDI data are sent and received by the devices. On each channel, a device can transmit messages that are independent of the other channels. When sending MIDI data, the transmitting device "stamps" on each message the channel on which the information was sent so that the receiving device will assign it to the correct channel.

One of the aspects of MIDI that is important to understand and remember is that MIDI messages do not contain any information about audio. In the traditional world of MIDI production, MIDI and audio signals are always kept separate. Think of MIDI messages as the notes that a composer would write on paper; when you record a melody as MIDI data, for example, you "write" the notes in a sequencer but you don't actually record their sound. While the sequencer records the notes, it is up to the synthesizers and sound modules connected to the MIDI system to play back the notes received through their MIDI interfaces. The role of the sequencer in the modern music production process is, in fact, very similar to that of the paper score in the more traditional compositional process. You "sketch" and write (sequence) the notes of your composition on a sequencer, then you have your virtual musicians (synthesizers, samplers, etc.) play back your composition. This is the main feature that makes MIDI such an amazing and versatile tool for music production. If one is dealing only with notes and events instead of sound files, the editing power that is available is much greater, meaning that one is much freer to experiment with one's music. This distinction has become a bit more blurred nowadays since the introduction of software synthesizers.

Every device that needs to be connected to a MIDI studio or system must have a MIDI interface. The MIDI standard uses three ports to control the data flow: IN, OUT, and THRU. The connectors for the three ports are the same: a five-pin DIN female port on the device and a corresponding male connector on the cable. While the OUT port sends out MIDI data generated from a device, the IN port receives the data. The THRU port is used to send out an exact copy of the messages received from the IN port. Nearly all professional electronic musical instruments, such as synthesizers, sound modules, and hardware sequencers, have built-in MIDI interfaces. The only exception is the computer, which usually is not equipped with a built-in MIDI interface and, therefore, must be expanded with an internal or external one. Nowadays, the computer (along with the DAW running on it) is the central hub of both your MIDI and audio data, becoming the main tool for your writing and arranging tasks. While the synthesizers, samplers, and sound generators in general may be referred to as the virtual musicians of the twenty-first-century orchestra, the computer can be seen as its conductor.

Depending on the type of MIDI devices and interface you get for your computer and DAW, there are three main MIDI configurations available: DAW-based, star network (SN), or daisy-chain (DC). The DC setup is mainly used in very simple studio setups or live situations where a computer may not be involved; it utilizes the THRU port to cascade more than two devices to the chain. The most used nowadays in home and project studios is the DAW-based configuration (Figure 1.1). In this situation, the main computer provides the power for both the MIDI sequencing/audio recording and the sound generation. For complex production, the main computer must be very powerful in order to handle all the real-time processing required. Usually you will have a single MIDI controller connected directly to a USB port on the computer. While this setup can seem simple, it is very common. Favoring the simplicity of the production environment allows you to focus more on the music and less on the gear. Do not be deceived, though, by the streamlined connections, since this particular setup relies on the power of the main

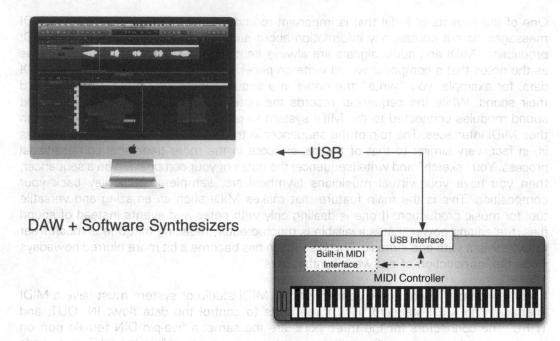

DAW + Software Synthesizers

Figure 1.1 DAW-based setup (Courtesy of Apple Inc.)

computer. If the computer is really powerful in terms of CPU speed, then it will be enough to run all your soft synths and the MIDI/audio tracks.

For an advanced and flexible MIDI studio, a multi-cable (or multiport) interface provides more flexibility, as it allows you to take full advantage of the potential of your MIDI devices. By using a multi-cable interface all the devices connect to the computer in parallel; therefore, the MIDI data won't experience any delay. This configuration, involving the use of a multi-cable MIDI interface, is referred to as a star network. One of the big advantages of the star network setup is that it allows one to use all 16 MIDI channels available on each device, as the computer is able to redirect the MIDI messages received by the controller to each cable separately, as shown in Figure 1.2.

It is important to note that in the modern digital studio setup the presence of external hardware gears has been reduced to a minimum due to the increasing power, flexibility, and cost-effectiveness of personal computers. This has in fact changed how we design and conceive the modern home and project studio. So, while, in the past, a complicated MIDI network was almost an essential part of your working environment due to the large use of hardware synthesizers, nowadays things have changed drastically. The explosion of the software synthesizers market (sound generator engines that run on your computer instead than on dedicated hardware platforms) has had the main advantage, among many others, of reducing cable clutter and simplifying your setup immensely (more on this

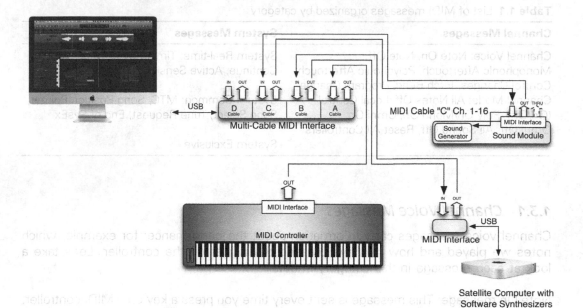

Figure 1.2 Star network setup with additional satellite computers (Courtesy of Apple Inc.)

later). Since the computer is usually not equipped with a built-in MIDI interface, if you want to have it connected to a MIDI device through the IN and OUT ports, you will need to expand its I/O with an external MIDI interface (which usually connects to the computer through a USB interface).

In order to exploit fully the creative power offered by the MIDI standard, it is crucial to precisely know and identify the MIDI messages that are available to us. While you may be familiar with some of the most common messages (e.g. Note On, Note Off), there are many others (CC#11, CC#73, CC#74, etc.) that are essential if you are trying to bring your MIDI productions to the next level. Let's take a look first at the main categories of the MIDI standard.

1.3 MIDI Messages and Their Practical Applications

The messages of the MIDI standard are divided into two main categories: channel messages and system messages. Channel messages are further subdivided into channel voice and channel mode messages, while system messages are subdivided into real-time, common, and exclusive messages. Table 1.1 illustrates how they are organized.

Table 1.1 List of MIDI messages organized by category

Channel Messages	System Messages
Channel Voice: Note On, Note Off, Monophonic Aftertouch, Polyphonic Aftertouch, Control Changes, Pitch Bend, Program Change	System Real-time: Timing Clock, Start, Stop, Continue, Active Sensing, System Reset
Channel Mode: All Notes Off, Local Control (On/Off), Poly On/Mono On, Omni On, Omni Off, All Sound Off, Reset All Controllers	System Common: MTC, Song Position Pointer, Song Select, Tune Request, End of SysEx
	System Exclusive

1.3.1 *Channel Voice Messages*

Channel voice messages carry information about the performance; for example, which notes we played and how hard we pressed the trigger on the controller. Let's take a look at each message in this category in detail.

Note On message: This message is sent every time you press a key on a MIDI controller. As soon as you press it, a MIDI message (in the form of binary code) is sent to the MIDI out of the transmitting device. The Note On message includes information about the note you pressed (the note number ranges from 0 to 127 or C2 to G8), the MIDI channel on which the note was sent (1–16), and the velocity-on, which describes how hard you pressed the key and ranges from 0 to 127 (with a value of zero resulting in a silence).

Note Off message: This message is sent when you release the key of the controller. Its function is to terminate the note that was triggered with a Note On message. The same result can be achieved by sending a Note On message with its velocity set to 0, a technique that can help to reduce the stream of MIDI data. It contains the velocity-off parameter, which registers how hard you released the key (note that this particular information is not used by most MIDI controllers at the moment).

Aftertouch (pressure): This is a specific MIDI message that is sent after the Note On message. When you press a key of a controller, a Note On message is generated and sent to the MIDI OUT port. This is the message that triggers the sound on the receiving device. If you push a little bit harder on the key after hitting it, an extra message, called aftertouch, is sent to the MIDI OUT of the controller. The aftertouch message is usually assigned to control the vibrato effect of a sound, but, depending on the patch that is receiving it, it can also affect other parameters, such as volume, pan, and more.

There are two types of aftertouch: polyphonic and monophonic. Monophonic aftertouch affects the entire range of the keyboard, no matter which key or keys triggered it. This is the most common type of aftertouch, and it is implemented on most (but not all) controllers and MIDI synthesizers available on the market. Polyphonic aftertouch allows you to send an independent message for each key. It is more flexible as only the intended notes will be affected.

Pitch bend: This message is controlled by the pitch-bend wheel on a keyboard controller. It allows you to raise or lower the pitch of the notes being played. It is one of the few MIDI data that do not have a range of 128 steps. In order to allow a more detailed and accurate tracking of the transposition, the range of this MIDI message extends from 0 to 16,383. Usually, a sequencer would display 0 as the center position (non-transposed), 18,191 fully raised and –8,192 fully lowered.

Program change: This message is used to change the patch assigned to a certain MIDI channel. Each synthesizer has a series of programs (also called patches, presets, instruments, or, more generically, sounds) stored in its internal memory; for each MIDI channel we need to assign a patch that will play back all the MIDI data sent to that particular channel. This operation can be done by manually changing the patch from the front panel of the synthesizer, or by sending a program change message from a controller or a sequencer. The range of this message is 0 to 127. As modern synthesizers can store many more than 128 sounds, nowadays programs are organized into banks, each bank storing a maximum of 128 patches. In order to change a patch through MIDI messages it is, therefore, necessary to combine a bank change message and a program change message. While the latter is part of the MIDI standard specification, the former changes depending on the brand and model of MIDI device. Most devices use CC#0 or CC#32 to change bank (or sometimes a combination of both), but you should refer to the synthesizer's manual to find out which MIDI message is assigned to bank change for that particular model and brand.

Control changes (CCs): These messages allow you to control certain parameters of a MIDI channel. There are 128 CCs (0–127); that is, the range of each controller extends from 0 to 127. Some of these controllers are standard and are recognized by all the MIDI devices. Among the most important of these (because they are used more often in sequencing) are CC#1, 7, 10, and 64. CC#1 is assigned to modulation. It is activated by moving the modulation wheel on a keyboard controller. It is usually associated with a slow vibrato effect. CC#7 controls the volume of a MIDI channel from 0 to 127, while number 10 controls its pan. Value 0 is pan hard left, 127 is hard right, and 64 is centered. Controller number 64 is assigned to the sustain pedal (the notes played are held until the pedal is released). This controller has only two positions: on (values > 64) and off (values < 63). While the four controllers mentioned above are the most commonly used, there are other controllers that can considerably enhance the MIDI rendition of acoustic instruments and the control that you have over the sound of your MIDI devices. Table 1.2 lists all 128 controllers, with their specifications and their most common uses in sequencing situations.

Among the 128 control change (CC) messages available in the MIDI standard, there are a few that can be particularly useful in a sequencing and music production environment. In particular, certain CC messages can be extremely helpful in improving the realism of

Table 1.2 MIDI control change (CC) messages

Controller #	Function	Usage
0	Bank select	It allows you to switch bank for patch selection. It is sometimes used in conjunction with CC#32 in order to send the bank number higher than 128
1	Modulation	It sets the modulation wheel to the specified value. Usually this parameter controls a vibrato effect generated through a low-frequency oscillator (LFO). It can also be used to control other sound parameters such as volume in certain sound libraries
2	Breath controller	Can be set to affect several parameters but is usually associated with aftertouch messages
3	Undefined	
4	Foot controller	Can be set to affect several parameters but is usually associated with aftertouch messages
5	Portamento value	Controls the rate used by Portamento to slide between two subsequent notes
6	Data entry (MSB)	Controls the value of either registered (RPN) or non-registered (NRPN) parameters
7	Volume	Controls the volume level of a MIDI channel
8	Balance	Controls the balance (left and right) of a MIDI channel. It is mostly used on patches that contain stereo elements (such as stereo patches). 64 = center, 127 = 100 percent right and 0 = 100 percent left
9	Undefined	
10	Pan	Controls the pan of a MIDI channel. 64 = center, 127 = 100 percent right and 0 = 100 percent left
11	Expression	Controls a percentage of volume (CC#7)
12	Effect controller 1	Mostly used to control the effect parameter of one of the internal effects of a synthesizer (for example, the decay time of a reverb)
13	Effect controller 2	Mostly used to control the effect parameter of one of the internal effects of a synthesizer
14–15	Undefined	
16–19	General purpose	These controllers are open and they can be assigned to aftertouch or similar messages
20–31	Undefined	
32–63	LSB for control 0–31	These controllers allow you to have a "finer" scale for the corresponding controllers 0 through 31
64	Sustain Pedal	Controls the sustain function of a MIDI channel. It has only two positions: off (values between 0 and 63) and on (values between 64 and 127)

Table 1.2 *continued*

Controller #	Function	Usage
65	Portamento on/off	Controls if the Portamento effect (the slide between two consequent notes) is on or off. It has only two positions: off (values between 0 and 63) and on (values between 64 and 127)
66	Sostenuto on/off	Similar to the sustain controller, but holds only the notes that are already turned on when the pedal was pressed. It is ideal for the "chord hold" function, where you can have one chord holding while playing a melody on top. It has only two positions: off (values between 0 and 63) and on (values between 64 and 127)
67	Soft pedal on/off	Lowers the volume of the notes that are played. It has only two positions: off (values between 0 and 63) and on (values between 64 and 127)
68	Legato footswitch	Produces a legato effect (two subsequent notes without pause in between). It has only two positions: off (values between 0 and 63) and on (values between 64 and 127)
69	Hold 2	It prolongs the release of the note (or notes) playing while the controller is on. Unlike the sustain controller (CC#64), the notes won't sustain until you release the pedal but instead they will fade out according to their release parameter
70	Sound controller 1	Usually associated with the way the synthesizer produces the sound. It can control, for example, the sample rate of a waveform in a wavetable synthesizer
71	Sound controller 2	Controls the envelope over time of the VCF (voltage-controlled filter) of a sound, allowing you to change the shape of the filter over time. It is also referred to as "resonance"
72	Sound controller 3	Controls the release stage of the VCA (voltage-controlled amplifier) of a sound, allowing you to adjust the sustain time of each note
73	Sound controller 4	Controls the attack stage of the VCA (voltage-controlled amplifier) of a sound, allowing you to adjust the time that the waveform takes to reach its maximum amplitude
74	Sound controller 5	Controls the filter cutoff frequency of the VCF, allowing you to change the brightness of the sound
75–79	Sound controllers 6–10	Generic controllers that can be assigned by a manufacturer to control the non-standard parameters of a sound generator
80–83	General purpose controllers	Generic button-switch controllers that can be assigned to various on/off parameters. They have only two positions: off (values between 0 and 63) and on (values between 64 and 127)
84	Portamento control	Controls the amount of Portamento
85–90	Undefined	

Table 1.2 *continued*

Controller #	Function	Usage
91	Effect 1 depth	Controls the depth of effect 1 (mostly used to control the reverb send amount)
92	Effect 2 depth	Controls the depth of effect 2 (mostly used to control the tremolo amount)
93	Effect 3 depth	Controls the depth of effect 3 (mostly used to control the chorus amount)
94	Effect 4 depth	Controls the depth of effect 4 (mostly used to control the celeste or detune amount)
95	Effect 5 depth	Controls the depth of effect 5 (mostly used to control the phaser effect amount)
96	Data increment (+1)	Mainly used to send an increment of data for RPN and NRPN messages
97	Data increment (−1)	Mainly used to send a decrement of data for RPN and NRPN messages
98	Non-registered parameter number (NRPN) LSB	Selects the NRPN parameter targeted by controllers 6, 38, 96, and 97
99	Non-registered parameter number (NRPN) MSB	Selects the NRPN parameter targeted by controllers 6, 38, 96, and 97
100	Registered parameter number (RPN) LSB	Selects the RPN parameter targeted by controllers 6, 38, 96, and 97
101	Registered parameter number (RPN) MSB	Selects the RPN parameter targeted by controllers 6, 38, 96, and 97
102–119	Undefined	
120	All sound off	Mutes all sounding notes regardless of their release time and regardless of whether the sustain pedal is pressed
121	Reset all controllers	Resets all the controllers to their default status
122	Local on/off	Enables you to turn the internal connection between the keyboard and its sound generator on or off. If you use your MIDI synthesizer on a MIDI network, most likely you will need the local to be turned off in order to avoid notes being played twice
123	All notes off	Mutes all sounding notes. The notes that are turned off by this message will still retain their natural release time. Notes that are held by a sustain pedal will not be turned off until the pedal is released
124	Omni mode off	Sets the device to omni off mode
125	Omni mode on	Sets the device to omni on mode
126	Mono mode	Switches the device to monophonic operation
127	Poly mode	Switches the device to polyphonic operation

MIDI sonorities when used to reproduce the sounds of acoustic instruments. Let's take a look at the CC messages (and their functions) that are particularly helpful in these types of applications. In order to tackle so many control changes without being overwhelmed, they can be organized according to their function and simplicity. Here, we will start with the most basic and most commonly used messages, and end with the more advanced ones.

1.3.2 Most Commonly Used Control Changes

Among the most used CCs, there are four that, in one way or another, you will use even for the most basic sequencing projects. These CCs are volume (CC#7), pan (CC#10), modulation (CC#1), and sustain (CC#64). While their names and functions are basically self-explanatory, their advanced use can bring your projects and your MIDI orchestration techniques to another level. Let's take a look at each message individually.

Volume (CC#7) enables you to control the volume of a MIDI channel directly through the sequencer or MIDI controller. Like most of the MIDI messages, it has a range of 128 steps (from 0 to 127), with 0 indicating basically a mute state and 127 full volume. Keep in mind that this is not the only way to control the volume of a MIDI track (more on this later), but it is certainly the most immediate. Think of CC#7 as the main volume on the amplifier of your guitar rig. It controls the overall output level of a MIDI channel. Also keep in mind that, as is the case for most MIDI messages, the message is sent to a MIDI channel and not to a MIDI track, and you have one volume control per MIDI channel and not per track. Therefore, if you have several tracks (i.e. drums) sent to the same MIDI channel and MIDI cable, they will all share the same volume control. The more advanced sequencing techniques involving the use of CC#7 will be discussed later in the book.

Pan (CC#10) controls the stereo image of a MIDI channel. The range extends from 0 to 127, with 64 being panned in the center, 0 hard left, and 127 hard right. As for CC#7, this message is sent to the MIDI channel and not to a specific track.

Modulation (CC#1) is usually assigned to vibrato, although in some cases it can be assigned to control other parameters of a MIDI channel. For example, certain software synthesizers and libraries use CC#1 to control the volume and sample switch of the instruments. This controller is a very flexible one and can, in fact, be used to manipulate several parameters that do not necessarily relate to vibrato. The way modulation affects the sound depends on how the synthesizer patch is programmed.

Sustain (CC#64) is usually associated with the sustain pedal of a keyboard controller. By pressing the sustain pedal connected to your controller you send a CC#64 with value 127; by depressing the pedal, you send a value of 0. Whenever the MIDI channel receives a CC#64 with value 127 it will sustain the notes that were pressed at the moment the control message was sent, until a new message (this time with a value of 0) is sent to the same MIDI channel. The overall effect is the same as you would obtain by pressing the sustain pedal on an acoustic piano.

1.3.3 Extended Controllers

In addition to the basic controllers described previously, there is a series of extended controllers that allow you to manipulate other parameters of a MIDI channel in order to achieve a higher degree of flexibility when controlling a MIDI device. These are the messages that you will take more advantage of when trying to take your sequencing, MIDI orchestration, and arranging skills to a higher level. They are particularly suited to adding more expressivity to such acoustic parts as string, woodwind, and brass tracks, as these instruments usually require a high level of control over dynamics, intonation, and color. Let's take a look at the extended MIDI controllers that are available under the current MIDI specifications.

Breath controller (CC#2): This controller can be set by the user to affect different parameters; it is not tied to a specific operation. It is usually set to the same parameter controlled by aftertouch. Generally, you will find it programmed to control modulation, volume, or vibrato. Breath controller is found mostly in MIDI wind controllers, where the amplitude of the controller is commanded by the pressure of the airflow applied to the mouthpiece.

Foot controller (CC#4): As in the case of the previous MIDI message, CC#4 can be assigned by the user to a series of parameters, depending on the situation. It can control volume, pan, or other specific parameters of a synthesizer. It is a continuous controller with a range of 0 to 127.

Portamento on/off (CC#65) and portamento time (CC#5): These give you control over the slide effect between two subsequent notes played on a MIDI controller. While CC#65 allows you to turn the portamento effect off (values 0–63) or on (values 64–127), with CC#5 you can specify the rate at which the portamento effect slides between two subsequent notes (0–127).

Balance (CC#8): This controller is similar to pan (CC#10). It controls the balance between the left and right channels for MIDI parts that use a stereo patch, while pan is more often used for mono patches. It ranges from 0 to 127, where a value of 64 represents a center position, 0 hard left, and 127 hard right.

Expression controller (CC#11): This particular controller is extremely helpful and is often used to change the volume of a MIDI channel. While you might recall that CC#7 controls the volume of a MIDI channel, expression allows you to scale the overall volume of a MIDI channel by a percentage of the value set by CC#7. In practical terms, think of CC#7 as the main volume on the amplifier for your guitar, and CC#11 as the volume on your guitar. They both, in fact, have an impact on the final volume of the part (MIDI channel), but CC#11 allows you to fine-tune the volume inside the range set by CC#7. To clarify further, think about the following examples. If you set CC#7 of a MIDI channel to 100 and CC#11 for the same channel to 127, you will get a full volume of 100. Now think what happens if you lower CC#11 to 64 (128 divided by 2); now, your overall volume will be 50 (100 divided by 2). Thus, expression can be extremely useful if used in conjunction with CC#7. A practical application would be to do all your volume automation with the expression controller and use CC#7 to raise or lower the overall volume of your MIDI tracks. We will discuss the practical application of the expression controller in the following chapters.

Sostenuto on/off (CC#66): CC#66 is similar to CC#64. When sent by pressing a pedal, it holds the notes that were already on when the pedal was pressed. It differs, though, from the sustain message because the notes that are sent after the pedal is pressed won't be held, as they are in the case of CC#64. It is very useful for holding chords while playing a melody on top.

Soft pedal on/off (CC#67): This controller works exactly like the pedal found on an acoustic piano. By sending a CC#67 to a MIDI device/part it lowers the volume of any notes played on a MIDI channel while the pedal is pressed. Soft pedal is off with values ranging from 0 to 63, and on with values from 64 to 127.

Legato footswitch (CC#68): This controller enables you to achieve a similar effect to the one used by wind and string players when playing two or more subsequent notes using a single breath or bow stroke. The legato effect achieved creates a smoother transition between notes. CC#68 achieves a similar effect by instructing the synthesizer to bypass the attack section of the voltage-controlled amplifier's (VCA) envelope of the sound generator and, therefore, avoid a second trigger of the notes played.

Hold (CC#69): CC#69 is similar to the sustain controller #64. While the latter sustains the notes being played until the pedal is released (values between 0 and 63), CC#69 prolongs the notes played by simply lengthening the release part of the VCA's envelope of the sound generator. This creates a natural release that can be used effectively for string and woodwind parts to simulate the natural decay of acoustic instrument sounds.

1.3.4 *Coarse versus Fine*

All controllers from 0 to 31 have a range of 128 steps (from 0 to 127), as they use a single byte of data to control the value part of the message. While most controllers do not need a higher resolution, for some applications there are other controllers that would greatly benefit from a higher number of steps in order to achieve a more precise control. For this reason, the MIDI standard was designed to have coarse and fine control messages. Each controller from 0 to 31 has a finer counterpart in controllers 32 to 63. By combining two bytes of data (least significant byte (LSB) and most significant byte (MSB)), the values have a much greater range. Instead of the coarse 128 steps, the finer adjustments use a range of 16,384 steps (from 0 to 16,383), achieved by using a 14-bit system (2^{14} = 16,384). While this function is a valuable one, most often you will be using the traditional coarse setting, as not all MIDI devices are, in fact, programmed to respond to the finer settings.

1.3.5 *Control Your Sounds*

The controllers analyzed so far are targeted to generic parameters that mainly deal with pan, volume, and sustain. There is a series of controllers, though, that can go even further and give you control of other effects present on your MIDI synthesizer, such as reverb, chorus, tremolo, detune, attack, and release time. Through the use of such powerful MIDI messages you can achieve an incredible realism when sequencing acoustic instruments. Let's take a look at this series of controllers.

Effect controllers 1 and 2 (CC#12 and 13): These two controllers allow you to change the parameters of an effect on a synthesizer. They are usually associated with the parameters of a reverb, such as reverb decay and size.

Sound controller 1—sound variation (CC#70): This controls the generator of the waveform on a synthesizer. One of the most common applications is the control of the sampling frequency of the generator, thereby altering the speed and "pitch" of the sound.

Sound controller 2—timbre/harmonic intensity (CC#71): This controls the shape of the voltage-controlled filter (VCF) of a synthesizer over time. It enables you to alter the brightness of the patch over time.

Sound controller 3—release time (CC#72): This controls the release time of the voltage-controlled amplifier (VCA) of a sound generator, giving you control over the release time of a patch on a particular MIDI channel. This message is very useful when sequencing acoustic patches, such as strings and woodwind, as it enables you to quickly adjust the patch for staccato or legato passages.

Sound controller 4—attack time (CC#73): This controller is similar to the one just described above. The main difference is that CC#73 gives you control over the attack parameter of the VCA of your sound generator. This is particularly indicated when sequencing acoustic instruments, where different attack values can help you re-create more natural and realistic results. We will discuss specific techniques related to the practical applications of controllers later in this book.

Sound controller 5—brightness (CC#74): CC#74 controls the cutoff frequency of the filter for a given patch and MIDI channel. By changing the cutoff frequency you can easily control the brightness of a patch without having to tinker with the MIDI device directly. Once again, this message gives you extreme flexibility and control over the realism of your MIDI instruments.

Sound controller 6—decay time (CC#75): This is often used to control, in real time, the decay parameter of the VCA's envelope. Note that sometimes, depending on the MIDI device you are working with, CC#75–79 are not assigned to any parameter and, therefore, are undefined.

Sound controllers 7, 8, and 9—vibrato rate, depth, and delay (CC#76, 77, and 78): Using these controllers you can vary, in real time, the rate and depth of the vibrato effect created by the sound generator of a synthesizer. You can effectively use these MIDI messages to control the speed and amount of vibrato for a certain patch and MIDI channel. These controllers can be particularly useful in improving the realism of sequenced string, woodwind, and brass instruments for melodic and slow passages.

Sound controller 10—undefined (CC#79): CC#79, like a few other controllers, is not assigned by default to any parameter. It can be used by a manufacturer to control, if needed, a specific parameter of a synthesizer.

Effects depth 1 (CC#91): Controllers 91 through 95 are dedicated to effects parameters often associated with general MIDI devices. Through these messages you can interact with the depth of such effects. Think of these controllers almost as effect send levels for each MIDI channel. CC#91 is specifically targeted to control the depth of the built-in reverb effect of a synthesizer. While, for most professional productions, the use of built-in effects is fairly limited, for quick demos and low-budget productions the ability to quickly control your reverb depth can sometimes be useful.

Effects depth 2 (CC#92): This enables you to control the depth of a second onboard effect of your synthesizer. It is often associated with the control of the depth of a tremolo effect, if available on your MIDI device.

Effects depth 3 (CC#93): this controls the send level of the built-in chorus effect on your synthesizer.

Effects depth 4 (CC#94): This enables you to control the depth of a generic onboard effect of your synthesizer. It is often associated with the control of the depth of a detune effect, if available on your MIDI device.

Effects depth 5 (CC#95): This enables you to control the depth of a generic onboard effect of your synthesizer. It is often associated with the control of the depth of a phaser effect, if available on your MIDI device.

1.3.6 Registered and Non-registered Parameters

In addition to the messages discussed so far, there is a series of parameter messages that expands the default 128 CCs. These extended parameters are divided into two main categories: registered (RPN) and non-registered (NRPN). The main difference between the two categories is that the former includes messages registered and approved by the MIDI Manufacturers' Association (MMA), while the latter is open and doesn't require manufacturers to comply with any particular standard. The way both categories of messages work is a bit more complicated than the regular CCs, which we just discussed, but they provide users with an incredible amount of flexibility for their MIDI productions. Let's learn how they are constructed and used.

Table 1.3 RPN list

Parameter # CC#101	CC#100	Function	Comments
0	0	Pitch-bend sensitivity	MSB controls variations in semitones LSB controls variations in cents
0	1	Channel fine-tuning	Represents the tuning in cents (100/8,192), with 8,192 = 440Hz
0	2	Channel coarse tuning	Represents the tuning in semitones, with 64 = 440Hz
0	3	Tuning program change	Tuning program number (part of the MIDI tuning standard rarely implemented)
0	4	Tuning bank select	Tuning bank number (part of the MIDI tuning standard rarely implemented)
0	5	Modulation depth range	Used only in GM2 devices
0	127	Reset RPN/NRPN	Reset the current RPN or NRPN parameter

RPN messages—CC#101 (coarse), 100 (fine): RPN messages allow for up to 16,384 parameters to be controlled, as they use the coarse–fine system explained earlier by using a double 7-bit system. As you can see, this allows for a huge selection of parameters that can eventually be controlled in a synthesizer. At the moment, though, only seven parameters are registered with the MMA: pitch-bend sensitivity, channel fine-tuning, channel coarse tuning, tuning program change, tuning bank select, modulation depth range, and reset. RPN messages are based on a three-step procedure. First, you have to send CC#101 and 100 to specify the desired RPN that you want to control, according to the values shown in Table 1.3.

After specifying the RPN parameter, you then send the data part of the message through CC#6 (coarse) and, if necessary, CC#38 (fine). If you want to modify the current status of the parameter, you can either resend the entire message with the new parameter data or simply use the data increment messages CC#96 (11) and CC#97 (21).

NRPN messages—CC#99 (coarse), 98 (fine): A similar procedure can be used to send NRPN messages. The only difference is that these messages vary depending on the manufacturer and on the device to which they are sent. Each device responds to a different set of instructions and parameters, very much like system exclusives that are specific to a MIDI device.

1.3.7 Channel Mode Messages

This category includes messages that mainly affect the MIDI setup of a receiving device.

All notes off (CC#123): This message turns off all the notes that are sounding on a MIDI device. Sometimes it is also called the "panic" function, since it is a remedy for "stuck notes": MIDI notes that were turned on by a Note On message but that, for some reason (data dropout, transmission error, etc.), were never turned off by a Note Off message.

Local on/off (CC#122): This message is targeted to MIDI synthesizers. These are devices that feature keyboard, MIDI interface, and internal sound generator. The "local" is the internal connection between the keyboard and the sound generator. If the local parameter is on, the sound generator receives the triggered notes directly from the keyboard, and also from the IN port of the MIDI interface (Figure 1.3). This setting is not recommended in a sequencing/studio situation, as the sound generator would play the same notes twice, thereby reducing its polyphony (the number of notes the sound generator can play simultaneously) by half. It is, though, the recommended setup for a live situation in which the MIDI ports are not used. If the local parameter is switched off (Figure 1.4), the sound generator receives the triggered notes only from the MIDI IN port, which makes this setting ideal for the MIDI studio. The local setting can also usually be accessed from the "MIDI" or "General" menu of the device, or can be triggered by CC#122 (0–63 is off; 64–127 is on).

Poly/mono (CC#126, 127): A MIDI device can be set as polyphonic or monophonic. If set up as poly, the device will respond as polyphonic (able to play more than one note at the same time); if set up as mono, the device will respond as monophonic (able to play only one note at a time per MIDI channel). The number of channels can be specified by the user. In the majority of situations we will want to have a polyphonic device, in order to take advantage of the full potential of the synthesizer. The poly/mono parameter

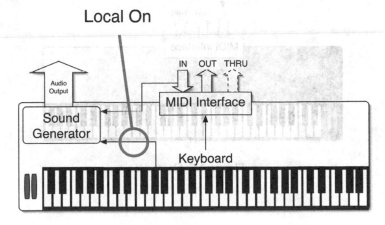

Figure 1.3 Local on

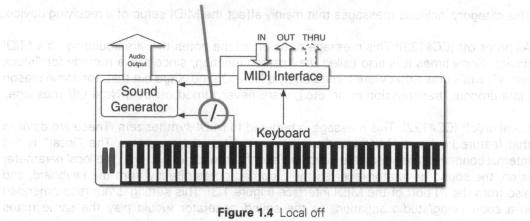

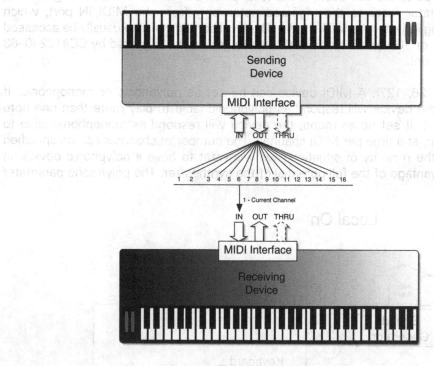

Figure 1.4 Local off

Figure 1.5 Omni on

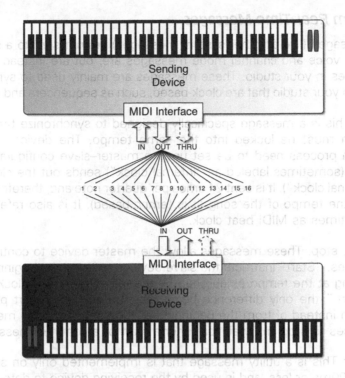

Figure 1.6 Omni off

is usually found in the "MIDI" or "General" menu of the device, but it can also be selected through CC#126 and CC#127, respectively.

Omni on/off (CC#124, 125): This parameter controls how a MIDI device responds to incoming MIDI messages. If a device is set to omni on, it will receive on all 16 MIDI channels but it will redirect all the incoming MIDI messages to only one MIDI channel (the current one) (Figure 1.5). If a device is set to omni off, it will receive on all 16 MIDI channels, with each message received on the original MIDI channel to which it was sent (Figure 1.6). This setup is more often used in sequencing, as it enables one to take full advantage of the 16 MIDI channels on which a device can receive. Omni off can also be selected through CC#124, while omni on is selected through CC#125.

All sound off (CC#120): This is similar to the "all notes off" message, but it doesn't apply to notes that are being played from the local keyboard of the device. In addition, this message mutes the notes immediately, regardless of their release time and whether the hold pedal is pressed.

Reset all controllers (CC#121): this message resets all controllers to their default states.

1.3.8 System Real-Time Messages

Real-time messages (like all other system messages) are not sent to a specific channel as the channel voice and channel mode messages are, but are instead sent globally to the MIDI devices in your studio. These messages are mainly used to synchronize all the MIDI devices in your studio that are clock-based, such as sequencers and drum machines.

Timing clock: This is a message specifically designed to synchronize two or more MIDI devices, which must be locked into the same tempo. The devices involved in the synchronization process need to be set up in a master–slave configuration, where the master device (sometimes labeled as "internal clock") sends out the clock to the slave devices ("external clock"). It is sent 24 times per quarter note and, therefore, its frequency changes with the tempo of the song (it is tempo-based). It is also referred to as MIDI clock, or sometimes as MIDI beat clock.

Start, continue, stop: These messages allow the master device to control the status of the slave devices. "Start" instructs the slave devices to go to the beginning of the song and start playing at the tempo established by the incoming timing clock. "Continue" is similar to "start," the only difference being that the song will start playing from the current position instead of from the beginning of the song. The stop message instructs the slave devices to stop and wait for either a start or a continue message to restart.

Active sensing: This is a utility message that is implemented only on some devices. It is sent every 300ms or less, and is used by the receiving device to detect if the sending device is still connected. If the connection were interrupted for some reason (e.g. the MIDI cable were disconnected), the receiving device would turn off all its notes to avoid causing "stuck" notes.

System reset: This restores the receiving devices to their original power-up conditions. It is not commonly used.

1.3.9 System Common Messages

System common messages are not directed to a specific channel, and are common to all receiving devices.

MIDI time code (MTC): This is another syncing protocol that is time-based (as opposed to the MIDI clock, which is tempo-based). It is mainly used to synchronize non-linear devices (such as sequencers) to linear devices (such as tape-based machines). It is a digital translation of the more traditional SMPTE code used to synchronize non-linear machines. The format is the same as SMPTE. The position in the song is described in hours:minutes:seconds:frames (subdivisions of one second). The frame rates vary depending on the format used. If you are dealing with video, the frame rate is dictated by the video frame rate of your project. If you are using MTC simply to synchronize music devices, it is advisable to use the highest frame rate available. The frame rates are 24, 25, 29.97, 29.97 drop, 30, and 30 drop.

Song position pointer: This message tells the receiving devices to which bar and beat to jump. It is mainly used in conjunction with the MIDI clock message in a master–slave MIDI synchronization situation.

Song select: This message allows you to call up a particular sequence or song from a sequencer that can store more than one project at the same time. Its range extends from 0 to 127, thus allowing for a total of 128 songs to be recalled.

Tune request: This message is used to retune certain digitally controlled analog synthesizers that require adjustment of their tuning after hours of use. This function does not apply to most modern devices, and is rarely used.

End of system exclusive: this message is used to mark the end of a system exclusive message, which is explained in the next section.

1.3.10 *System Exclusive Messages (SysEx)*

System exclusives are very powerful MIDI messages that allow you to control any parameter of a specific device through the MIDI standard. SysEx are specific to each manufacturer, brand, model, and device and, therefore, cannot be listed here in the same way that we have the generic MIDI messages described so far. In the manual for each of your devices there is a section in which all the SysEx messages for that particular model of device are listed and explained. These messages are particularly useful for parameter editing purposes. Programs called editors/librarians use the computer to send SysEx messages to connected MIDI devices in order to control and edit their parameters, making the entire patch editing procedure much simpler and faster.

Another important application of SysEx is the MIDI data bulk dump. This feature enables a device to send system messages that describe the internal configuration of that machine and all the parameters associated with it, such as patch/channel assignments and effects setting. These messages can be recorded by a sequencer connected to the MIDI OUT of the device and played back at a later time to restore that particular configuration, making it a flexible archiving system for the MIDI settings of your devices.

1.4 **Principles of Orchestration**

Having reviewed the MIDI standard and its messages, it is time now to review the principles of orchestration in order to gain an overall comprehensive view of the variables involved in the acoustic realm of arranging. This is crucial in order to bring your MIDI orchestration skills to a higher level. Writing and sequencing for MIDI instruments is no different than writing and orchestrating for acoustic ensembles. All the rules that you must follow for an acoustic ensemble apply to a MIDI sequence. Often one of the biggest problems that I encounter when assessing some of my colleagues' and students' MIDI productions is the lack of classical and traditional orchestration skills that no matter how sophisticated your MIDI equipment is, should never be overlooked. In the following paragraphs, Richard DeRosa will guide you through the main concepts and terms that constitute the fundamentals of orchestration.

1.4.1 Composition

The first stage of a musical creation is, of course, composing. This occurs most often at a piano. The piano provides the ultimate palette since it encompasses the complete range of musical sound and is capable of demonstrating the three basic textures of musical composition: monophonic texture (melody), polyphonic texture (multiple melodies occurring simultaneously), and homophonic texture (harmony).

Arranging

The second stage of the process is arranging. At this point, the factors under consideration may be the addition of an introduction or an ending, as well as any transitional material. Some new melodic ideas may be added that would serve as counterpoint to the original melody. Sometimes there are harmonic concerns such as the need to choose a specific key or create a modulation.

Orchestration

The third and final stage requires the assignment of individual parts to various instruments. This process involves an acute awareness of instrumental color, weight, balance, and intensity, as well as physical practicalities. The ultimate artistic goal is to flesh out the mood and expression of the composition.

There are composers who are comfortable with all three procedures and many prefer to do all three when time permits. Today, composers are usually obligated to create a mock performance of the orchestrated composition/arrangement so the client has an idea of what the final result will be with real instruments. Of course, economically, many situations dictate the need to have the final product exist in the MIDI format. Whatever the case, it will help greatly to have a traditional understanding of orchestration in order to enhance the MIDI representation.

Traditional Orchestration

Orchestration, in the traditional sense, is the process of writing for the instruments of a full orchestra. There are four distinct groups within the ensemble, characterized as follows: strings, woodwinds, brass, and percussion. The broader definition of orchestration essentially means to write for any combination of instruments. There are the wind ensemble, concert band, marching band, woodwind quintet, brass quintet, string quartet, jazz ensemble, also known as a big band, and a multitude of chamber groups that are comprised of a variety of instruments.

1.4.2 Range

Every instrument has a certain span of notes that it can play. This is referred to as the range of the instrument. Usually the bottom of the range is fixed but, quite often, the upper end of the range can be extended in accordance with the abilities of the individual performer. Within an instrument's range there is a portion that is used most commonly because it is the most flexible in ability and expression. This is referred to as the practical range.

It is most beneficial when the orchestrator can stay within this range without compromising the artistic integrity of the composition and/or arrangement. This is especially true for the commercial composer (music for recording sessions or live concerts) because of the time constraints involved in rehearsing music. In today's world there is very little time to prepare music, for several reasons. Studio time is very expensive (there is the cost of the studio, the engineer, an assistant, the musicians, the conductor, etc.), so it is advantageous to the producer(s) and the music creator(s) to be as expedient as possible. Even if budget is not a factor, there may be time constraints imposed by scheduling (recording musicians have other gigs to get to, a recording studio has another session booked after yours, a group of live musicians may be on a strict rehearsal schedule due to union rules, or there is only a limited amount of time that may be devoted to your piece as there are other pieces that need to be prepared as well).

1.4.3 Register and the Overtone Series

Within each instrument's range there are registers. They are divided into three general areas—low, middle, and high—and each register usually offers a different color, mood, or expression. The overtone series plays an important role in the natural production of a tone (Figure 1.7). Each single note produces its overtone series based on the following intervals, from low (first overtone) to high. The first overtone is at the interval of the octave, then the perfect fifth, the perfect fourth, the major third, two consecutive minor thirds, and finally a series of major and minor seconds.

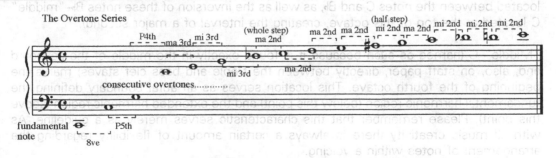

Figure 1.7 Overtone series

The overtone series is most helpful to any orchestrator since it acts as a guideline with regard to clarity and resonance for notes sounding simultaneously in harmony. It can also inform the arranger/orchestrator as to the use of extended harmonic color tones versus fundamental tones. It is most helpful to locate the overtone series at the piano starting from the lowest C. There is a total of eight Cs (outlining seven octaves) on the 88-key piano. The first two Cs up to the third C establish the contrabass and bass registers. If the parameters of the overtone series are not respected, the end result could be an unclear or a muddy presentation (Figure 1.8).

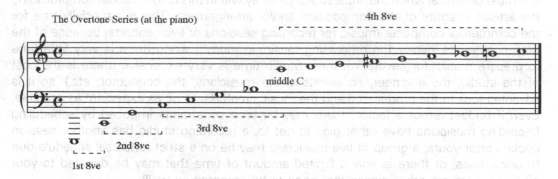

Figure 1.8 Overtone series at the piano

Using the series as a guide, it can be determined that within the first octave only an interval of a perfect octave will be resonant. Henceforth, any other interval within this first octave (contrabass register) will be unclear. Moving to the second octave, it can be determined that, in addition to the interval of a perfect octave, the perfect fifth and perfect fourth intervals may be used. Only in the third octave can the orchestrator begin to arrange chord voicings. In accordance with the overtone series, at this point the major and minor thirds are heard clearly (other intervals emerge when considering non-adjacent notes; there is the tritone located within the notes E–B♭; there is also the minor seventh located between the notes C and B♭, as well as the inversion of these notes B♭–"middle" C located at the top of this octave, creating the interval of a major second).

"Middle" C (named as such because it is found exactly in the middle of the keyboard and, also, on staff paper, directly between the treble and bass clef staves) marks the beginning of the fourth octave. This location serves as an aural boundary defining the fundamental harmonic region (below this point) and the extended harmonic region (above this point). Please remember that this characteristic serves merely as a guideline. As with all music creativity there is always a certain amount of flexibility regarding the arrangement of notes within a voicing.

The chord in Figure 1.9 is derived from most of the notes of the overtone series as it is heard from the note C. It creates a dominant thirteenth with an augmented eleventh.

Notice that there is only one chord tone in the first octave (in this case, the root). In the second octave, along with the root, there is the fifth of the chord (this is why instruments in the bass register play roots and fifths. This is apparent in music such as a Sousa march, ragtime, and Dixieland jazz). In the third octave, the chord color increases with the emergence of the third and seventh. In the fourth octave, the upper or extended chord tones emerge. The D is heard as a major ninth, the F♯ is heard as the augmented eleventh and the A is heard as the major thirteenth of the chord.

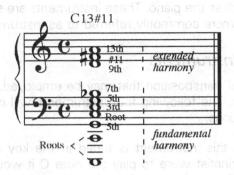

Figure 1.9 Overtone series chord

In jazz, there is an abundant usage of extended harmony and it is imperative that the arranger/orchestrator understand how to organize harmonic structures within the guidelines provided by the overtone series. Modern classical music (found often in film scoring) also demands careful attention to this principle.

1.4.4 Transposition

Many instruments require the need to transpose. This is because instruments are made in various sizes and, owing to the inherent physical qualities, larger objects will sound lower than smaller ones. The following analogy should help the reader to understand this aspect. Imagine blowing into three empty bottles: small, medium, and large. The effect will be that the smallest bottle will offer the highest pitch. The medium-sized bottle will offer a pitch lower than the small bottle but its pitch will also be higher than the largest bottle. The same principle applies to instruments of any type: drums, saxophones, string instruments, etc.

Using the saxophone family as an example, there are four types of saxophone that are used most commonly today. They are the soprano sax, alto sax, tenor sax, and baritone sax. Each saxophone has the same set of "fingerings" (button combinations needed to create various pitches). This makes it quite easy for a sax player to perform on any of the saxes. The problem occurs for the orchestrator because he or she cannot simply write the notes from the piano and give them to the saxophonist, or the result will be that the music will be heard in a different key. This situation becomes worse if two different saxophonists (i.e. an alto and tenor sax) were to play the same set of written notes. The result would be that each saxophone is heard in its respective key, creating an undesirable effect of polytonality (music heard simultaneously in two different keys).

1.4.5   Concert Instruments

It is important to know that there are also instruments that do not require any transposition. In this case the instrument sounds as written. This also means that the instrument sounds at the same pitch as the piano. These instruments are said to be in concert pitch with the piano and are more commonly referred to as instruments in C.

1.4.6   Transposing Instruments

There are three types of transposition that may be employed. They are pitch, register, and/or clef transposition. The following four instruments will demonstrate the need for these various transpositions:

Clarinet in B♭: Because this instrument is tuned in the key of B♭, it requires a pitch transposition. If the clarinetist were to play the note C it would actually sound as the note B♭ on the piano. To compensate for this differential, the clarinetist would need to play one whole tone higher. In other words, the clarinetist would play a D so that it sounds in unison with the C played on the piano (an illustration is provided in Chapter 4).

Double bass: Many low instruments (and high ones too) require a register transposition. This is a bit more arbitrary as it supports the need to simplify the note-reading process. The bulk of the bass's range exists well below the bass clef staff and would require an excessive amount of leger lines. To bypass this problem, it was decided that all bass parts should be written an octave higher. As a result, the reading of music becomes much simpler since most of the notes now lie within the bass clef staff (an illustration is provided in Chapter 3).

Guitar: The guitar is similar to the double bass as it uses a register transposition of an octave. However, guitarists read only treble clef (at least we must assume so in accordance with tradition). Most of the notes found on the guitar lie within the bass clef staff, so a conversion to treble clef is necessary (an illustration is provided in Chapter 2).

Baritone saxophone: This instrument actually requires the use of all three transpositions. Pitched in the key of E♭, it sounds a major sixth lower, along with a register transposition of an octave, when it plays the note C. Since saxophonists read only in treble clef, the clef transposition must take place as well (an illustration is provided in Chapter 4).

1.4.7   Weight, Balance, and Intensity

As an instrument passes through its range, the dynamic contour (volume of sound) can change. Some instruments are thicker and fuller in the bottom range. Others become thinner near the bottom. The opposite effect can occur in other instruments. Some instruments are naturally softer than others and vice versa. As a result, the balance may be affected and it is the job of the orchestrator to know how to maintain good balance with the instruments. The intensity of an instrument's timbre is determined by how rapidly the vibrations move within the sound, as well as the physical energy required by the performer.

Using a violin, a flute, and a trumpet to demonstrate the difference, imagine that all three instruments were, in unison, playing a high concert E located on the third leger line above the treble staff. The violinist uses the least amount of energy since there is no breath support needed. The sound of the string actually becomes thinner since there is less string length. (This is why many violins are needed in an orchestra in order to maintain balance with the stronger woodwind and brass.) The flautist needs to support the note with a significant amount of breath. The timbre of this note is quite bright but not shrill and it penetrates rather nicely. The trumpeter needs an exorbitant amount of energy to play this note and many trumpet players cannot even accomplish this. To capture specific pitches on a brass instrument, beyond the specific valve or slide positions, the performer must use air in combination with embouchure (the mouth's position against the mouthpiece) support. The trumpet in this register is exceedingly powerful and would certainly outbalance the other two instruments.

It is imperative that the orchestrator keeps in mind the physical endurance of the performers. Excessively long musical passages without places to breathe, or constant playing in the high register, will be detrimental to the performance and, in some cases, may be impossible to execute.

1.4.8 *Hazards of Writing at the Piano*

As mentioned previously, the piano offers the best palette from which to create. However, there are some distinct pitfalls that can lead a novice orchestrator astray. The piano offers only one timbre, so the blend and balance of pitches within a harmonic structure occurs naturally. If the orchestrator is writing for a string section, then the transfer from the piano to the string section will be fairly true. The same can be said for a section of trombones or any other type of instrumentation that offers a monochromatic timbre. However, if the performing group were a woodwind quintet (flute, oboe, clarinet, bassoon, and French horn), the final presentation of sound would be very different from what is heard initially at the piano. In this scenario, the orchestrator must be keenly aware of the distinct timbral color differences in addition to the factors of weight, balance, and intensity.

The factors of breathing and endurance are missing when playing a piano. This of course is not a concern if the performers are playing instruments that do not require breath support, but if the instrumentation is for a brass group these factors are of critical importance.

Technical difficulties are different on other instruments. Wide leaps in quick succession may be difficult on brass instruments. Complex melodies with difficult chromatic tones can be more challenging for string players. (In general, it is more difficult for string players to play in tune since the pitches on string instruments are not fixed as they are on the piano.) A piano piece is not so easily performed on the harp since the harp is set up as a diatonic instrument. In order to play chromatically, the harpist must implement pedal changes. The pedals are found at the bottom on either side of the strings and obviously two pedal changes on one side would take longer than changing two pedals simultaneously on opposite sides of the strings. Mute changes in brass instruments and equipment changes in the percussion section are other factors that must be considered.

Tonal capability varies according to instrument. The orchestrator cannot simply write loud or soft on the part and expect the performer to produce the desired effect. The orchestrator must know the general capability of an instrument before assigning a specified dynamic level. It is analogous to trying to force a car to go faster than the capability of its engine or trying to lift an amount of weight beyond what is humanly possible.

There is no sustain pedal in an orchestra. The orchestrator must be keenly aware of this and establish sustain within the instruments (more commonly referred to as "pads") by actually writing out the specific duration of each sustained passage.

Voice-leading is not as apparent at the piano. The orchestrator should not simply "take the notes from top to bottom" and assign them accordingly to the corresponding instruments. This may seem logical but, in fact, can expose a novice orchestrator since the end result may produce unfavorable musical results in the form of incorrect or awkward voice-leading. For example, an oboe and a clarinet have two distinct sounds. If the orchestrator were to simply assign in an obvious way the notes for these instruments from the original sketch conceived at the piano, each instrument might expose an unfavorable or unmelodic resolution. This phenomenon only becomes apparent as a result of the two distinct timbres. At the piano, the voice-leading (or lack thereof) goes unnoticed since all of the notes are colored with the same timbre.

The physical context of performance at the piano is ultimately quite different from the reality of performance by a multitude of individuals on various instruments. At the piano the monophonic texture (melody) is usually played in the right hand and the homophonic texture (harmony) is played in the left hand. Imagine hearing the melody played only by the violins, flutes, and trumpets, never by the French horns, cellos, or bassoons. This would deny the wider spectrum of expression that music is capable of. Quite often, the wonderful aspect of polyphonic texture (multiple melodic ideas) is omitted because of the physical constraints of performance at the keyboard. Sometimes, when using polyphony, the lines to be played by different instruments may cross. This is quite acceptable and even interesting, especially when each line is represented by a distinct instrumental color. At the piano, this possibility might not be considered since it is usually awkward physically to cross the melodic lines.

Generally, in the case of a complex orchestration, it will probably be impossible to "perform" all of the parts at the piano. For the music arranger/orchestrator, the piano is viewed primarily as a tool that can guide the writing process. It is sometimes easier for non-pianists to adopt this line of thinking and more difficult for the pianist to think outside of his or her normal performance domain. For the modern music writer, the advantage of realizing the process via MIDI helps that writer to think beyond the limited scope of what the piano keyboard has to offer.

1.5 **Final Considerations**

The secret to a professional-sounding and realistic contemporary production is the right combinations of traditional orchestration techniques and advanced sequencing tools and skills. Only by mastering the two aspects of modern production are you able to bring the quality of your work to another level. In order to do so, in this chapter, we first analyzed the basic tools and concepts that form the backbone of a MIDI studio, for you to have a full grasp of the environment in which you will be working. This is a crucial aspect of contemporary MIDI orchestration since having a fully functional, versatile, and flexible project studio is a key element. Knowing the tools that are available to you as a MIDI orchestrator is extremely important in order to be able to achieve any sonority or effect you want in your composition. We don't want to settle only for the known features of your sequencer or your MIDI studio, but instead we want to explore more advanced techniques, commands, and controllers in order to be able to create the perfect MIDI rendition of your scores. You should always keep a copy of the extended MIDI controllers list handy in your studio. It would be a bit intimidating to memorize them all, so have a copy of Table 1.2 available next to you or on your computer. Always remember that the tools that you use to create your music are exactly that: tools. Try not to be caught in the middle of a technological struggle that would get in the way of your creativity. The purpose of this book is precisely to help you avoid technological "breakdowns" that would stop your creativity. My advice: learn the tools, set your studio in an efficient and streamlined way and then let the creative process flow!

Most people are not musicians and therefore listen to music from a different perspective. Primarily, music speaks to the listener more from the standpoint of emotion and expression than technique. It is most important for the music creator to respect this and remember that live musicians will also experience music this way. The orchestrator must respect the physical laws of sound as they relate to the overtone series, along with the range, registers, and tonal colors of the various instruments as there is a direct connotation to expression. Technique is only what is used to facilitate and enhance the overall presentation.

In the next chapters we are going to analyze specific sections of the modern large ensemble. For each chapter we will analyze the traditional orchestration techniques first, in order to give you the necessary background you need to understand the ranges transposition, styles, and combinations of each instrument that is part of a specific section. Once you are familiar with the more traditional techniques, you will be ready to move on to learn and experiment with the advanced MIDI techniques used to render your parts within the realm of a MIDI studio, using extended MIDI controllers, automation, mixing techniques, and more.

1.6 Summary

Almost all contemporary music productions are based on a hybrid combination of acoustic instruments and MIDI parts. To be able to master both aspects of the production process is crucial for the modern composer and orchestrator. To understand the MIDI standard is extremely important for a successful and smooth production process. The MIDI standard is based on 16 independent channels on which MIDI data are sent and received by the devices. On each channel a device can transmit messages that are independent of the other channels. Every device that needs to be connected to a MIDI studio or system must have a MIDI interface. The MIDI standard uses three ports to control the data flow: IN, OUT, and THRU. While the OUT port sends out MIDI data generated from a device, the IN port receives the data. The THRU port is used to send out an exact copy of the messages received from the IN port. Depending on the type of MIDI interface (single cable or multi-cable) you get for your computer and sequencer, you can have, respectively, two main MIDI configurations: daisy-chain (DC) or star network (SN). The DC setup is more limited and it is mainly used in simple studio setups or live situations where a computer is (usually) not used. The SN setup allows you to take full advantage of the potential of your MIDI devices. One of the big advantages of the star network setup is that it allows one to use all 16 MIDI channels available on each device, as the computer is able to redirect the MIDI messages received by the controller to each cable separately.

The messages of the MIDI standard are divided into two main categories: channel messages and system messages. Channel messages are further subdivided into channel voice and channel mode messages, while system messages are subdivided into real-time, common, and system exclusive messages. Channel voice messages carry information about the performance; for example, which notes we played and how hard we pressed the trigger on the controller. In this category, the control change messages are among the most used when it comes to MIDI orchestration and sequencing because of their flexibility and comprehensive coverage. These messages allow you to control certain parameters of a MIDI channel. There are 128 CCs (0–127); that is, the range of each controller extends from 0 to 127. Among the 128 control change (CC) messages available in the MIDI standard, there are a few that can be particularly useful in a sequencing and music production environment. In particular, some CC messages can be extremely helpful in improving the realism of MIDI sonorities when used to reproduce the sounds of acoustic instruments. Among the most used CCs, there are four that, in one way or another, you will use even for the most basic sequencing projects. These CCs are volume

(CC#7), pan (CC#10), modulation (CC#1), and sustain (CC#64). In addition to these basic controllers, there is a series of extended controllers that allow you to manipulate other parameters of a MIDI channel in order to achieve a higher degree of flexibility when controlling a MIDI device. They are particularly suited to adding more expressivity to such acoustic parts as string, woodwind, and brass tracks, as these instruments usually require a high level of control over dynamics, intonation, and color. All controllers from 0 to 31 have a range of 128 steps (from 0 to 127), as they use a single byte of data to control the value part of the message. While most controllers do not need a higher resolution, for most applications there are other controllers that would greatly benefit from a higher number of steps in order to achieve more precise control. For this reason, the MIDI standard was designed to have coarse and fine control messages. Each controller from 0 to 31 has a finer counterpart in controllers 32 to 63.

The channel mode messages include messages that affect mainly the MIDI setup of a receiving device, such as all notes off (CC#123), local control on/off (CC#122), poly on/mono on (CC#126, 127), omni on/off (CC#124, 125), reset all controllers (CC#121), and all sound off (CC#120).

Real-time messages (like all other system messages) are not sent to a specific channel, as the channel voice and channel mode messages are, but instead are sent globally to the MIDI devices in your studio. These messages are used mainly to synchronize all the MIDI devices in the studio that are clock-based, such as sequencers and drum machines. System common messages are not directed to a specific channel, and are common to all receiving devices. They include MTC, song position pointer, song select, tune request, and end of system exclusive.

System exclusives are very powerful MIDI messages that allow you to control any parameter of a specific device through the MIDI standard. SysEx are specific to each manufacturer, brand, model, and device and, therefore, cannot be listed here in the way that we have the generic MIDI messages described earlier. These messages are particularly useful for parameter editing purposes.

1.7 Exercises

Exercise 1.1

Connect the MIDI equipment shown in Figure 1.10 in a DAW-based setup.

Exercise 1.2

Connect the MIDI equipment shown in Figure 1.11 in a star network setup.

DAW + Software Synthesizers

Figure 1.10 Exercise 1.1

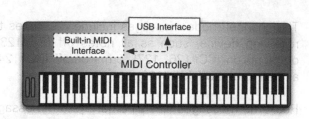

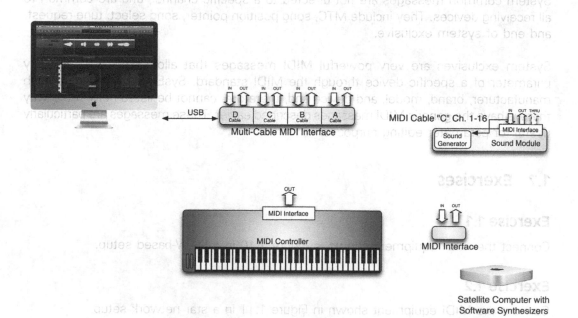

Figure 1.11 Exercise 1.2

Exercise 1.3

Give a brief description and practical application of the following MIDI CCs:

CC#	Description	Application
1		
2		
7		
10		
11		
64		
65		

Exercise 1.4

Give a brief description of the following MIDI messages:

MIDI message	Description
Note On	
Monophonic aftertouch	
Polyphonic aftertouch	
Pitch bend	
Control change	
Program change	

Exercise 1.5

Take a phrase from a simple, popular melody and write it in the treble and bass clefs.

Exercise 1.6

Write the key signatures for the B♭ and E♭ instruments in accordance with their relationship to each of the concert keys. Begin in the concert key of C major.

Exercise 1.7

Transpose the keyboard melody written in Figure 1.12 up a whole step for clarinet in B♭.

Figure 1.12 Exercise 1.7

Exercise 1.8

Take the left-hand piano part in Figure 1.13 and transpose it an octave higher for the double bass so that instrument sounds in unison with the piano.

Figure 1.13 Exercise 1.8

Exercise 1.9

Take the piano accompaniment written in Figure 1.14 and transpose it an octave higher for guitar so both instruments sound in unison.

Figure 1.14 Exercise 1.9

Exercise 1.10

Take the piano melody written in Figure 1.15 and transpose it an octave and a major 6th higher so a baritone saxophone can be heard in unison.

Baritone sax

Figure 1.15 Exercise 1.10

2 Writing and Sequencing for the Rhythm Section

2.1 Introduction: General Characteristics

The rhythm section is the foundation of most contemporary popular music productions. In more-traditional ensembles (orchestra or band), rhythmic momentum is usually supplied through melodic counterpoint along with various harmonic accompaniment patterns heard in the strings and winds. The percussion section in such ensembles is usually less active, supplying more of a color with an occasional rhythmic, perpetual motion for reinforcement when desired. The marching band is perhaps the earliest example of the relationship changing significantly between the rhythm section and the winds (and strings). Here, the percussion section supplies a continuum of rhythm and support upon which the other melodic and harmonic instruments rely. This is also true of the jazz style, in which the rhythm section (piano, bass, drums, and guitar) supplies the rhythmic and harmonic foundation for the melodic instruments. It is usually moving in perpetual motion, generating a vibrant groove that becomes infectious for the listener to the point of wanting to dance. Popular music styles, from the latter part of the twentieth century into the twenty-first, can incorporate complex textures of expanded rhythm sections that demand an internal discipline from the players in order to avoid mayhem. Previously, orchestral composers controlled these actions through carefully scripted parts. In other cases, particularly with non-Western styles (African, as well as Latin styles from various places in the Caribbean and South America), there is an aural tradition handed down through the generations by which a certain protocol is to be followed with regard to the performance of a rhythmic idea or style. Today, however, many rhythm section players create parts or embellish on rather simple guidelines that a composer or arranger may present. This is most notable in jazz, country, rock, folk, and most American-style popular music that depends substantially on a rhythm section as part of its musical texture.

2.2   The Components of the Rhythm Section

The rhythm section is rather complex in its composition and musical function. It consists of a wide tonal palette of instruments, designed to generate perpetual rhythmic motion while providing a harmonic foundation as it creates a tapestry of rhythmic, harmonic, and melodic ideas that work simultaneously. Essentially, these instruments may be categorized into three large groups. In no particular priority, they are:

- Group 1: keyboards (acoustic and electric piano, vibes, marimba, etc.).
- Group 2: strings (bass, guitars, banjo, mandolin, etc.).
- Group 3: percussion (drum set, congas, timbales, shaker, triangle, tambourine, etc.).

The percussion element is outstanding since the various tones of the instruments are considered to be non-defined with regard to pitch. This helps to establish a sonic canvas that is more distinctive for the listener as this group of instruments can play continuously without muddling the other harmonic and melodic components. Of course, it is essential that these instruments maintain a balance concerning volume.

2.3   The Role of the Essential Instruments

Rhythm sections can vary in the number of performers and instruments in accordance with a particular style of music. Some instruments are more important than others. Professional rhythm players understand the function of each instrument and how the instruments should interact. A common rule to be followed is: the more players there are, the more self-discipline is needed. Careful lateral listening is essential and precision is most important. To achieve rhythmic precision, each player must be cognizant of the beats but, more importantly, the subdivision of each beat. In general, the fastest-moving or most dense part is the one that broadcasts the subdivision, such as a hi-hat playing sixteenth notes. Conversely, a simple constant part, such as a cowbell playing each beat, can also act as a "handle" for the entire section. Concerning harmony, when there are multiple chord instruments (keyboards and guitars), the players must find a place on the sonic canvas that enables their individual voices to be heard or felt without obscuring the overall texture. This harmonic "comping" (slang for accompanying) may often be problematic with unprofessional rhythm sections. Here again it is most helpful to use the overtone series as a guide with regard to register and clarity. Remember that the first two octaves of the piano are essentially the bass registers. This also means that chords sound muddy here and consequently this register should be avoided when creating harmony. The third and fourth octaves offer the best tone and support for harmonic structures. The fifth octave may also be used in conjunction with one of these lower octaves, but the top two octaves are not really useful in this regard. Even though the range of the guitar and, in particular, the keyboard extends beyond either side of the third and fourth octaves, professionals stay primarily in this preferred area when comping. To create perspective for the listener, each instrument finds a different register, allowing both instruments to be heard simultaneously in a clear manner. When two comping instruments need to play in the same register, quite often the rhythm is the element that creates contrast. For example, a guitarist could be strumming in a more rapid fashion while the keyboardist is playing long, sustained chords, or the keyboardist might be playing a syncopated rhythmic figure using full chords while the guitarist is playing a more transparent idea using arpeggios. These are just two ways to establish contrast. It is wise to listen to rhythm sections with a keen ear for each instrument within the section to become aware of the various possibilities. (The music scores in Appendix A will also be useful to get you started in this endeavor.)

2.3.1 The Bass

The bass (acoustic or electric) is probably the most fundamental instrument since it provides rhythmic propulsion as well as the harmonic foundation. Its low tones also add "bottom," supporting the overall texture. For contemporary playing, the strings are plucked with the fingers, adding a percussive quality. As for the piano, the primary register for the bass to be heard is in the first two octaves. There are some times when the bass will venture beyond this fundamental register, but that is more for melodic effect or for a somewhat floating quality. (Chapter 3, on the strings, discusses the bass in more specific detail.)

There is a distinct difference between the acoustic bass and the electric bass. Visually they are quite different since the electric bass looks more like a guitar and should be thought of as such. This difference should be respected aurally as the two instruments each have a distinct set of advantages (and disadvantages). They are outlined below:

Acoustic bass	Electric bass
Warmer, deeper tone	More treble in tone
Less percussive	More percussive
Less agile	More agile (particularly for multiple stops)
Less accurate with intonation	Fretboard helps to "lock in" the pitch
Arco is an option	Arco is not an option
No amplification needed	Needs amplification
Has four strings (may have C extension)	May have five strings (offering a low B)

Many professional bass players can play both instruments, but more commonly there are those who have dedicated themselves to mastering one over the other. Further, many players specialize in a particular style of music, so it is important to keep this in mind when hiring a bassist. In either case, the bassist knows his or her function in the rhythm section as the provider of the harmonic foundation via the bass line along with supporting the drummer rhythmically.

2.3.2 *The Drum Set (and Auxiliary Percussion)*

The drummer supplies most of the rhythmic momentum but, in particular, helps to intensify the rhythmic feeling by advertising the subdivisions within each beat (e.g. in 4/4 time there are four beats within each bar, but the feeling or "groove" between each beat can differ; the subdivision of this space usually incorporates notes in eighth, sixteenth, or triplet rhythms). Since the drummer potentially uses four limbs, it is possible to play more than one of these subdivisions. In any case, there is usually one predominant subdivision broadcast through the drummer's beat pattern or general playing. The vast amount of tone color and the ability to add significant power also make this instrument the most dramatic in the group. Another distinctive characteristic is that the multitimbral drum set is essentially considered to be a non-defined pitch (i.e. while cymbals and drums do have pitches, in context within a piece of music, they appear to lack any distinctive pitch). The drums themselves (snares, toms, bass drum) are "tuned" in a sense of being higher and lower, but they are not thought of as contributing in a melodic or harmonic way. (The only exception would be timpani, which are actually tuned to specific pitches and, of course, are not considered to be part of the rhythm section.) A non-defined pitch quality enables the drummer to be active without obscuring the precise pitches heard from the other members of the section.

Of course, it is also possible to expand the drum area of the rhythm section with some percussion. Most commonly, Latin percussion instruments are added: cowbell, claves, congas, timbales, maracas, guiro, etc. Then there is also the more traditional tambourine or triangle. Usually there is a specific rhythmic pattern assigned to a percussion instrument when playing perpetually. Sometimes a percussion instrument may be heard on a regular basis, but only occasionally (e.g. a triangle sounding on the upbeat of three during the first bar of a four-bar phrase, or a tambourine emerging with a pattern on the chorus of a tune). All of these colors can be very effective in accordance with style if not overused. Using a food analogy, they should be thought of as seasoning for the main dish.

2.3.3 *Keyboards and Guitars*

The homophonic texture (harmony) is supplied usually by piano or guitar. This group can also expand to several guitars and keyboards of different timbre such as the mandolin, banjo, electric piano, accordion, vibes, or marimba. There are certainly other instruments as well, but the point is made. Generally, there are keyboards or guitars. Although their function, musically, is similar, because the instruments are physically so different their practical application must be considered. Outlined below are some of the considerations:

Keyboard	Guitar
Can use all 10 fingers	Can only use four fingers to create voicings
Cluster harmony is possible	Cluster harmony is difficult if not impossible
Transposition is more difficult	Transposition is easier (planing technique; capo)

Wider range of notes	Smaller range of notes
Repeated notes are difficult	Repeated notes are easy (on two strings; with a pick)
Strumming is not possible	Strumming is easy and quite characteristic
Sustaining arpeggios is easy	Sustaining arpeggios is harder or impossible

Consequently, some musical ideas work better when assigned to one versus the other:

Musical idea	Instrument choice
Large chords (many notes)	Piano
Fast percussive strumming	Guitar
Extended range	Piano
Harmonics	Guitar
Melody with accompaniment (e.g. ragtime)	Piano
"Planing" harmonic technique	Guitar
Sustaining arpeggios with many chords	Piano
Repeated notes in quick succession	Guitar

2.4 Color and Style

The vast array of instruments found in a rhythm section offers an arranger many choices. The choice of instrument affects the music's color and style. Outlined below are several of the more commonly used instruments, along with their most stereotypical effects.

2.4.1 Basses

Acoustic (double bass): a warm, deep and woody tone that works very well in jazz, folk, country, and sometimes light rock.

Electric (with frets): the classic electric bass sound most commonly heard in rock, Motown, funk, and studio music.

Electric (fretless): A smoother sound that is most commonly heard in jazz-rock fusion. Jaco Pastorius used this bass.

2.4.2 Guitars

Electric (solid body): Capable of a wide range of timbre, especially with the use of the chorus, overdrive, and fuzz-box effects. It can also produce a clean sound for more-traditional contemporary styles.

Electric (hollow body): produces a warmer sound that works very well in the jazz style.

Steel string acoustic: the most typical of the acoustic guitars and most commonly used in folk, country, and bluegrass music.

Nylon string acoustic: These strings provide a mellow and warm sound. Used extensively in Brazilian bossa nova music and in mellower fusion jazz, in particular by Earl Klugh. It is great for melodies and for delicate arpeggiated passages, and also for classical music. This instrument is usually played with the fingers instead of a pick, and vibrato should be used in the left hand when playing sustained notes.

Twelve-string: This guitar has a very metallic and big sound due to the extra strings, some of which are tuned in octaves. Its timbre is unique and is used most commonly for country and folk music.

2.4.3 Keyboards

Piano: The most classic of the keyboards. A distinction should be made between the grand piano and the upright piano. For beauty, the grand is better, but the upright can also be used, especially for a honky-tonk or saloon-style piano effect.

Rhodes electric: A unique sound offering a bright twinkle of sound in the higher register as well as a thick, lush texture in its middle register. (The low register is rather muddy and probably should be avoided.) This sound became very popular in the 1970s. Since then, many other electric pianos have been based on the Rhodes sound, offering slight variations and characteristics. It is used extensively in fusion jazz and studio music productions.

Wurlitzer electric: A predecessor of the Rhodes, it is cleaner, thinner, and darker in timbre with little or no twinkle. It is used in many rock bands.

Clavinet: A gritty, nasal-sounding keyboard that is used primarily in funk music, it is somewhat similar to the harpsichord. The harpsichord is a keyboard instrument from the Baroque era that produces a distinctly metallic and staccato sound. It is rather soft but many synthesized simulations add volume. This sound is excellent for music productions where a distinctly formal elegance is desired. (It is very often used in conjunction with a string section.)

Organ: This is a world unto itself. There are few keyboardists who feel as comfortable playing the organ as they would a piano. This instrument is far more complex and requires a great amount of dexterity, not only with the hands but with the feet, too. Most organs have multiple keyboard decks and a wide array of drawbars that offer various tones and frequencies, similar to using equalization. There are also pedals that play deep bass tones. Organs are mostly heard in jazz, gospel, rock, and church music. There are many types of organ, but the most classic one for popular music is the Hammond B-3. Its sound is heard through a very large Leslie speaker cabinet, which helps identify its unique sound. There are still a few organists who are so dedicated that they will perform in live situations with this instrument (it is quite large and difficult to transport) but, for convenience, there are many smaller electric keyboards that, with the best of intentions, attempt to capture the sound of this wonderful instrument. Of course, for most keyboardists who are non-organists, this is also a welcome opportunity regarding the technique of performance, since the pedals and multiple keyboard decks are absent. In general, the organ timbre is very valuable as a thickener of the rhythm section sound. Unlike more conventional keyboards, whose notes decay uncontrollably, the organ can sustain a note indefinitely. This feature provides the sense of having a string section (with more colorful timbres) as it can create a lush pad of sound. This feature is also used with great effect, mainly at climax points, where the tension is greatest. The timbral array of organs is vast, so the reader is encouraged to explore them beyond the scope of what this book can offer.

2.4.4   Mallet Percussion

Vibes: Made of metal bars, this instrument offers a liquid sound ranging from dark and creamy to shimmery and bright. It is most commonly used in jazz and is also very effective for film music. The mallets can be softer or harder as they affect the mood and expression of the sound.

Marimba: Made of wooden bars, this instrument offers a stereotypically African, as well as Mexican, flavor, but is also quite effective in film music and some contemporary music productions. The sound is warm and rich in the lower to middle registers. When the player rolls softly on the bars using soft mallets, the effect is very mesmerizing as it offers a floating and ethereal quality.

Xylophone: also made of wooden bars, but much brighter and more brittle than the marimba, this instrument is used mostly in orchestral music.

Orchestra bells: Similar to a celeste, they offer a sparkle of sound. Not too distinct in terms of clarity of pitch, they are best used as a highlighter when doubled with other instruments such as woodwind or other keyboards. They are used mostly in music for film and theater.

2.4.5   Drum Set

There are many variations of sound within the basic drum set. The same drum set can sound quite different depending on the drummer who tunes it, according to personal taste.

Here are some general guidelines for adjusting the sound:

Bass drum: A larger shell will produce a deeper sound, but smaller shells can still sound deep as well. The standard sizes are 22, 20, and 18 inches. For a deeper sound with less "ring," loosen the heads and dampen them with felt. For more "thud," fill the inside of the shell with a pillow.

Snare drum: Perhaps the most diverse in sound, the snares (metal strands on the bottom head) can be tightened or loosened with a control knob. Tight snares will produce a more brittle and crisp sound. Loosen the snares and the drum will "breathe" and be more legato. Also be aware of the tension of both the top and bottom heads. The looseness and crispness will be affected accordingly. Another factor is the width of the drum and the material of the shell itself. A wider shell will offer a deeper and thicker sound and a wood shell, in contrast to a metal shell, will offer a warmer sound with less "crack."

Toms: There are many different-sized tom-toms that can be tuned as desired. The head will often determine the timbre. A thick head will produce more of a thud (in conjunction with a "loose" tuning), whereas a thin head with a tighter head tension will produce a more brittle and defined pitch. Muffling may also be applied to take out the "ring."

Cymbals: These metallic plates are like wine in their variety; what is best depends on taste regarding each musical situation. There are several broad categories to consider:

Ride cymbal: Used primarily for a lighter, more transparent beat. This cymbal should be thick enough to speak clearly when played with sticks, but be thin enough to breathe and to avoid a "clangy" sound. The overtones from the ring of the cymbal should not overbear the articulation of the stick sound. Types of ride cymbal range from the standard 20-inch size with a dome to flat-rides (fewer overtones) to "sizzle" ride cymbals, which use rivets to create the effect.

Crash cymbal: The larger the cymbal, the deeper the sound (the best range is 14–18 inches in diameter). These cymbals should not be thick.

Hi-hat cymbals: Played with the foot, they are used primarily to "clap" on beats 2 and 4 in jazz, but can be quite conversational in rock and funk by opening and closing while playing on them with the stick. The marking (+) indicates that the cymbals are in the closed position and (o) indicates the open position. The hi-hat pattern found in disco music from the 1970s demonstrates this technique most abundantly.

Striking material: Drumsticks in particular have a significant effect on the overall sound of the drum set. Wood-tip sticks are generally warmer than plastic tips and the thickness of the handle as well as the bead can also affect the sound. (If you don't like the sound of the drummer's cymbals, see whether a different stick might be better.) Brushes (traditionally made of wire and now also made with plastic) are also very effective; most commonly used in jazz and country music, they have become increasingly popular for light rock styles as well. Mallets (made of hard felt) offer a more exotic sound on the drums and are great for cymbal swells or a quasi-timpani effect on the toms.

2.4.6 Auxiliary Percussion

The percussion world is vast and the reader is advised to consult books that are devoted exclusively to this topic. Here are some of the most common instruments that are used to color music in general:

Conga drums: Used in Cuban music, there are usually at least two and sometimes three drums, offering different pitches. Played with the hands, they are more shallow in sound than "rock" toms, but deeper than bongos.

Bongo drums: Smaller than conga drums, there are only two: one high and one low. They are also played with the hands and offer a high-pitched sound that is crisp and exciting.

Timbales: Made of metal shells, they are somewhat similar to bongos but larger and are played with sticks. They are also louder than bongos.

Cowbell: A metallic sound that is muffled to achieve a more staccato sound. It is quite loud and its stereotypical, constant quarter-note pattern (in Latin music) serves as a guidepost for the more complex rhythms that embrace it.

Agogo bells: Similar to the cowbell, but much lighter and higher in pitch, the two bells offer high and low pitches.

Claves: Two rosewood sticks that offer a crisp, robust, and distinct pitch when struck together. The clave patterns (2–3 and 3–2) are an essential fundamental rhythm on which all Latin music is based.

Maracas/shakers: Essentially containing beads or some other similar substance, these instruments create a unique "ch" sound that can be crisp or legato. They are very effective in intensifying the subdivision within a groove and also very subtle in the overall ensemble as the sound is not overbearing.

Triangle: Although not found in Latin rhythm sections, this instrument is very effective for Brazilian music and for film music. One typical rhythmic pattern involves playing constant eighth notes while ringing and dampening (similar to the open and closed hi-hat) the metal with one hand while the other strikes using a small metal beater. Like the shaker, it is effective in adding intensity with subtlety (small triangles work best).

Tambourine: Also not found in Latin music, but used commonly in folk and rock in America and in Brazil (where it is referred to as a pandeiro). The metal disks inside the wood frame offer a distinct "jingle," which can be short or long when rolled. The head on one side of the wood enables the player to use it also as a drum. This instrument is not as subtle as the shaker or triangle and should not be used in a constant manner.

2.5   Writing Parts for the Rhythm Section

This task has always been challenging for the arranger. There is no particular way to write the parts and, depending on style, it is sometimes best to write a suggestion of what you want and let the player create the part based on your input. This can sometimes mean simply communicating to the player using English (e.g. swing feel in "2." Play "time" with ride cymbal for 24 bars). Other times you may want to indicate the melodic elements played by the horns so that the rhythm player knows what exactly to accompany. In doing so, he or she will create a part that will work well with the scribed parts. It is usually more necessary to write specific parts in a funk style, particularly when the rhythm section must interact with a large ensemble such as a big band. Ultimately, you should have a balance of communication using English and music to outline clearly what are your needs; and don't forget to use dynamics. If you can consolidate the part with a minimum of page turns, it will be easier for the performer as well.

2.5.1   Bass Parts

The contemporary bassist can read chord symbols (Cma7—Bmi7b5—E7—Ami) in addition to the traditional note-writing method. In most professional circumstances, the symbols are used while written-out notes are used only when a specific part is desired. In some instances, both are used simultaneously; this allows the arranger and the performer a choice of either method. If the arranger wants more control, the written part is there for the bassist; conversely, the bassist may create more of a sense of freedom, embellishing upon the arranger's original idea and ultimately enhancing the music. When writing a specific part, remember that the bass reads in bass clef. (On specific occasions, when the bassist is required to play in the extreme upper register, the treble clef may be used.) Also remember that the bass requires a register transposition; it is written one octave higher from where it sounds in order to avoid an excessive amount of leger lines.

2.5.2   Drum Set Parts

For many arrangers, drum set parts are the hardest to write. Quite often the part is not even written in drum notation. Professional arrangers know that, if every hand and foot stroke were dictated, most drummers could probably not sight-read the part perfectly and, if done so, it would probably not sound natural or as good as the drummer's instinctive playing. It is best to understand that you are providing a part that serves as a guide for the drummer. When provided with enough general information (regarding feel, style, and dynamics) and specific information (horn cues), the drummer will be able to create sound that will effectively interact with the band. When specific notation is desired, the conventional notehead represents the drums (bass, snare, toms) and an x notehead is used for cymbals (ride, crash, hi-hat).

2.5.3   Guitar Parts

As with bass parts, most guitar parts are written with chord symbols. However, do not discount the guitar as a melodic instrument. It is excellent for playing a melody or countermelody, especially when doubled with other instruments in the band. Sometimes the guitar may play a line with the bass but sound an octave higher. In order to write specific notes, the arranger must understand that the guitar has six open strings, tuned from the bottom string (E–A–D–G–B–E) to the top string. They sound at the piano starting in the second octave and ending in the fourth octave (Figure 2.1).

The open strings of the guitar

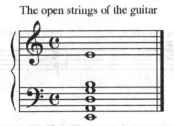

Figure 2.1 The "open" strings of the guitar

Since only the top string is actually within the treble clef of the grand staff, one would assume that the parts would be written in the bass clef. Unfortunately for the arranger, guitarists read their written music only in treble clef. As a result, the guitar part must be transposed (Figure 2.2).

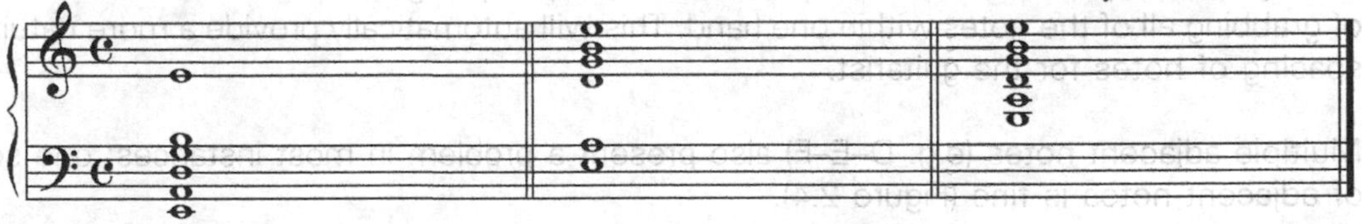

The open strings of the guitar as they sound at the piano.	The open strgs of the gtr as written using the register transposition	The open strgs of the gtr as written solely in treble clef

Figure 2.2 The "open" strings of the guitar as written

There is always a register transposition (written up an octave from where the notes sound on piano) to avoid excessive leger lines. Notice that the low E, when transposed as such, is now in the third octave of the piano. The next step is to convert (or transpose) this E, as well as the A string, to treble clef. You will notice that the E now sits below the third leger line below the treble staff. (Imagine how many leger lines you would need to use if there were not a register transposition!) Many guitarists have dealt with an arranger's negligence regarding the register transposition so they are used to accommodating the mistake, but it helps to be aware of the matter. In other cases, they must transpose when they are reading a master rhythm part (discussed at the end of this chapter).

The guitar part may also incorporate written notes in conjunction with chord symbols to provide musical options. Be careful when writing out a specific chord voicing. Remember that most chord voicings on piano do not simply transfer to the guitar since the physical properties of the instruments are so different. A wise approach is to write no more than four notes in a voicing and to keep them arranged in a more "open" position instead of a "closed" position. Simply take the second note from the top of a "closed" voicing and drop it down an octave; voicing becomes "open" and consequently more conducive to the setup of the guitar (Figure 2.3).

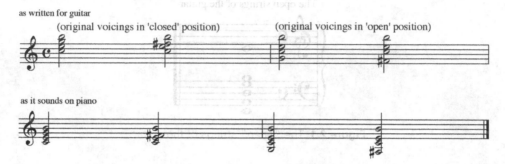

Figure 2.3 Voicings transposed for the guitar versus piano

The reader should be aware that this technique is a general guideline. When the voicing is nearer to the bass register, lowering a note down an octave would create a muddy sound. Simply do the opposite in this case: take the second note from the top and raise it an octave. Another general guideline to follow, from the standpoint of playing guitar voicings at the piano, is to use only the index and pinky fingers from each hand instead of grabbing all of the notes within one hand. This will automatically provide a more natural spacing of notes for the guitarist.

Multiple adjacent notes (e.g. D–E–F) also present a problem in most instances; one set of adjacent notes is fine (Figure 2.4).

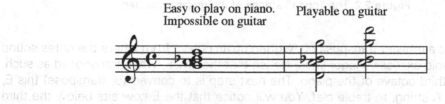

Figure 2.4 Creating open voicings for the guitar

It should also be noted that sharp keys, as with all string instruments, are more favorable technically than flat keys because of the tuning of the open strings. Of course, there are many great jazz guitarists who can play incredibly complex harmony in keys such as A♭ and D♭ major, but it will be of significant advantage to try to use "friendlier" keys where possible.

2.5.4 Keyboard Parts

Keyboard parts are usually written using the grand staff. Sometimes, to consolidate the part, one stave may be used (usually if the part contains chord symbols only). The presence of chord symbols with written material can aid the performer in deciphering the mass of notes that can occur on a given part. In large ensemble settings, it is most helpful for the pianist, when supplied with chord symbols, to have the lead horn or string lines cued in the part. This enables the pianist to accompany with a clearer context.

2.5.5 Mallet Parts

Vibes, marimba, xylophone, and orchestra bells fall into this category. (Within the contemporary-style rhythm section, only vibes and possibly marimba would be found.) The performer would use two mallets for fast melodies and four for chord voicings. Similar to guitar voicings, wider spaces between the notes conform slightly better to the manner in which the mallets are held. Clusters are possible, but they are not as natural to the confines of the hand.

2.6 Sample Scores for the Rhythm Section

The following music examples should provide some context regarding the rhythm section's function, orchestration, and part-writing. The scores are available at the end of the book, in Appendix A.

Score 2.1 is in a funk style. The score also includes a concert sketch of the horn parts for reference. The rhythm parts are quite specifically notated with the guitar and bass parts transposed as the player should see them (written an octave higher than the actual sound). Many times, experienced rhythm players create a part that may ultimately sound like this written material, but the notation here will help to illustrate the complexity of the rhythm section and how to create parts within the MIDI domain.

Analysis: The bass and drum parts are based on a two-bar phrase. The two guitar parts are written in one-bar phrases. In guitar 1, the written pitch F indicates the top note of the voicing. The chord symbol allows the player to view a simpler part with fewer pitches to read. The rhythm is provided to control the activity so that the part does not become too dense. This concept is maintained throughout the rest of the part. In guitar 2, contrast is provided through a muting technique. The guitarist "picks" the notes while the base of the hand rests against the string, resulting in the muted effect. The single note suggests, more than using a full chord, a percussive droning. The piano part offers melodic commentary, playing with the horns and filling in between them.

Score 2.2 is in a contemporary European-jazz style (not fusion). The bass and drums do not adhere to any specific or constant idea. Instead, it is encouraged that a freer approach be taken to allow for more spontaneous interaction. Overall, the feeling is legato but with an abundant sense of "dancing" syncopation within. The parts are written to provide enough context for the performer, after which they evolve into more of a guideline; this helps to consolidate the part and to encourage the player to embellish upon the original ideas.

Analysis: In the first eight bars, the piano plays a melodic ostinato in the fifth and sixth octaves with the sustain pedal left down so that the notes ring into each other indefinitely. The guitar provides some occasional melodic commentary and the echo effect carries it into infinity. The bass provides bottom, completing the wide spectrum of sound, but moves very slowly. The drummer uses only cymbals to achieve a transparent, "washy" effect. The drums themselves are inappropriate for the mood at this point. In bars nine to 16, the rhythm section now supports the main theme and provides a more fundamental sound. The piano part now moves more slowly, but offers a very important ingredient to the composition: a harmonic ostinato pattern that works in counterpoint to the melody. The voicings here must be written out to capture the essence of the composition's flavor. The bass notes, though not as essential, are written out to present a clear idea of the interaction and general activity level. The part becomes busier than the first eight bars, dancing around the less active harmonic ostinato pattern, but still not outlining any defined beat pattern. The guitar is still used primarily for ambient color. The written notes must be respected, but the player can interpret regarding the timbre (the chorus and echo effects). The drum part is written to provide a sense of activity level as well as orchestration (to control the player's use of the equipment, in this case to avoid becoming overindulgent prematurely). The written notes in the first two bars offer a clear suggestion and the repeat signs, rather than dictating to the drummer to repeat exactly, are merely indicating that embellishment is expected within the established context. As the piece unfolds and improvisation occurs, the parts become more general (see Score 2.3).

Score 2.4 (Example 2.64) illustrates an example of a master rhythm part. It may take more careful preparation by the arranger, but it can be very helpful to the rhythm section, provided that the players are of professional quality and are adept at reading a multiple-staff part as well as transposing the register where necessary. In most cases, the bass and guitarist will need to transpose an octave higher when reading a master rhythm part to establish a proper unison with other instruments. Such is the case here.

This music offers three distinct grooves (a quasi-Brazilian funk-samba style switching into a jazz "swing" feel and then into more of a Cuban-style Latin pattern). In the Brazilian section, the piano and bass parts are written exactly; the bass clef should be read by the bassist and also played by the pianist's left hand, thickening the ostinato pattern.

The chord symbols above the piano voicings help the pianist to understand better the harmonic intention. There is no specific drum part other than stylistic direction and some small requests (the most important one being the placement of the bass drum on the "back half" of the bar), and it is expected that the drummer will embellish within the context. The "swing" section is written very loosely as the players need to have the freedom that is appropriate to the style. Here, the players are instinctively more aware of their responsibilities and begin to incorporate more improvisation within their accompaniment roles. The first and second endings delineate rhythmic figures (more commonly referred to as "hits") using "slash notation." The drummer provides the "glue" that connects the figures, creating some fills in conjunction with his or her general beat pattern (referred to as "time playing"). The voicings here can be more generic so chord symbols are used, but in conjunction with a specific "top note" that helps make the chord progression more melodic. The hits themselves deconstruct the full swing energy, creating tension and, ultimately, a transition into the Cuban-style section. As the style demands, the parts once again become more strict. Here, the guitar and the piano are unified rhythmically, but the octaves in the guitar provide a nice contrast to the full harmony in the piano. (If, for some reason, the arranger wanted a thicker sound, the guitarist would now have the piano part available to play the full voicing.)

2.7 Sequencing the Rhythm Section: an Introduction

In the previous paragraphs you learned how the generic term "rhythm section" can cover a wide range of sonorities and instruments. This holds particularly true for the modern rhythm section, where synthesizer and effected guitars and basses can extensively expand the color palette of your productions. The second part of this chapter will discuss the techniques and tools that are available in your project studio in order to get the most out of your MIDI gear when sequencing for the instruments of the modern rhythm section. As for any other section of your virtual orchestra, the choice of sonorities and libraries is crucial when it comes to the rhythm section. In addition, the choice of the right MIDI controller to perform and program the parts plays a very important role. To each instrument you can then apply a series of specific sequencing techniques that can help you to bring your productions to a higher level. These important topics are followed by a step-by-step guide through each aspect of sequencing for the rhythm section. The mix stage will be discussed on an instrument-by-instrument basis first and as a rhythm section as a whole at the end of the chapter. If your MIDI gears are up and running, let's move on!

2.7.1 The Keyboard

The keyboard is, in most cases, the main MIDI controller used to input parts into a sequencer. In this category we also find a series of important instruments that form the harmonic backbone of the rhythm section, such as acoustic piano, electric piano, organ, and synthesizer. Before getting into the specifics of how to deal with each of these individual instruments in a MIDI environment, let's take a look at the most common options that are available in terms of MIDI keyboard controllers.

Every MIDI-based studio needs some type of controller in order to be able to generate and send out MIDI messages to other devices. You have already learned that both the keyboard controller and MIDI synthesizer are MIDI devices capable of transmitting MIDI messages. A good controller is indispensable for your studio if you are serious about MIDI. Keep in mind though that keyboards are not the only MIDI controllers available. If you are not a keyboard player there are other options available. In addition, remember that one does not exclude the others. Keyboard controllers can vary depending on their features and available options. The action of the key, for example, can be either weighted, semi-weighted, or synth-like. Weighted action features a response similar to the one of an acoustic piano. It is ideal for sequencing classical and jazz piano parts, but can be a bit of a problem when trying to sequence fast synth parts. In addition, some weighted action controllers do not have pitch-bend and modulation wheels or any other assignable faders and knobs. A semi-weighted controller (lighter than a fully weighted action but heavier than a synth keyboard) provides more flexibility for a wider range of sequencing styles. If you absolutely cannot renounce the real feel of a piano keyboard and the ability of controlling your MIDI studio at the same time, the ultimate solution is to use, as your main MIDI controller, an acoustic piano able to send MIDI data. While there are different brands and models on the market, one of the most successful combinations of acoustic piano and MIDI controller is the Disklavier by Yamaha.

Make sure that the keyboard controller of your choice has pitch-bend, a modulation wheel and at least one (the more the better) slider assignable to any MIDI control change (CC). This feature is extremely important since it allows you to achieve a higher control in terms of the expression of certain instruments such as strings, woodwind, and brass. It can also help in creating an automated mix without using the mouse. Have at least one keyboard controller in your studio capable of transmitting aftertouch messages. This feature will give you much more control over the expressivity of a performance. While devices that respond to polyphonic aftertouch are harder to find, the majority of them are monophonic aftertouch-ready. The size of the controller (in terms of key range) also plays a very important role when sequencing for advanced projects. In general, the range goes from a small two-and-a-half-octave (30 keys) extra-light and portable controller, through a more grown-up version that features four or five octaves (49 or 61 keys), to a full 88 keys (seven and a third octaves). For a serious project studio, it is highly recommended that you use a keyboard with 88 keys, the main reason being that when you sequence for strings, woodwind, and brass you might use a library with key-switching technology. This option allows you to change the sonority assigned to a MIDI channel through MIDI Note On messages assigned to keys that are out of the range of the instrument you are sequencing (usually placed in the very low or very high range). For this reason, you need a controller with a wide key range in order to be able to access the key switches. In a complete project studio you would want to use a multicontroller environment where you can sequence parts using one main keyboard controller along with one or more keyboards or alternate controllers, as shown in Figure 2.5.

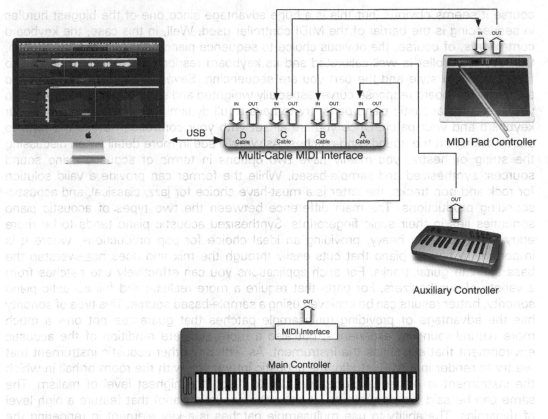

Figure 2.5 A real example of a MIDI project studio built around a main 76-key keyboard controller to sequence orchestral parts, a small 25-key keyboard controller for key switches and CCs and a MIDI pad controller used to sequence drums and percussion parts (Courtesy of Apple Inc., Roland Corporation U.S., and M-Audio)

2.7.2 The Acoustic Piano

With your main keyboard controller you will be sequencing mostly keyboard parts and, in most cases, at least at the beginning, mainly piano parts. "Piano" refers to both acoustic and electric piano. The term "synthesizer parts" refers principally to sonorities and parts that do not fall into any acoustic category. The synthesizer (including some of the most used synthesis techniques) will be discussed later in this chapter. For now, the discussion will focus on the sequencing techniques used in inputting both acoustic and electric piano parts. While acoustic and electric pianos are very similar in terms of the controller techniques involved in the sequencing process, there are a few differences worth noticing when it comes to the advanced use of CCs and mixing.

Let's start by analyzing the acoustic piano. This is one of the easiest parts to sequence since we have the advantage of using a keyboard MIDI controller to input the notes. Of course it seems obvious, but this is a huge advantage since one of the biggest hurdles in sequencing is the barrier of the MIDI controller used. Well, in this case, the keyboard controller is, of course, the obvious choice to sequence piano parts. Make sure, though, that your controller is well calibrated and its keyboard response curve well balanced to fit your playing style and the part you are sequencing. Several controllers allow you to adjust the keyboard response curve, especially weighted and semi-weighted ones. When you sequence acoustic piano parts, try to have a full dynamic response from both your keyboard and your patch. Once you are all set with your controller, now is the time to start working on the sounds and patches. As we will see in more detail when discussing the string orchestra, you mainly have two options in terms of acoustic piano sound sources: synthesized and sample-based. While the former can provide a valid solution for rock and pop tracks, the latter is a must-have choice for jazz, classical, and acoustic-sounding productions. The main difference between the two types of acoustic piano sonorities lies in their sonic fingerprints. Synthesized acoustic piano tends to be more edgy and mid-range heavy, providing an ideal choice for pop productions, where it is important to have a piano that cuts easily through the mix and does not overstep the bass and the guitar tracks. For such applications you can effectively use patches from a variety of synthesizers. For parts that require a more realistic and full acoustic piano sonority, better results can be achieved using a sample-based source. This type of sonority has the advantage of providing multisample patches that guarantee not only a much more natural-sounding experience, but also a more accurate rendition of the acoustic environment that surrounds the instrument. As with any other acoustic instrument that we try to render in a MIDI studio, the acoustic interaction with the room or hall in which the instrument is placed is crucial in order to reach the highest level of realism. The same can be said for instruments (such as acoustic piano) that feature a high level of dynamics. The ability to use multisample patches is a key element in rendering the complex interaction between all the different dynamic levels of the instrument. The professional options available in terms of libraries for acoustic piano are endless and, as is the case for the majority of sample-based libraries, constantly changing. When choosing an acoustic piano library, make sure that it is versatile enough to cover different styles, ranging from pop and rock (edgy and clear with a punchy sonority) to classical and jazz (mellow and warm). Extremely popular and flexible are dedicated acoustic piano plug-ins based on a mix of sampling and physical modeling (PM) synthesis that allows you to achieve astonishing results. These plug-ins (Figure 2.6) provide not only multisample sonorities based on studio recordings of popular pianos such as Steinway, Yamaha, and Bösendorfer, but also give you control over any possible parameter, such as keyboard response curve, dynamic range, timbre, and soft pedal. These plug-ins are fairly CPU-intensive, so if you plan to use one make sure that you have a powerful computer.

The choice of the acoustic piano sound that you will use in your sequence is extremely important. As a general rule, use a smoother, mellower, and warmer sound for classical and jazz parts, and an edgier and brighter sonority for pop and rock tracks. If the piano is particularly exposed (no matter which style you are sequencing), a mellower sound will in most cases provide a more realistic rendition. Once you have found the sonority that best matches the part, it is time to start sequencing.

Figure 2.6 The Ivory II piano module by Synthogy: a virtual acoustic piano plug-in with a variety of options and controllers that allow you quickly to "create" the piano of your dreams
(Courtesy of Synthogy)

2.7.3 Sequencing and Mixing Techniques for the Acoustic Piano

As mentioned earlier, using a keyboard controller to sequence piano parts gives you a great advantage. Make sure that you have a sustained pedal connected. Try not to quantize the piano parts that you sequence unless absolutely necessary. If you do quantize them, avoid drastic settings such as 100 percent strength. As we will see in more detail when discussing drums and percussion, 100 percent strength will translate into a very stiff and mechanical performance. Use less-extreme quantization settings, such as 80 percent strength, instead. This will provide a smoother and more natural feel and will improve the overall realism of the part. As in any other sequencing situation, avoid as much as possible copying and pasting repetitive passages. Play the parts all the way through instead. Variation is extremely important in any sequencing situation, so artificially repeated phrases or passages should be avoided.

While the use of the sustain pedal (CC#64) is crucial for a convincing MIDI piano part, there are a few other controllers in the MIDI specification that can help you achieve the ultimate acoustic piano rendition. Two of the most useful are CC#66 (sostenuto) and CC#67 (soft pedal). The sostenuto pedal works in a similar way to the sustain controller. The main difference is that sostenuto, when pressed (values between 64 and 127), holds only the notes that are already turned on when the pedal was pressed. This function is useful to give the impression of the left hand (chords) sustaining longer than the right hand (melody or solo). The soft pedal option works pretty much the same way as in an acoustic piano. This controller has only two positions: on (values between 64 and 127) and off (values between 0 and 63). When in the on position, it lowers the volume of the notes that are played. This controller can be used to increase the dynamic range of a part or passage, creating a smooth and natural effect. While most synthesizers will respond to sostenuto, not all synthesizers and samplers respond to soft pedal as a default. In general, sample-based patches are more flexible in terms of MIDI controller programmability, giving you the option to assign any controller to any parameter of a patch. The drawback of this approach is that some standard MIDI CCs need to be assigned manually or, in some rare cases, won't work at all.

At the mix stage there are three main aspects to consider when dealing with the acoustic piano: panning, equalization, and reverb/ambience. When working on the panning of the acoustic piano, you have two main options: performer point of view or audience point of view. If you use the former, the low register of the piano will be panned to the left of the stereo image and the high register more toward the right. This panning is usually clearer and slightly easier to mix with other instruments. If you choose the latter, then your options are wider. You can reverse the stereo image completely (low register to the right and high register to the left), but you can also reduce the width of the stereo image and use a more conservative approach where the low and high registers are not so widely spread. Try to avoid extremely wide and extremely narrow panning settings unless required by a particular style of music, for example solo piano pieces. In general, a stereo spread between 75 and 85 percent is a good starting point for several genres and styles. Use a light equalization to correct problematic frequencies embedded in the samples or annoying frequency interactions (such as masking) occurring with the other instruments of the mix. Remember that the acoustic piano covers the entire audible range and therefore it often interferes with the other instruments of the arrangement. Table 2.1 lists some of the key frequencies that can be used to improve the sound of an acoustic piano.

Table 2.1 Key frequency ranges of the piano for equalization

Frequencies	Application	Comments
100–200Hz	Boost to add fullness	This frequency range effectively controls the powerful low end of a mix
1.4–1.5kHz	Boost for intelligibility	
5–6kHz	Boost to increase attack	A general mid-range frequency area to add presence and attack
10–11kHz	Boost to add sharpness	High-range section that affects clarity and sharpness
	Cut to darken piano overall sonority	
14–15kHz	Cut to reduce sharpness	

In terms of reverberation, a convolution reverb is preferable to a synthesized one. Convolution reverbs have the advantage of producing acoustic responses that are sampled from real environments, making them the perfect choice for any sample-based acoustic sonorities such as piano or strings. The amount and length of the reverberation largely depend on the style and instrumentation of the production. For classical composition, you can use a longer reverberation time between 2.3 and 3.2 seconds. For jazz productions, you may use a more intimate and smaller reverb with a length between 2 and 2.5 seconds. For punchier and more direct sonorities, such as those required for pop and rock productions, you can shorten the length even further (between 1.7 and 2 seconds). These are generic settings that should be used as a starting point. Feel free to experiment and make sure that the overall mix is always balanced and as clear as possible.

2.7.4   *The Alternate Keyboards: Electric Pianos and Organs*

The majority of the techniques discussed for the acoustic piano apply also to the electric piano. The same CCs can be used to achieve a higher degree of control over the performance. One of the main advantages of sequencing for the electric piano over its acoustic version is sonic versatility. The electric piano can adapt to a variety of situations and styles. In addition, its sounds and colors are almost endless. Synthesizers have done a great job in emulating a variety of electric piano timbres and colors, from the Wurlitzer to the Fender Rhodes and the Hohner Pianet. Physical modeling synthesizers can effectively generate convincing and smooth electric piano patches. Software plug-ins, such as the EVP88 found in Logic Pro X, FM8 from Native Instruments, and Lounge Lizard EP-4 by Applid Acoustics System, combine sample-based and PM techniques to generate extremely versatile and expressive timbres. When sequencing the electric piano, keep in mind that this instrument (in its most basic form and in its variations) has a more restricted dynamic range and therefore that the keyboard response of the MIDI controller can be adjusted to re-create the real response of the original models. A gentle exponential curve may be used to re-create a more realistic velocity response. Electric piano sonorities can cover a wide range of timbres, from fat and heavy to light and edgy. Their sonic features can easily be adjusted through either layering techniques or equalization. If you feel that the electric piano you want to use in your production lacks some low frequencies, you can layer a second electric piano with such sonic features, or you can brighten up a dark and muddy electric piano patch with the addition of a thin one. Electric pianos have the advantage of being extremely flexible and easily layered with any other timbres. When it comes to pan and equalization, the same principles apply as for the acoustic piano. If you want to reach a more realistic sound, use a more conservative panning setting with a spread between 60 and 70 percent. Use the equalizer settings listed in Table 2.1 to correct any frequency imbalance due to masking or frequencies overlapping with the other instruments of the arrangements. In terms of reverberation, use a shorter reverb than the one used for the acoustic piano. Start with a conservative 1.8 seconds and move up if necessary. Do not add too much reverb to the electric piano, especially if the patch also has a chorus effect. Too much reverb would decrease the definition and attack of the hammers. Keep the electric piano as dry as possible to obtain a punchier effect.

Another extremely popular keyboard-based instrument of the contemporary rhythm section is the organ, and in particular the Hammond B-3 and C-3. The sonorities of this instrument are extremely flexible and adaptable to a variety of musical styles and situations. Both wavetable and sample-based/PM sound-generating techniques can create astonishing clones of the original instruments. In particular, you should have in your studio at least one software plug-in, such as the Logic Pro's EVB3 or the Native Instrument's B4 II. These two software organs' emulators are particularly flexible and realistic. They not only emulate the original tone-wheel generators, but also virtually re-create the sonic imperfections that gave this instrument its legendary sonic signature. When sequencing for the Hammond organ, keep in mind that, because of the original nature of the instrument, the keyboard is not dynamic. This means that in a synth patch (if well programmed) the velocity information of the MIDI message is not used to control the loudness of each note. The dynamic of the performance is usually controlled through CC#11 (expression). Ideally, you would have a controller that has an expression pedal input in the back (it looks like a regular quarter-inch jack). The use of CC#11 can help you add a realistic touch to the Hammond's parts since most B-3 synthesizers and emulators imitate the tonal changes that occur at different dynamic levels. If you don't have a keyboard controller that allows you to connect such a pedal, you can use one of the assignable sliders to send CC#11. To take full advantage of the flexibility offered by this instrument, you should work with a multicontroller environment. The original B-3 had two keyboard layers (called manuals) plus the pedal board. Each of the three layers features independent settings in terms of colors and timbres. You can use two keyboard controllers by assigning them to the two main manuals, each with independent drawbar settings. The same can be said for the pedal board. You can even buy a pedal board MIDI controller for the ultimate B-3 emulation. Most B-3 patches can also be played from a single keyboard controller with a minimum of a 76-key range. The ideal controller, though, would have 88 keys, since the keyboard can be split into two manuals and the pedal section, as shown in Figure 2.7.

As shown in Figure 2.7, the lower section of the controller is dedicated to the drawbar preset switches. Using these lower notes as triggers you can change drawbar presets while playing. Although this split setup can work for the majority of the sequencing

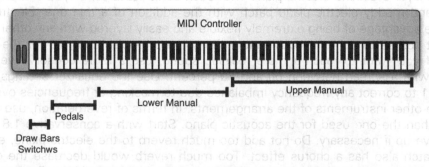

Figure 2.7 The two manuals plus the bass pedal section controlled using a split keyboard. Notice how the lower register is dedicated to control the presets for the drawbars

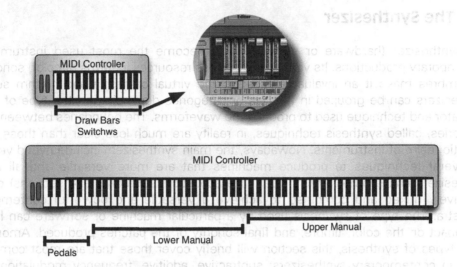

Figure 2.8 A B-3 two-controller setup. The second smaller controller is used to trigger the drawbars presets (Courtesy of Apple Inc.)

situations, it will limit the range of each manuals (in a B-3 organ, each of the two manuals has a five-octave range). You should use at least a separate small controller for the drawbar switches, as shown in Figure 2.8.

In most B-3 emulators you can also assign CCs to the single drawbars. This allows you to change and morph seamlessly between presets and timbres. Use a separate external controller to program the CCs that will activate the drawbars.

At the mix stage you can pan the organ in the center of the stereo image or mildly left or right. Because of the nature of the original instrument, the stereo spread is usually not very wide. You can play, though, with different pan settings for the two manuals and the pedal section. In this case, pan the two manuals left and right and keep the bass pedal board centered. When working with B-3 sounds, it is preferable to avoid the use of equalizers and instead use the drawbars to control the sonic features of the patch. This will give you a much more realistic effect. Each drawbar controls the amplitude of the partials that, added together, form the final sonority. The drawbars to the left control the lower partial (lower frequencies). The drawbars on the left side control lower partials (lower frequencies), while the ones on the right side control higher-frequency partials. Use the drawbars to shape the frequency range of the sound. You will be amazed how effective this technique can be and how natural your patch will sound. Try to keep the organ patch on the dry side when it comes to reverberation. A wet sonority will give the impression of a large stadium or church, taking all the punch and definition off your parts. Use a reverberation time between 1.5 and 1.7 seconds as a starting point. The addition of some chorus, saturation, vibrato, and Leslie effect can increase the realism of an average patch. Remember, especially, that chorus and Leslie can reduce the definition and edge of the overall sound, so try to be conservative when using such effects.

2.8 The Synthesizer

The synthesizer (hardware or software) has become the most used instrument in contemporary productions. Its variety and endless resources in terms of color, sonorities, and timbres make it an invaluable asset to the virtual orchestra and rhythm section. Synthesizers can be grouped in families or categories depending on the type of sound generator and technique used to produce the waveforms. The boundaries between these categories, called synthesis techniques, in reality are much less clear than those in the acoustic realm of instruments. Nowadays, the main synthesizers include hybrid versions of several techniques to produce machines that are more versatile and all round. Synthesizers and synthesis techniques (like acoustic instruments and sections) can be effectively mixed and combined to achieve a versatile and powerful contemporary orchestra. The type of synthesis used by a particular machine or software can have a big impact on the color, timbre, and final sonority of the patches produced. Among the many types of synthesis, this section will briefly cover those that are most commonly found in contemporary synthesizers: subtractive, additive, frequency modulation (FM), wavetable, sampling, PM, and granular. The goal of this section is not to provide a complete guide to synthesis and its many features, subjects that would take up a manual of their own, but instead to render an overall picture of the aforementioned types of synthesis available on the market for the modern composer and producer in order for you to be able to make the right decision when selecting sounds and devices for your projects. For more specific and detailed information on synthesis, read *Sound Synthesis and Sampling* by Martin Russ and *Computer Sound Designing: Synthesis Techniques and Programming* by Eduardo Miranda, both published by Focal Press, UK.

2.8.1 Hardware and Software Synthesizers

Since the 1950s, and up to six or seven years ago, the word "synthesizer" meant for most of the composers and producers some sort of hardware component (sound module, keyboard synthesizer, drum machine, etc.) that was able to generate waveforms resembling original acoustic sonorities or create completely new waveforms such as pads and leads. As hardware synthesizers become more and more demanding in terms of hardware power, and increasingly complicated to program, a new breed of synthesizers started to become more popular: the software synthesizer. This new tool, based on the combination of a computer and a software application, takes advantage of two main factors: first, that in recent years computers have become more and more powerful, and, second, that hardware synthesizers have become mainly a combination of internal processor (CPU) and software written for that particular type of processor. The main difference between hardware and software synthesizers is how the sound is generated. While the former utilize a dedicated CPU to produce the sounds, the latter take advantage of the CPU of a computer and specially written software to create and process the waveforms. The advantages of software synthesizers are many. After the initial investment in the computer, the price of a software synthesizer is much lower than its hardware counterpart. In addition, the continuous evolution of algorithms, raw waveforms, and patches can be easily integrated with new revisions of the software. However, the upgrade procedure of hardware synthesizers is much more problematic and limited. Software synthesizers also have the advantage of a seamless integration with sequencers and other music software. Through the use of plug-ins and standalone versions, it is nowadays possible to create, sequence, record, mix, and master an entire project completely inside a computer, without the audio signal leaving the machine. This provides a much cleaner overall signal. Of course, one major drawback of this type of synthesis is that it requires very powerful computers in order to be able to run several plug-ins at the same time on top of the MIDI and audio sequencer. Software synthesizers today are the standard for every contemporary production in both studio and live situations. The availability of portable computers that are almost as powerful as their desktop counterparts contributed greatly to the spread of software synthesizers. An example of the increasing interest in the development of software synthesizer is the fact that even major hardware manufacturers, such as Korg, Roland, and Yamaha, are developing new synthesis engines based on software-only algorithms. More and more companies are bringing back vintage hardware synthesizers as software versions, making available the warm sonorities of the old analog devices with the power and the possibilities of the new software-based technology. The types of synthesis described in the following pages are not platform-dependent, meaning that the actual way the waveform is produced does not basically change from hardware to software. The software approach opens up possibilities that are usually not conceivable at affordable prices on hardware platforms.

A software synthesizer is usually available in two formats: as a plug-in or as standalone. The former requires a host application in order to run. Without this application the synthesizer is not able to launch. When buying a software synthesizer, make sure that your platform (sequencer and operating system) supports that particular format. Among the most widely used formats there are VST (Virtual Studio Technology), AU (Audio Units) and AAX (Avid Audio eXtension) for Apple's iOS, and Dxi, VSTi, and AAX for Windows platforms. Through the use of a "wrapper," certain formats that are not available natively for a certain application can be used. A wrapper is a small application (a sort of software adapter) that is able to translate one plug-in format into another in real time.

An alternative to the plug-in format is the standalone application. A standalone software synthesizer is a separate independent application that does not need a host in order to run. This option has the advantage of being slightly less demanding in terms of CPU power since it doesn't require an entire sequencer to run. One of its drawbacks is a lower integration in terms of routing and interaction with other components of your virtual studio. Most software synthesizers are available as both plug-in and standalone versions. The choice between formats really depends on how the software synthesizer will be used. For live settings, a standalone version may be preferable, since these usually are more stable. For a MIDI/audio studio situation, where the sequencer becomes the central hub of both MIDI and audio signal, a plug-in system may be better, as it is more flexible in terms of signal routing.

2.8.2 Synthesis Techniques

The modern synthesizers (both hardware and software) available on the market use one or more synthesis techniques that have been developed during approximately the last 40 years. Each approach to synthesis has a peculiar way of synthesizing original and replica waveforms. Each synthesis technique features a unique sonority and sonic print. The choice regarding which machine, module, or plug-in to buy or to use in a particular project depends mainly on the type of sonority and patches you need. For example, you would hardly choose an analog synthesizer to program a realistic string ensemble, as you wouldn't probably choose a sampler to program a multistage complex synthesized pad. So that you can make an educated decision when choosing a synthesized patch or sound, the most common types of synthesis available to the modern MIDI arranger are described below.

2.8.3 Analog Subtractive Synthesis

A synthesizer is a device capable of generating electrical soundwaves through one or more voltage-controlled oscillators (VCOs). A VCO is an electronic sound source (analog or digital) capable of generating simple or complex waveforms depending on the level of sophistication of the oscillators used. The analog subtractive synthesis constitutes one of the oldest types of synthesis that were made commercially available and marketed in the 1960s. The "sound source" or "generator" section of a subtractive synthesizer is based on one or more oscillators generating a basic and repetitive geometrically shaped waveforms such as sine waves, triangular waves, square waves, sawtooth waves, pulse waves, and noise. The other two main areas of a synthesizer based on the subtractive approach are the "control" section and the "modifiers" section. The former comprehends the keyboards, pitch bend, modulation wheel, foot pedals, etc., while the latter includes the filter section that is used to alter the basic waveform generated by the oscillators. The name "subtractive" comes from the process involved in creating more complex waveforms starting from a basic cyclic soundwave. After being generated, the simple waveform is sent to the modifiers section, where a series of high-pass and low-pass filters (the number and complexity of which vary depending on the sophistication of the synthesizer) "subtract" (or, better, remove) harmonics, producing, as a result, more interesting and complex wave shapes. The altered sound is then sent to the amplifier section called VCA (voltage-controlled amplifier), where the amplitude of the waveform is amplified.

The VCA can be altered through the use of an envelope generator (EG). This is a multistage controller that allows the synthesizer to control over time the amplitude of a waveform. The most basic EG has four stages: attack, decay, sustain, and release. In more modern synthesizers, the EG can be divided into more stages, allowing higher flexibility. Some modern synthesizers, for example, feature a hold stage inserted between the attack and the sustain. Others feature a second stage for each main section of the EG, such as attack 1 and 2, or sustain 1 and 2. In addition to controlling the VCA, a second EG can be assigned to control how the filters change over time, enabling the synthesizer to produce even more complex waveforms. In order to be able to introduce some variations in the repetitive cycle of subtractive synthesizer, one or more auxiliary oscillators are introduced. These low-frequency oscillators (LFOs) have a rate that is much lower than the one used to generate the waveforms. They can be assigned to control several parameters of the synthesizers, such as the filters section or the pitch of the main oscillators. Usually the LFO can be altered by changing its rate (the speed at which the LFO changes over time) and its attack time. The signal flow and the interaction among the sections and components of a subtractive synthesizer can be seen in Figure 2.9.

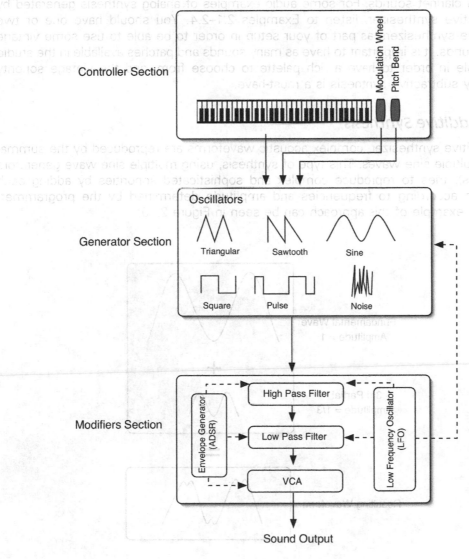

Figure 2.9 Components and sections of a subtractive synthesizer

When you think of subtractive synthesis, the first machines that may come to mind are the legendary Minimoog, the Prophet 5 by Sequential Circuits, and the Juno series by Roland. Subtractive synthesis is particularly suited for low and deep analog basses and edgy and punchy analog leads. These are among the sounds and patches that made subtractive synthesis famous and that are still largely used in contemporary and dance productions. Another area in which subtractive synthesis is capable of producing original and interesting sonorities is rich and thick pads. Subtractive synthesis, because of its limited basic waveforms, is usually not suited to re-creating acoustic instruments such as strings and woodwind. It can somehow be effective, however, in producing synthesized brass and clarinet sounds. For some audio examples of analog synthesis generated by a subtractive synthesizer, listen to Examples 2.1–2.4. You should have one or two subtractive synthesizers as part of your setup in order to be able to use some vintage analog sounds. It is important to have as many sounds and patches available in the studio as possible in order to have a rich palette to choose from, and the vintage sonority offered by subtractive synthesis is a must-have.

2.8.4 Additive Synthesis

In an additive synthesizer, complex acoustic waveforms are reproduced by the summation of multiple sine waves. This type of synthesis, using multiple sine wave generators (oscillators), tries to reproduce complex and sophisticated sonorities by adding each waveform according to frequencies and amplitude determined by the programmer. A graphic example of this approach can be seen in Figure 2.10.

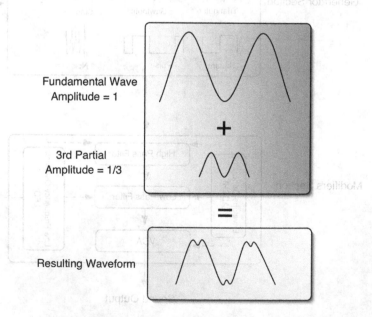

Figure 2.10 Example of basic additive synthesis

In this example, a simple sine wave (the fundamental) of amplitude 1 and frequency 1 is added to a sine wave (third harmonic) of amplitude 1/3 and frequency 3 (the numbers are used simply to illustrate the examples and do not have any reference to real frequencies and amplitudes). The result is a more complex waveform than the two original ones compiled from the addition of sine wave components. One of the main drawbacks of the additive approach is that a large series of oscillators is needed to produce complex waveforms and therefore synthesizers based on the additive approach require a very powerful sound engine. Legendary commercially available synthesizers based on additive synthesis include the Synclavier, produced by New England Digital in the mid-1970s, and the Kawai K5. Also, in the case of additive synthesis there are three main sections, as we have seen in the case of subtractive synthesizers: controller, generator, and modifier.

Additive synthesizers are usually able to produce convincing thin sounds mainly related to vibrating strings such as guitars and harps. They are not particularly effective in reproducing percussive sonorities with fast transients such as drum sounds. In recent years, a new series of software synthesizer based on an evolution of additive synthesizers has emerged. These applications use the concept of resampling to resynthesize waveforms that are fed to the software as audio files. In other words, a resampler analyzes a complex waveform provided as an audio file and then tries to synthesize it by reconstructing its complex wave through the summation of sine wave components. Examples 2.5–2.8 present some patches generated through additive synthesis.

2.8.5 *Frequency Modulation Synthesis*

Another answer to the need for synthesizing more complex and sophisticated waveforms came from yet another innovative approach to synthesis, called frequency modulation (FM). This technique involves a multioscillator system. In FM, each oscillator influences and changes the output of the others. In its most simple setup, FM synthesis uses two oscillators: one is called the modulator and the other the carrier (advanced FM synthesizers use three or more oscillators to generate more complex waveforms). The modulator changes and alters the frequency of the carrier by constantly modulating the basic frequency at which the carrier operates. In a multioscillator setup there are usually several matrices available that control the flow of the signal path among modulators and carriers according to predetermined algorithms. One of the most commercially successful FM synthesizers ever made was the Yamaha DX7, which was released in the early 1980s and has undergone many successful revisions and improvements over the years. The sonorities that FM synthesis can produce are usually "glassy," bell-like tones with good fast transients and a string-like quality. FM synthesizers are considered best at generating patches such as electric pianos, harps, synth electric basses, bells, and guitars. One of the drawbacks of FM synthesis is that it is fairly hard to program. As with the other synthesis techniques analyzed so far, you should have at least one FM synthesizer available in the studio. There are a few software synthesizers based on FM synthesis, such as the classic FM8 by Native Instruments, which is based on the engine of the original DX7, but with new controls and matrices available. Listen to Examples 2.9–2.12 to get an idea about the quality of the timbres generated by FM synthesis.

2.8.6 *Wavetable Synthesis*

The synthesis techniques described so far use basic waveforms (mostly geometrically shaped such as sine, square, and triangular) combined with complex filters and algorithms to produce more-sophisticated sonorities, whereas wavetable (WT) synthesis starts with complex waves instead. These soundwaves are usually sampled from both acoustic and synthesized instruments and are stored in tables in the ROM (read only memory) of the synthesizer, to be recalled and used to build complex patches. This approach became more and more popular among synthesizer manufacturers in the mid- and late 1980s, when the price of memory started decreasing and the size of the chips started increasing. The waveforms are accessed through a variety of tables that can be altered and reshuffled in real time, thereby allowing the creation of ever-changing patches.

A generic wavetable synthesizer stores presampled short waveforms that can be reproduced (played back) sequentially or randomly, thus creating the illusion of a constantly evolving sound. Usually a WT software synthesizer features two oscillators, each capable of "browsing" through a series of waves. A table can have up to 256 waveforms stored in it. What makes WT particularly appealing is the way you can move from waveform to waveform. A complex matrix of assignable LFOs allows us to modulate any aspect of the synthesizer, such as the position of the oscillator inside the table, or the start, end, and mid-point of the oscillator inside an individual waveform. The initial set of waves available with the machine can be increased by sampling your own sounds, making the WT practically endless in terms of creative sound design. Wavetable synthesizers are particularly appreciated for their overall flexibility, comprehensive lists of patches, and excellent sound quality. I particularly like WT synthesizers for evolving pads, leads, and synth basses. Listen to Examples 2.13–2.16 to compare patches generated with wavetable synthesis.

2.8.7 *Sampling*

A sampler (hardware or software) takes the approach of wavetable synthesis to an entirely new level. Using a similar approach to that seen in the wavetable technique, instead of being limited to a table of small samples that are stored by the manufacturer in the ROM of the machine, it stores the sample in RAM (random access memory), which can be erased and refilled with new samples at the user's will. When you turn on a sampler, it usually does not contain any samples in RAM; you have to either load a bank of samples from a hard disk (HD) or CD-ROM or sample your own waveforms. Saving to HD the samples and the changes made to the banks will allow you, at a later time, to reload the settings for future sessions. Samplers are the best options to reproduce acoustic instruments. The amount of RAM available on a device or your computer is strictly related to the amount and length of samples that you can load and use at the same time. The samples that are recorded and stored in the memory are mapped to the keyboard in order to have the full extension and range of the instrument that was sampled. The higher the number of samples that form a particular patch the more accurate the patch will be, since for every key that doesn't have an original sample assigned the machine will have to create one by interpolating the digital information of the two closest samples available.

As in the case of wavetable synthesizers, the sampled waves can be altered through a filter section similar to the one found in subtractive synthesizers.

Software samplers have practically replaced their hardware counterpart in any professional production environment. Their ability to take advantage of large memory sizes and (for certain types such as Kontakt (Native Instruments) and Mach5 (MOTU), to mention just two) their unique feature of being able to stream the samples directly from the hard disk, has basically put an end to the limitations created by the RAM-based architecture of hardware samplers. The ability to edit and program a software sampler with a touch of the mouse, through clear and comprehensive graphic interfaces, has greatly contributed to the success of software samplers. This type of approach to sound creation constitutes the core of the modern MIDI and audio studios. The majority of acoustic instruments that are sequenced nowadays are recorded using sample-based devices. Among the most used sampling applications are Mach5 by Mark of the Unicorn, Kontakt by Native Instruments, and Logic Pro ESX24 by Apple. In Examples 2.17–2.20 you can evaluate some sample-based patches.

2.8.8 *Physical Modeling Synthesis*

With physical modeling (PM) synthesis (a relatively new sound-generation technique), the synthesizing process involves the analytical study of how a waveform is produced and of all the elements and physical parameters that come into play when a certain sound is produced. The sound-producing source, not the resulting waveform, is the key element here. Physical modeling synthesis is based on a series of complex mathematical equations and algorithms that describe the different stages of the sound-producing instrument. These formulas are derived from physics models that are designed through analysis of acoustic instruments. Behind PM lies the principle of how a vibrating object (strings, reeds, lips, etc.), the medium (air), and an amplifier (the bell of a trumpet, the body of a piano, the cone of a speaker in an amplifier) all interact together to produce a particular sonority. Extremely complicated mathematical models that describe how the several elements of any sound-producing instrument interact together are stored in the synthesizer, which generates the waveform, calculating in real time the conditions, relations, and connections between each part involved in creating a particular sound. The number of algorithms present in a PM synthesizer determines its versatility and sound-generating power. A diagram representing the various stages in which PM would dissect the process involved in the production of a violin waveform is shown in Figure 2.11.

PM requires an incredible amount of CPU power to process in real time the calculations necessary to generate the waveforms, power that only recently has become available to every musician with the introduction of more powerful CPUs and computers. The strengths of PM are many and they all affect the final users. The programmer is presented with parameters that are usually fairly easy to understand. No longer do you have to guess which parameter (or parameters) will have an impact on the way a sound will change when sending a higher-velocity value or how another sound will change when using a certain vibrato technique. With PM sounds are programmed using real parameters

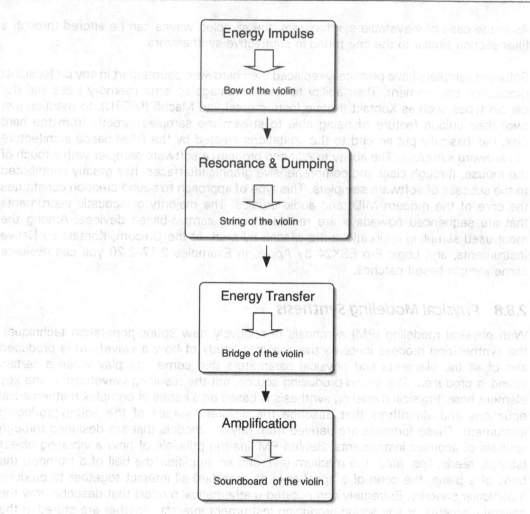

Figure 2.11 Example of physical modeling model for a violin sound

that apply to real acoustic instruments. Physical modeling synthesizers require much less RAM than wavetable devices and samplers. Physical modeling can calculate the waveform for the entire range of the instrument without requiring a multisample setup. In order to slim down the amount of calculation that a PM synthesizer needs to do in real time, the original algorithms are often simplified and translated in sequence of traditional preprogrammed filters. Other tricks such as cycle loops are often used to simplify even further the computational process. One of the most active manufacturers in developing accessible and marketable solutions for musicians based on PM synthesis has been Yamaha, which in the mid-1990s introduced one of the first commercially available and relatively affordable PM synthesizers, the VL1, which had a two-note polyphony limit. As in the case of other synthesis techniques, PM is progressively taking advantage of the software synthesizer revolution. The extreme flexibility and adaptability of software synthesizers are the perfect companions for a type of synthesis such as PM. As we saw earlier, the number of algorithms preprogrammed in a PM synthesizer determines the capabilities of the machine. While most hardware devices are based on a pretty closed architecture that makes software upgrades rather complicated, limiting the possibilities of expendability in terms of new models, a software synthesizer allows for a much higher degree of flexibility owing to its intrinsic open architecture. One of the more intriguing and fascinating aspects of PM is related not to the reproduction of real acoustic instruments, but instead to the unlimited possibilities in generating morphed instruments, which are constructed by mixing and matching models and stages from several sources. Imagine having an "energy generation" module of a violin fed into the "resonance and damping" section of a drum and then sent to an "amplifier" stage of a trumpet. The resulting sonority will be something totally new, yet extremely musical and inspiring! Physical modeling synthesis is particularly effective in reproducing wind instruments and in particular brass sonorities. A combination of sampling and PM, is often adopted to produce extremely realistic acoustic instruments (such as woodwind or brass instruments). This technique, referred to as sample modeling, offers the advantage of combining the flexibility of PM with the realism and practicality of sampling. Listen to Examples 2.21 and 2.22 to compare respectively some "acoustic" instruments first, and some "morphed" instruments programmed with PM.

2.8.9 *Granular Synthesis*

Even though the concept on which granular synthesis (GS) is based can be traced back to as early as the end of the nineteenth century, its practical use in synthesis was only developed around the late 1940s and early 1950s. The main idea behind GS is that a sound can be synthesized starting from extremely small acoustic events called "grains." These grains are usually between 10 and 100ms in length. Following this approach, granular synthesizers start from a waveform (it can be any type of waveform, such as FM, wavetable, or sampled) that is "sliced" to create a "cloud," which is made of hundreds or even thousands of grains. A completely new and complex waveform is then assembled by reordering the grains according to several algorithms, which can be random-based or functional. One of the advantages of GS is that the sounds created from such small units can constantly change and evolve depending on the grain-selection algorithm. Since the grains forming the sound cloud are so small and so many, their repetition is not perceived as a loop (as in the case of wavetable synthesis or sampling), but instead as "quantum" material that constitutes the basic substance of which sounds are made. Grains and resulting waveforms can be processed using more standard modifiers such as envelopes generators and filters. Usually GS is particularly suited for the creation of sophisticated and fresh-sounding synthesized pads and leads (Audio Examples 2.23 and 2.24). One of the drawbacks of GS is the CPU power required to calculate the grains and the clouds in real time. Granular synthesis synthesizers have become extremely popular in the last few years, especially in contemporary electronic music, dance, and sound design. Granular synthesis patches are a constant inspiration and they can really open new horizons in terms of colors and textures.

As you can see from this brief overview of the most popular synthesis techniques available on the market, each one of them has strengths and weaknesses that make it the ideal choice for a certain style, texture, and sonority. To understand their fundamentals and to be familiar with their overall colors and textures is key to producing versatile, original projects. When buying or considering using a particular synthesizer (software or hardware), think which texture and sonority you are looking for, keep in mind the strengths of the synthesis techniques just analyzed, and make a decision based on facts and goals. While the synthesis techniques we learned can be specifically and individually found in devices on the market, it is more likely that you will have to choose between products that will incorporate several approaches to synthesis at the same time. In fact, most of the devices (and applications) available nowadays feature three or more techniques that are integrated with each other, bringing their flexibility and their power to higher levels.

2.9 Sequencing for the Guitar: Overview

Another extremely versatile instrument of the rhythm section is the guitar. As you learned in the first half of this chapter, this instrument can vary in terms of texture, sonic features, and colors. In the MIDI domain, its multifaceted nature is often hard to translate in a realistic way. While, for example, nylon guitars and steel string guitars can be rendered fairly accurately when played finger-style, they are extremely hard to reproduce when strummed. When sequencing for guitar there are three main aspects that you have to keep in mind in order to achieve successful results: the sound (patch) you are using, the sequencing techniques involved, and the mixing methods. Each stage plays an important role in rendering a guitar part in a MIDI setting. Let's analyze each stage to learn the techniques and options that will give the best results.

When it comes to guitar sounds and patches that are commonly used in contemporary productions, there are three main types: the nylon string (sometimes also called classical), the steel string (folk), and the electric guitar, with all its variations, such as clean, saturated, and distorted. In general, as for most of the acoustic sounds that we want to render using MIDI technology, sample-based libraries provide the most realistic and flexible technology. The advantage of using this sound-generation approach for the guitar is that it allows us to take advantage of multilayer patches where different velocities can trigger different samples. In the case of the guitar, it means being able to reproduce all the different colors and shades of the instruments when played with different intensities. Usually higher velocities (such as higher than 120) are used to trigger special samples that re-create slides, pull-offs, bends, etc., thus increasing the level of variation and control of the patch. PCM-wavetable synthesizers can do a decent job in rendering guitar parts, especially if you use machines that are fairly recent and that implement the latest technologies. A newer approach to guitar synthesis is offered, sometimes with successful results, by PM. As we learned previously, this technique involves the reduction to mathematical formulas and algorithms of the way a sound is produced. In the case of the guitar, a PM synthesizer has to take into consideration the vibrating string (or strings), whether the string was plucked (with a pick or with fingers) or strummed, and where the pick-up is positioned in relation to the strings. Listen to Examples 2.25–2.27 to hear the difference between a PCM-wavetable guitar patch, a sample-based guitar patch, and a PM guitar patch, respectively.

The nylon string guitar is rendered extremely well by using samples (and multisamples) of the original instruments. Only through sample-based technology can all the nuances of the original instrument be re-created. PCM-wavetable synthesis also does quite well, even though the limitations in terms of stored waveforms of this synthesis technique will limit the rendering of some of the more subtle nuances of this instrument. Physical modeling also does a decent job, not necessarily in accurately reproducing the original tone, but mainly in creating alternative sonorities that are related to it but do not mimic it. The steel string guitar sound can be rendered extremely well by sample libraries, especially when used in conjunction with key-switching and multisample techniques. As for the nylon guitar, wavetable can be pretty effective, whereas PM usually does not achieve acceptable results. Electric "clean" guitars can be reproduced very well by all three synthesis techniques, with, as usual, sample-based being the clear favorite. Distorted guitar can be rendered extremely well by sample-based libraries right out of the box. Wavetable and PM can get very close to the real sonority if either re-amped (more on this later in this chapter) or filtered through a guitar amp simulator such as Guitar Rig by Native Instruments, Amplitube by IK-Multimedia, or Amp Farm by Line 6. Among the available guitar libraries, the excellent Chris Hein Guitars and Real Guitar (by Music Lab) can be recommended.

2.9.1 Sequencing Techniques for the Guitar

How you sequence guitar parts in your MIDI studio varies greatly depending on the type of controller you use. Assuming that you are not a guitar player (otherwise why would you need to use MIDI to record guitar parts?), the common choice for inputting your guitar parts is the keyboard controller (the other controllers, including the guitar/bass to MIDI controller, will be covered later). Well, let's face it, using a keyboard to sequence a guitar part is probably as unnatural as it gets. The two instruments have practically nothing in common. Voicing, dynamics, sonorities, and fingering are almost completely unrelated. So, where should you start? The first thing that you have to keep in mind is that the chord voicings of piano and guitar are different. If you play a MIDI guitar patch from a keyboard controller and use the regular piano voicing, the result will be exactly that: your guitar parts won't sound at all like a guitar. So your first assignment is to go out and buy a simple guitar voicing book (there are literally hundreds available on the market). You can also use the Internet as a wonderful resource to look up guitar chord voicings. For instance, www.chordbook.com and www.all-guitar-chords.com are great websites with an extensive database of guitar voicings and their audio renditions. In order to understand fully what a difference it makes to use the right voicing, compare the two diagrams in Figure 2.12.

On the left there is a C-9 chord with a typical piano voicing, while on the right the same chord is voiced for guitar. Listen to Examples 2.28 and 2.29 to compare the two different

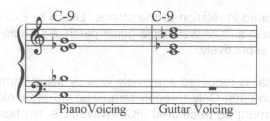

Figure 2.12 A C-9 chord voiced for keyboard and for guitar

voicings. In general, a "drop two" voicing is as ideal a starting point for voicing a chord as a guitarist would do. This voicing is particularly indicated for guitars since it avoids close small intervals between the notes of the chord being played. To create a "drop two" voicing start from the closed position one (for example C E G B from a CMaj7 chord) and drop the second note from the top (G in this case) an octave lower. Now the "drop two" voicing will be G C E B (Figure 2.3).

Usually, arpeggio passages are definitely easier to render, even from a keyboard controller. Set your controller to a more gentle response to render the soft passages of a finger-picking style. Use the sustain pedal to re-create the natural sustain of the picked strings during the arpeggio passages. Try to avoid the repetition of arpeggio styles from the low to the high register. Create variation instead by alternating the arpeggio notes inside the same voicing. Take full advantage of multisamples by using a wide range of velocities. This will inject life into the guitar parts by triggering the highest number of different samples possible. If available to your library, use key switches to insert slides, pull-offs, and other colors that are specific of the guitar vocabulary. Some libraries offer specific performance-related noises that are peculiar to the guitar's sonority. These usually include string noises, fingers sliding on the strings, frets buzz, etc. Use them in your sequenced performance to bring a guitar passage to life. If you don't have access to such noises, it is very easy to create your own. Ask a guitarist friend to stop by your studio one day and record some of the noises that can be produced on the guitar. Save them in a folder as part of your "bag of tricks." Next time you sequence a guitar part, add some of them to the guitar sound and you will see how your part will come alive! Of course, keep in mind that different types of guitars make different types of noises. If you plan to build your own small guitar noises library, use Table 2.2 as a reference.

Listen to Examples 2.30 and 2.31 to compare a guitar part sequenced using piano voicing and without noises and a similar part sequenced using guitar voicings with the addition of performance noises, respectively.

Strumming guitar parts are the hardest to render effectively using a keyboard controller. In general, everything learned so far for arpeggio guitar parts applies also to strummed parts, especially in terms of voicing. The most effective way to create convincing strummed guitar parts is to use a guitar-to-MIDI converter (more on this in a moment). If you have to use a keyboard controller, use the function called "flam," available in most MIDI sequencers. This option allows you automatically to distance each note

Table 2.2 Summary of guitar performance-related noises that can be used in a MIDI rendition to infuse life into a virtual performance

Type of noise	Guitar	Comments
Fingers sliding on strings	Any type of guitar with specific sonic characteristic for each string type: nylon, steel and electric	Particularly effective on steel and somehow on nylons
Fret noise	Any type of guitar	Very effective on steel and electric
Hum noise	Electric guitar only	Effective to re-create grounding problems, hum, or pick-up noises especially if used intermittently
Guitar body noises	Nylon and steel strings guitar	Particularly effective on quiet and solo passages

played simultaneously in a chord by a certain number of ticks (subdivisions of a beat). This will create the typical strumming effect of an acoustic guitar where each string (and note of the chord) is played a fraction of a second later then the previous one. The amount of flam applied gives you control over the distance of each note. For a more realistic effect you can alternate positive and negative flam amounts. This will create the illusion of alternating down and up strums as if you would do on a real instrument. In Figure 2.13 you can see how the original pattern sequenced from a keyboard controller (top screen) has been split into down strums and up strums (lower screens). After the split, positive flam values were assigned to the down strums (lower left screen) and negative values to the up strums (lower right screen). Make sure to use the sustain pedal to sustain the notes of each chord all the way through the chords. This will re-create the natural ringing behavior of the strings on an acoustic guitar.

If your sequencer does not provide an automatic flam option you can manually create a similar effect by moving each note one by one. By creating more complex rhythmic patterns and using the same technique explained above you can create convincing guitar parts fairly quickly.

Solo guitar parts are usually easier to sequence even if using a keyboard controller, especially if you take advantage of CCs such as modulation, pitch bend, and sustain. Solo lines can really come to life when sequenced using a high dynamic range. Soft passages can bring out the intimacy of a nylon guitar samples, while louder dynamic levels are ideal for electric distorted parts. As mentioned earlier, the addition of acoustic noises is particularly important when sequencing solo parts. Single lines usually bring forward the small nuances of finger slides, pull-off, harmonics, and fret noises.

If, after trying the above techniques, you still feel that your MIDI guitar parts are not up to speed, well, then it is time to do some shopping and purchase a guitar-to-MIDI converter (Figure 2.14).

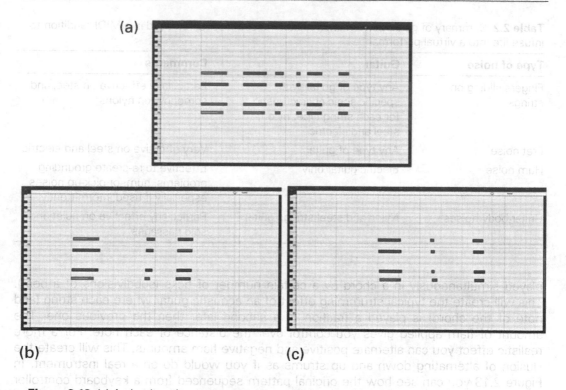

Figure 2.13 (a) A simple guitar pattern sequenced with a keyboard controller. The same pattern has been split into two different tracks alternating; (b) down strums with positive flam; and (c) up strum with negative flam (Courtesy of Apple Inc.)

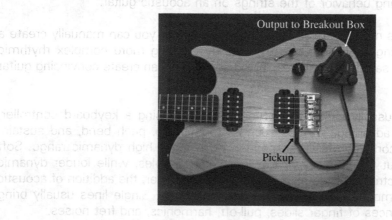

Figure 2.14 Example of a guitar-to-MIDI converter pick-up: the Yamaha G50 (Courtesy of Yamaha)

A guitar-to-MIDI converter (also known as a guitar MIDI controller or GMC) allows a regular acoustic or electric guitar to be connected to a MIDI system and output MIDI messages and notes to any MIDI device, including a sequencer. This technology has been around for many years and is constantly being perfected by companies like Roland, Yamaha, and Fishman. Even though the models vary from manufacturer to manufacturer, the principle on which this type of controller is based is simple: a pick-up (divided in six segments, one for each string) is mounted next to the bridge of the guitar (Figure 2.14). The pick-up detects the frequency of the notes played on each single string by analyzing their cycles, then this information is passed to a breakout unit that converts the frequencies into MIDI Note On and Off messages. From the unit a regular MIDI OUT port sends the messages to the MIDI network. The pick-up can also detect bending of the strings, which is translated in pitch-bend messages. Even though it takes a bit of practice to adjust your playing style to this type of converter, you will enter a whole new world of sonorities and possibilities. Fishman has recently introduced a new series of MIDI pick-ups for guitar called "Tripleplay Wireless Guitar Controller." The advantage of the Fishman solution is that it is wireless and the technology used to translate the pitch to MIDI data is faster than any previous system. A GMC is an alternative controller that is worth having in the studio as either the main controller or an occasional substitute for the keyboard. The ideal practical applications of a GMC vary depending on your musical background. If you are an experienced guitarist then you can use it as your main MIDI controller. If you are not a guitarist you can still use it to sequence guitar parts taking advantage of all the benefits that MIDI has to offer such as editing, tempo slow-downs, and overdub. You can, for example, sequence a difficult guitar passage at a much slower tempo, or sequence a part in an easier key and transpose it later in the sequencer. As the price of these devices keeps getting lower it is definitely recommended that you invest in one. A GMC is ideal for sequencing all types of guitars and parts. For arpeggio and strumming parts it allows you to use genuine voicings, natural dynamics, and realistic rhythmic patterns. When sequencing solo lines you can achieve the ultimate expressivity without using cumbersome pitch-bend and modulation wheels.

A GMC can also be used in a more creative way. Since each string has an individual pick-up, by assigning different strings to different MIDI channels and patches, you can create a "one-man band" and generate interesting layers and ensemble combinations. You can record-enable multiple tracks (maximum of six, one per string), all receiving on different MIDI channels, and assign their outputs to separate devices and MIDI channels. By pressing the record button you will record separate parts on separate tracks. You might assign a low sustained pad to the low E and A strings, use the D and G strings for sound effects or tenor voice counterpoint, and the B and high E for the melody. The advantage of this approach is that you can capture a much better groove and "live feel" than sequencing the tracks separately. Experiment with these techniques and let your imagination run free; you will be surprised by the results.

New technologies have emerged recently to facilitate the rendition of guitar parts in the project studio environment. As loops are being used more and more to render drum parts, phrase-based techniques are also taking over the six-strings world. But these new techniques are not based on boring repetitions of looped phrases. They are instead based on elaborate software engines that give musicians a much higher degree of control over their performance. Real Guitar by MusicLab is a clear and successful application of this approach. The results can be astonishing, especially for background parts and accompaniments. This approach can be particularly useful when sequencing strumming parts for acoustic and electric guitars.

For more complex guitar parts (such as strumming or busy arpeggios), I recommend using a dedicated software synthesizer such as Strum GS-2 by Applied Acoustic System (AAS), Acoustic Guitars by Ilya Efimov Production, or Ample Guitar by Ample Sound.

2.9.2 *Mixing the virtual guitar*

MIDI guitar parts can be brought to life at the mixing stage using effects such as re-amping, saturation/distortion, delays, and chorus/flanger. The use of effects at the mixing stage can really improve the overall sound of a MIDI-sequenced guitar patch. One very useful technique is re-amping. This approach involves the rerecording on an audio track of the output of the synthesizer (hardware or software) that generates the patch of the MIDI guitar track through a guitar amplifier. The signal coming out of the amplifier is recorded with a microphone. After sequencing the part and making sure that all the notes are edited, quantized, and trimmed in the way you want, send the audio signal of the module playing the guitar part through a guitar amplifier. Record the signal of the amp through a good microphone onto an audio track in the sequencer. The warmth and the realism of the ambience captured by the microphone will add a great deal of smoothness and realistic texture to the virtual guitar patch. This technique can be used with any type of guitar patch, from classical to folk, from jazz to rock. It works particularly well with guitars that need medium to heavy distortion or saturation applied to their sound. Even though it is possible to apply such effects through a plug-in inside the sequencer, by using a real amplifier and the microphone technique you will get as close as possible to a real live situation. Listen to Examples 2.32 and 2.33 on the website to compare a stiff guitar part sequenced using a keyboard controller played directly from the MIDI module and a more convincing part sequenced using a guitar-to-MIDI controller with the sound run through an amplifier to which some saturation was applied.

When it comes to equalization the guitar is an extremely flexible instrument. In guitar solo passages it can provide a deep and full sonority, while when mixed within an

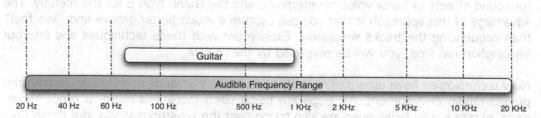

Figure 2.15 The fundamental frequency range of the guitar and of the bass

ensemble it can serve a more discrete rhythmic function. The fundamental tones of the guitar range from 60Hz to 1kHz (Figure 2.15), but keep in mind that the overtones of this instrument reach much higher frequencies.

When using equalization try to be gentle and not to overdo it unless absolutely necessary (to fix very obvious problems). To boost the low range of the guitar use the frequency range between 400 and 600Hz. This range controls the fullness of the guitar's low fundamentals. You can control the punch of a guitar sound by boosting or cutting the 2.8–3kHz range. The attack and presence can be effectively enhanced through a light boost of the frequency between 5 and 6kHz, while the range between 10 and 11kHz can be effectively used to enhance sharpness (boost) or darken the overall sonority (cut). The above equalization settings will serve as a starting point to get the most out of your virtual guitar parts and renditions.

To enhance further the realism of a sequenced guitar part you can also take advantage of a large number of plug-ins that fall into the category of amp or guitar box simulators. These virtual effects can inject new life into sampled and synthesized guitar patches, allowing you to reach new sonorities that you didn't know were possible. Pod Farm, AmpliTube, and Guitar Rig are a few of the several options available nowadays. Feel free to experiment with different settings and combinations of the most common guitar effects such as chorus, flanger, phaser, and delay. A very useful effect on guitar is a very short delay panned on the opposite side of the dry guitar signal in order to obtain a wider stereo image and a fuller (yet clear) guitar sound. Here's how it is done. First of all pan the dry guitar track between 75 and 100 percent on one side of the stereo image (let's say on the left side for the purpose of this example). Then create an auxiliary track, assign its input to an empty bus (e.g. bus 1) and insert on it a short delay (usually between 40 and 60ms). Pan the aux track so that it mirrors the guitar track panning (in this example to the right). Assign one of the send of the guitar track to the same bus on which the aux track is receiving (bus 1) and set the send knob fairly high, say between 80 percent and 90 percent. Your guitar now will sound much wider, fuller, and spread across the entire stereo image. The reason for this effect is that you end up with an exact copy of the dry guitar track (panned left) on the right channel, but the right channel is delayed by a few milliseconds, giving the impression of having a second guitarist performing the same part. This effect will improve the realistic rendition of sequenced guitar part, no matter which sound generator you use. To blend a virtual guitar part within a mix, add a little bit of chorus or flanger. These time-based effects use phase shifting to create a "watery" and "out-of-phase" effect that gives a warm and mellow color to the guitar sound. The basic principle on which these effects are based lies in the ability of a plug-in to delay (one or several times) the original input signal of an audio track by a very short amount of time (the signal can be displaced in time from a few milliseconds up to 1 or 2 seconds). The summation of the original (in-phase) signal and the delayed (out-of-phase) signals creates the typical chorus/flanger effect. Chorus effects can be used to enlarge the guitar sound, to add depth, and also (with extreme settings) to obtain completely original sonorities.

In terms of reverberation, you should use a convolution reverb on acoustic guitars (nylon and steel strings), as for all your mixes. This type of reverb will give the most realistic acoustic sonority, helping you and your virtual guitar parts to come across with more life. For background parts with a busy rhythmic signature use slightly shorter reverberation times, between 1.5 and 1.7 seconds. For solo guitar parts or up-front melodies you can afford usually to apply slightly longer reverberations, up to 1.9 seconds. If you are mixing a sampled or synthesized electric guitar patch you can definitely be more creative with the reverb setting. Don't be afraid to experiment with other reverb color such as plate, springs, or any other vintage type.

You will be amazed how realistic MIDI guitar parts can sound with a little bit of work and programming. One of the most crucial aspects of rendering a guitar part in a MIDI studio is how you treat the final sound. Think of your MIDI guitar part as a real guitar sound. Use the same effects that you would use on a regular guitar track and you won't be disappointed. You could use one or more actual real effect pedals on the output of a MIDI guitar patch. Chain, for example, an equalizer, a chorus effect, and a reverb. Not only you will get great realistic effects, but also a much warmer and fuzzy sonority, much closer to the real thing than the sterile and clean digital output of your synthesizer.

2.10   Sequencing for the Bass: Overview

Sequencing for the bass is, in general, a much easier task than sequencing for the guitar, even though there are many similarities between the two processes. Since the bass is primarily a monophonic (in some cases duophonic) instrument it can be sequenced more easily through the use of a conventional keyboard MIDI controller. The dynamic range and MIDI implementation of most keyboard controllers are adequate to render the overall dynamics of the bass, especially in the case of an electric bass. The bass used in the modern rhythm section is mainly electric or acoustic (double bass). The former is in general much easier to render effectively through a MIDI system, the main reason being that a typical electric bass is fretted and therefore it offers a more limited set of performance variations than its acoustic counterpart. This translates into a more straightforward, precise, and repetitive sound generation, all elements that make MIDI an ideal option for the creation and sequencing of bass parts. Fretless instruments (such as the electric fretless and the acoustic bass) present a much wider range of expressivity and variation (intonation, slides, vibrato, etc.). This can be particularly challenging to re-create through a series of samples or synthesized waveforms. Because of the fairly different nature of these two types of bass, sequencing techniques that apply to the fretted category will be analyzed first, followed by the fretless and acoustic types.

The electric bass (or bass guitar) can be convincingly reproduced using both wavetable synthesis and sample-based techniques. Very realistic electric bass patches are available from a variety of hardware and software synthesizers capable of creating punchy and deep electric basses that can sit nicely in a contemporary mix. Sample-based bass libraries are extremely versatile and they can often fool even the most seasoned producer if well mixed and equalized. There are several library options when it comes to bass patches.

Some libraries re-create vintage bass models, such as the excellent Trilian by Spectrasonics, or the versatile Scarbee Bass series by Native Istruments. Try to find samples that are well recorded and that give you the highest possible range of flexibility, such as patches played with both pick and fingers and recorded through real amplifiers. All these features will make your work much easier when sequencing bass parts. As for the guitar, a multisample and key-switch capable library will dramatically improve the rendition of a bass part.

For fretless and acoustic basses sample-based libraries are highly recommended. The complexity of the sonic structure of such instruments can only be rendered at its best by multisample libraries with a high number of samples for different velocity levels. This is particularly true for the acoustic bass, a very complex and sophisticated instrument. Several software plug-ins are also available that are targeted specifically to bass patches. These include Trilian by Spectrasonics, which features an incredible number of patches that cover a wide range of styles and sounds from electric to fretless, from acoustic to vintage analog synth, and the versatile Native Instruments Scarbee Bass series.

2.10.1   Sequencing techniques for the bass

The sequencing techniques adopted when rendering bass parts can be divided into two different types and they vary slightly depending on the type of part and bass on which you are working. The first type targets the actual inputting of the notes through a controller, while the second type focuses more on the overall groove and quantization of the part itself. For most fretted bass parts you can use your keyboard controller without compromising the global realistic rendition. There are some exceptions, such as very exposed parts like bass solos and high-register fills. For these exceptions and for fretless and acoustic parts use a guitar- (or bass-) to-MIDI converter. The quality of the rendition will increase exponentially when using such devices. If you are a guitarist you will find yourself at ease with this controller; if your principal instrument is not bass or guitar you can use all the tricks that are available in the MIDI domain (slowing down the tempo, transposition,

Figure 2.16 A Yamaha bass to MIDI converter in the 4-strings configuration (Courtesy of Yamaha)

quantization, editing, etc.) to come up with convincing bass parts without waiting 10 years to learn how to play the instrument! If you are a bass player you can use a bass MIDI controller for sonorities that wouldn't be available to you otherwise, such as synth bass and vintage electric basses. This system is similar to the GMC, but the pick-up is tailored to the bass strings configuration: four, five, and six strings (Figure 2.16).

Keep in mind, however, that because of the pitch-tracking system of the device the delay between the acoustic note and the MIDI note is inversely proportional to the frequency of the note. Therefore for very low notes the delay is noticeable and this might discourage some bass players. This is due to the fact the lower frequencies have a longer cycle and therefore the pick-up has to wait longer before detecting a full cycle. One way around this is to use a "piccolo bass," meaning a bass with strings that sound one octave higher. This is an alternative controller that is worth having in your studio as either your main controller or an occasional substitute for the keyboard. Listen to Examples 2.34 and 2.35 to compare a bass part sequenced using a regular keyboard controller and a MIDI bass/guitar controller. Using such a device to sequence bass parts gives you full control over note bending, slides, vibrato effects, etc. If you are not lucky enough to own one, make sure that you take advantage of the most common MIDI controller and messages to obtain a similar flexibility. You can somehow effectively re-create bends using the pitch-bend wheel from the keyboard controller. Program its range to not more than one whole step (one half-step is even better) to avoid artificial bends than would give away the MIDI nature of the production. Use aftertouch and modulation to add subtle vibrato touches in order to create variation. As with the guitar, you can add a touch of realism by inserting performance noises when sequencing for the bass. Consult Table 2.3 for a list of performance noises that can be used when sequencing electric, fretless, and acoustic bass parts.

Table 2.3 Summary of bass performance-related noises that can be used in a MIDI rendition to infuse life into a virtual performance

Type of noise	Bass	Comments
Fingers sliding on strings	Any type of bass, electric, fretless and acoustic	On the acoustic bass be careful to use string noises that are recorded from acoustic basses since their strings (guts' strings or steel flat) are very different from the ones used in fretted and fretless electric basses (usually rugged)
Fret noise	Electric fretted bass	It can be very effective for exposed bass parts such as solos or featured bass lines
Hum noise	Electric bass (both fretted and fretless)	Effective to re-create grounding problems, hum, or pick-up noises, especially if used intermittently
Bass body noises	Acoustic bass	On the acoustic bass the addition of very light wooden squeaks can improve the realism of a solo passage

Since bass parts are usually groove-based and they need to support the overall rhythmic groove of the rhythm section, a crucial aspect of the sequencing process is represented by quantization and groove quantization. With the exception of bass solos, quantize the bass parts immediately after you sequence them, since they will constitute the foundations of the rhythm section and you definitely don't want to build all your other tracks over a weak and unstable foundation. The problem with straight quantization (also referred to as "100 percent quantization"), however, is that usually it will stiffen up the natural performance groove that was played. To regain the groove lost due to straight quantization, use the "groove quantize" option on your sequencer. The principle behind groove quantize is simple: starting from templates, provided with your sequencer or that you can buy as expansion packs, you can apply several styles, such as "laidback," "push forward," or "shuffles," to your parts. The grooves control primarily the timing, velocity, and lengths of the notes in the MIDI tracks or parts you are quantizing (Figure 2.17).

According to the template you choose, the MIDI events will be shaped to give a more realistic rhythmic feel. Keep in mind that this type of quantization should not be used to correct rhythmic mistakes, but to create more realistic and natural-sounding performances. Therefore, quantize your part first with the regular quantization techniques and then apply the groove quantization to loosen up the stiffness of straight quantization. The groove quantization technique will be discussed in more depth in the next section, on sequencing for drums and percussion.

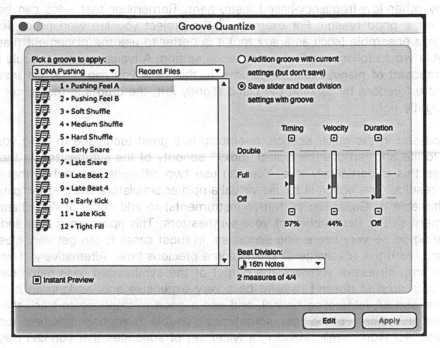

Figure 2.17 Example of groove quantization options in Digital Performer (Courtesy of MOTU)

Another effective technique to minimize the "drum machine" effect caused by straight quantization is to use a lower strength. Instead of applying 100 percent quantization strength, try to use 80 percent or 90 percent. This will make your bass parts sound rhythmically sound but it will also leave a bit of the human touch and feel of your performance. If you feel that 80 percent is still a bit too loose, try to quantize with a strength setting of 10 percent increments until you are satisfied with the groove and rhythmic feel.

2.10.2 Mixing the Virtual Bass

In terms of mixing the MIDI bass parts that you have sequenced, a few rules can be followed to obtain the most out of them. First of all, you should be quite conservative when panning any bass track. Low frequencies usually feel more natural if panned in the center of the stereo image. The so-called frequency placement approach is based on the fact that low frequencies are harder to place in space than high frequencies. The brain perceives sounds in space according to the difference in phase of the waveforms received by the two ears. Since low frequencies have longer periods, a small difference in phase is harder to perceive and therefore low frequencies are more difficult for the brain to place precisely in space. Thus, it is usually more natural for the ears to listen to audio material that features low-frequency instruments placed in the middle of the stereo image rather than panned hard left or right. Following this approach, you should avoid using extreme panning settings for instruments such as bass, bass drum, string basses, and any other low-frequency-based instrument. Remember that rules can be broken if it is for a good reason. For example, if the project you are working on features a virtual jazz ensemble (such as a jazz trio) it is better to use the placement that the live ensemble would follow in a live performance setting. A typical example would be a jazz trio composed of piano, bass, and drums. In this case you can place the instruments with more freedom by panning the piano slightly left, the drums in the center, and the bass slightly right.

As discussed in the guitar section, re-amping is a great technique to bring your virtual tracks to life and remove the typical "boxy" sonority of the synthesizer. In the case of the bass this is particularly true. You can use two different approaches that can give similar results. One option is to use virtual amplifier simulators (such as Amplitube 3 by IK Multimedia or Guitar Rig by Native Instruments) to add some analog character and excitement to the flat patches of your synthesizers. This approach is fast and capable of creating some very interesting sonorities. In most cases it can get very close to the real thing, with the advantage of saving some precious time. Alternatively, if you do not own an amp simulator software the output of the synthesized bass patch can be run into a bass amp (it doesn't have to be a very expensive one, as long as it has some character and an interesting sound), and use a good microphone to track the speaker response. In the case of the bass, sometimes a double-microphone approach (condenser plus dynamic) works well, providing a wider set of sonorities that you can play with at the mix stage. In this case you will record the two microphones on two different mono

Table 2.4 Key equalization frequencies that apply to the electric and acoustic bass

Frequencies	Application	Comments
60–80Hz	Boost to add fullness to both electric and acoustic bass	This range gives you a fuller sonority without over accenting the muddiness
200–300Hz	Cut to reduce muddiness	Use with discretion since by cutting too much you might reduce the fullness of the bass track but by boosting too much you might increase its muddiness
400–600Hz	Boost to add presence and clarity to bass	
1.4–1.5kHz	Boost for intelligibility	

audio tracks. For electric fretted bass you may like to feature more the character of the dynamic microphone, whereas for fretless and acoustic the condenser is better suited to pick up the subtleties of their more complex sonic ranges. Listen to Examples 2.36 and 2.37 to compare a MIDI bass part sequenced without and with the re-amping technique.

The use of equalization on bass tracks can drastically change how the low frequencies of a mix will sound (the fundamental frequencies of the bass are shown in Figure 2.15). Usually, in most contemporary productions, the bass needs to be deep and full, but not muddy. The goal is to create the perfect symbiosis between the bass drums and the bass in order to have a punchy and full, yet clean, low end. To learn more about key frequencies that typically apply to the bass, see Table 2.4.

The low register of the bass usually calls for very little reverb (or most likely no reverb at all). Instruments that cover mainly the low end of the frequency range, such as bass and bass drums, usually require much less (or no) reverberation, while on instruments that cover the mid- and high-frequency range, such as guitars, hi-hat (HH), cymbals, and snare drum, you can apply more reverb. This is due mainly to the fact that reverb in general contributes to add muddiness to the mix through the tail created by its reflections. Adding too much reverb to instruments that cover the low-frequency range will reduce the definition of the overall mix. This is true especially for the electric bass and the bass drums, which can quickly lose sharpness and clarity if too much reverb is added. There are some exceptions to these rules. Add some reverb to the acoustic bass if played in a small jazz ensemble. In this case, a little bit of reverb will add a nice room effect that will increase the realism of the sample patch (usually the reverb length will be between 1.5 and 1.7 seconds). Another exception is the use of reverb on bass solo lines that cover the higher register of the instrument. This approach works particularly well with fretless basses, which are more melodic in nature than their fretted cousins.

2.11   Sequencing for Drums and Percussion: Overview

In contemporary productions, drums and percussion form the core of the rhythm section in any modern music style. MIDI techniques and libraries have become more and more sophisticated in order to provide the composer and programmer with more flexible tools to render all the different nuances that these instruments are capable of producing. When it comes to drum sonorities, we have to distinguish between acoustic and electronic. Being versatile in terms of sonorities and drum kits is crucial nowadays. Your drum kit and percussion libraries should cover acoustic and electronic sonorities equally well. In general, it is much easier to find good-sounding electronic patches than acoustic ones. Pretty much any modern synthesizer is capable of providing convincing electronic kits that can be used for a wide variety of styles. Acoustic patches are harder to re-create in all their subtle nuances and variations. For acoustic sonorities, you should definitely choose the route of a professional multisample library (or software sample playback/plug-in). As shown earlier, multisample libraries can be extremely effective in re-creating acoustic sounds such as piano, guitar, and bass. This technology is particularly effective for drums. A snare, for example, features a totally different attack, release, and color depending on whether it is struck softly or hard. The multisample technique allows you to experience a similar response by assigning several different samples to different MIDI velocities. When you play a drum part, the sound will change depending on the velocity at which you play the notes on your controller (Figure 2.18).

There are several options available in terms of sample-based sound libraries, all providing sounds ranging from good to excellent. Pure sampled drum libraries are now being replaced by complete software packages that are real drum/rhythmic engines capable of extremely sophisticated sonorities and algorithms. Among those of particular interest are

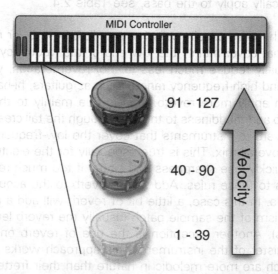

Figure 2.18 The multisample technique applied to a snare drum sound

Figure 2.19 OH and Room mic balance in Native Instruments "Studio Drummer"

BFD 3 by Fxpansion, Superior Drummer by Toontrack, Addictive Drums 2 by XLN Audio, and Battery by Native Instruments. They allow you not only to trigger multisampled sounds, but also to control the balance between overhead and room microphones of the kit that you are using in the sequence (Figure 2.19).

Having control over the actual microphone balance and over the ambience in which the drum kit is played will increase the realism of your drum tracks considerably. The software also gives you a much deeper control over the dynamic response of the samples to MIDI velocity, allowing you to fine-tune the response of the controller to the drum sound of your choice. New expansion kits are regularly produced to allow you to explore new styles and samples. These plug-ins usually require a more powerful CPU than a regular sample-based library for your sampler, but the results are definitely worth the extra power needed. Most libraries include some percussion instruments, such as bongos, congas, shaker, and tambourine, but in most cases you are required to buy expansion packs that cover all sorts of percussive instruments. You will need at least two good drum/percussion libraries or plug-ins to be able to mix several sources. This will help you render more natural and realistic rhythmic tracks. For quick sequences and productions you can use a MIDI sound module entirely dedicated to drum sounds in order to be able to find the most appropriate sonorities quickly. These modules are a good starting point for sequencing drum parts since they offer a comprehensive palette of sounds in one box. You might want to replace some of the patches later in the production, but this type of module offers a fairly good starting point.

Some of the most advanced drum plug-ins allow you also to use presequenced bundled patterns (either MIDI or audio loops) in sequences without actually sequencing anything on your own. The use of "precooked" material (be it MIDI patterns or audio loops) can be a valuable source of inspiration for temporary tracks or quick demos, but it can be limiting when it comes to delivering the final product. While the use of loops can be interesting for embellishing a production and to give a neat touch to a series of MIDI tracks, loops (by definition) are very repetitive and can flatten your sequence by making it boring and cold, especially if used as the primary source for the rhythmic tracks of a production. If you make the use of loops a constant in your productions you will soon have to adapt your creative work to the loops and not vice versa, as it should be! I would rather listen to a simple MIDI drums track that is groovy and original than to a short, sophisticated rhythm that keeps repeating over and over without variation. If a loop becomes an essential part of your project, then spice it up with some variations using MIDI parts, acoustic live recording of a few percussive elements (e.g. shaker, HH, cymbals), or, even better, a combination of the two. A good alternative to premixed stereo loops (which constitute the majority of the loop-based rhythmic material available) is the multitrack versions of the prerecorded drums and percussion parts offered by companies such as Discrete DrumsSonoma Wire Works. This type of drum track has the advantage of providing the original multitrack version of the loops, allowing a much higher degree of flexibility at both the editing and the mixing stage. Having separate tracks for each groove allows you to replace single instruments of the drum set to create a higher level of variety (you can change the HH or the bass drum pattern, for example, every so often). In addition, during mixing, you can apply different effects to each single track to achieve more realistic results. You can also use the multitrack sessions in a creative way by mixing and matching parts from different grooves (e.g. the HH part from a hip-hop groove with the bass drum and snare drum from a funk rhythm). No matter which technique or combination of techniques you use, always try to be creative and do not let the sequencer or the loops dictate your music.

2.11.1 *Sequencing Techniques for Drums and Percussion*

The first rule for re-creating convincing rhythmic parts with a sequencer is to try to avoid using your keyboard as a controller and use MIDI pads instead. One of the biggest problems with using a keyboard controller to sequence percussive instruments is the lack of a wide dynamic range (in terms of MIDI velocity data) required to render drums and percussion instruments adequately. The action of a keyboard gives you only around one inch to express 127 different velocities (from 1 to 127), making it very difficult for the performer to select the desired dynamic accurately. A MIDI pad (Figure 2.20) allows for much more control over the dynamic response of the controller and therefore a much more detailed handling of the multisample sounds.

The percussive MIDI controllers can vary depending on their size, style, and features. Nowadays you can find three main types of percussive controllers: drums, pads, and triggers. Drum controllers re-create the entire drum kit including bass drum and cymbals (Figure 2.21). Each pad of the kit has several velocity-sensitive zones that can be assigned to different notes and MIDI channels. Roland has mastered this technique with the V-Drum series. There are several configurations available, ranging from a full studio kit to a portable "live gig" setup.

MIDI pads are smaller controllers meant to complement and not substitute an entire drum kit. They usually feature four, six, or eight pads that can be played with either sticks or hands (to simulate percussive instruments such as bongos and congas). The MIDI channel and note assignment can be set for each pad individually. This type of

Figure 2.20 The Roland Octopad SPD-30 (Courtesy of Roland Corp.)

Figure 2.21 The Roland V-Drum TD 20KW (Courtesy of Roland Corp.)

controller is ideal for a small studio situation and for occasional use on some projects. Another creative way to program a percussion/drum controller is to use it as a MIDI "pitched" instrument by assigning different pads to different notes of a pitched instrument such as marimba, xylophone, or vibraphone. You can also assign certain pads to trigger loops or samples and use them as background layers for a one-man-band type of performance. As mentioned when discussing guitar/bass MIDI controllers, in this case you are not necessarily required to be a proficient drummer to use a MIDI pad. These devices are particularly appealing to non-drummers since you can take advantage of all the editing features of your sequencer to fix and improve your parts. Listen to Examples 2.38 and 2.39 to compare a similar drum part sequenced with a keyboard controller and with a MIDI drum pad controller, respectively.

If you can't afford a MIDI drum or a MIDI pad device, you can use a keyboard controller that has MPC style pads (Figure. 2.22). These pads are a hybrid solution between regular keyboard keys and the bigger MIDI pads. They provide enough sensitivity to allow for better dynamics than a keyboard. Using your fingers to trigger them is a good technique to get a fairly precise rhythmic feel.

Once you have chosen the samples and sounds that fit the style of your production and you have your controller ready, it is time to start sequencing. For drum parts, use separate tracks for each piece of the kit or of the percussion set. This means that, usually, toms, HH, and cymbals (sometimes it is good to separate the cymbals in separate tracks for each type, such as ride and crash) will each be on their own track. You could put bass drum and snare on the same MIDI track and record them on the same pass: it will give you a much more natural feel to sequence these two pieces together. Color code each track if the sequencer allows it. This will speed up the process at the mix stage. Sometimes, if you really feel inspired, it is worth putting the sequencer in loop record

Figure 2.22 MIDI Pads à la MPC on the AKAI MPK61 MIDI controller

and keep overdubbing on the same track in order not to lose the groove you have found. In this case you will have to separate the different pieces of the kit after recording. The main reason for having different tracks for each percussive instrument is that it will give you much more flexibility at the editing and quantization stage. When sequencing a drum part, start with the kick and snare; this will provide a solid foundation. Then move to the HH and cymbals and finish with toms and fills. After each pass make sure to quantize (if necessary) the part you just laid down, otherwise you will be building each pass over tracks that are not rhythmically solid. This approach will translate into a much more cohesive and solid drum part. Try to avoid as much as possible sequencing only four or eight bars and then copying and pasting the part over and over. Remember that for a successful and realistic MIDI rendition you have to avoid repetitions as much as possible. Play the part instead for either longer sections (16 or 32 bars) or, ideally, for the entire length of the piece. A simple and small variation in MIDI note velocity through the different sections of the sequence will be enough to add a little bit of life to your productions.

2.11.2 Quantization for Drums and Percussion

Quantization is a very important aspect of sequencing, but it becomes even more important, even vital, when applied to drum and percussion. To "quantize" literally means to limit the possible values of a quantity to a discrete set of values. This concept, applied to a MIDI sequencer, describes a function that allows you to correct rhythmically the position of MIDI data in general, and notes in particular, according to a grid determined by the "quantization value." In practice, when you quantize a part, a track, or a region of a sequence, you are telling the computer to move each MIDI data item to the closest grid point specified by the quantization value. One of the main problems is choosing the right value for the track (or section) you are quantizing. If your patterns are fairly simple, you will probably be able to get away with only one value (e.g. eighth notes). If your parts are more rhythmically sophisticated (e.g. a mix of eighth note triplets with straight eighth notes), then you will have to select manually each region with the value that best fits it. In general, the rule states that the quantization value needs to be equal (or in certain cases could be also smaller) to the smallest rhythmic subdivision present in the selected region. For example, if you recorded a percussion part where you have mixed rhythms such as eighth, sixteenth, and thirty-second notes, then you have to set the quantization value to thirty-second notes. When you quantize "straight" your part will, most likely, lose the natural feel and groove of the live performance, and you will get the "drum machine" effect, where the part sounds stiff and unnatural. To avoid such undesirable effects, you can use a couple of different options involving two slightly distinctive approaches. The first takes advantage of quantization parameters that can be adjusted at will to adapt to any part, while the second uses groove quantization to regain the original groove of a MIDI track. Let's take a look at these two options.

Any professional sequencer nowadays allows you to control at least three main parameters of the quantization function: sensitivity, strength, and swing. By accurately balancing these three parameters you can obtain much more fluid and natural quantized parts that are much more effective than the good old straight quantization (remember the dreadful "drum machine" effect!). Let's take a look at how you can effectively use them in your productions. By controlling the sensitivity of the quantization algorithm you can choose which events will be quantized and which will be left unquantized based on their position (relative to the grid) and not on their type. In most sequencers, the sensitivity parameter ranges from 0 to 100 percent (others allow you to choose the range in ticks). With a setting of 0 percent, no events will be quantized, while with a value of 100 percent all

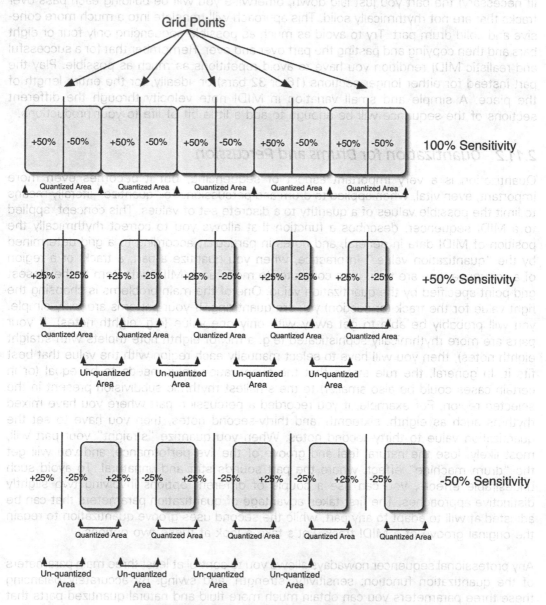

Figure 2.23 Quantization parameters: sensitivity

the events in the selected region will be affected. Any other value in between will allow you to extend or reduce the area around each grid point influenced by the quantization action. With a setting of 100 percent or with the regular quantization options, each grid point has an area of influence (a sort of magnetized area) that extends 50 percent before and 50 percent after the point (a total of 100 percent). Events that fall in these two areas will be quantized and moved to the closest grid point. By reducing the sensitivity, the area of influence controlled by the grid points is reduced. Therefore, at a sensitivity of 50 percent each point will "attract" only notes that are 25 percent ahead of 25 percent behind the grid points (a total of 50 percent). This setting is used mainly to clean up the events around the grid point and leave the natural rhythmic feel of the other notes. If you choose a negative value for the sensitivity parameter, you will achieve the opposite effect: only the events that were played further from the grid points will be quantized, leaving the ones that were fairly close to the click in their original position (Figure 2.23).

This setting is perfect to fix the most obvious mistakes but leave the overall natural feel of your performance intact. On a practical level, a sensitivity value between 250 percent and 280 percent can be used to fix major rhythmic mistakes but keep the overall feel and grove of the performance intact. Listen to Examples 2.40, 2.41, and 2.42 to compare different sensitivity settings.

Another parameter that you can use to improve the quantization of your parts is the strength option. While the sensitivity parameter has an impact on which events will be affected by the quantization, the strength allows you to control how much the events will be quantized (Figure 2.24). Any professional sequencer nowadays allows you to control at least three main parameters.

By choosing a value of 100 percent, the events will be moved all the way to the closest grid point. At the other extreme, if you choose a value of 0 percent their original position will not change. If you choose a 50 percent value, the events will be moved halfway between their original position and the closest grid point. This option gives you great control over the "stiffness" of the quantization. While with a 100 percent strength (usually the default) your parts will sound very rigid and mechanical, choosing a value between 50 percent and 80 percent will help maintain the original smoothness of the parts and at the same time correct the major timing mistakes. Listen to Examples 2.43–2.45 to hear the difference between different amounts of strength quantization.

Another parameter that you can use to control the rhythmic flow of your drum and percussion parts is the swing option. While the range of this function varies from sequencer to sequencer, in general it allows you to move every other note (based on the quantization value selected) closer to the next note (positive values) or closer to the previous note (negative values). For example, with the swing parameter set to 0 percent (Figure 2.25) you will get the same result as using a straight quantization, while with the swing parameter set to 100 percent you will get the same rhythmic feel of a triplet. What's interesting is that now you can choose any value in between to control how much your computer can swing.

With values below 50 percent, you will get a slightly looser feel than straight quantization, providing a more relaxed and slightly rounder rhythmic feel to the quantized part. If you use values between 50 percent and 80 percent, you can achieve a more natural swing feel, avoiding the unnatural and sometimes mechanical eighth note triplet feel (Audio Examples 2.46–2.48). Another trick that I like to use is to add a very small percentage of swing to some of the pieces of the drum kit (for example the HH), in order to have a slightly looser groove that is still in the straight note realm. Try to use 5 percent or 10 percent swing and you will notice how the groove sounds more relaxed! While the

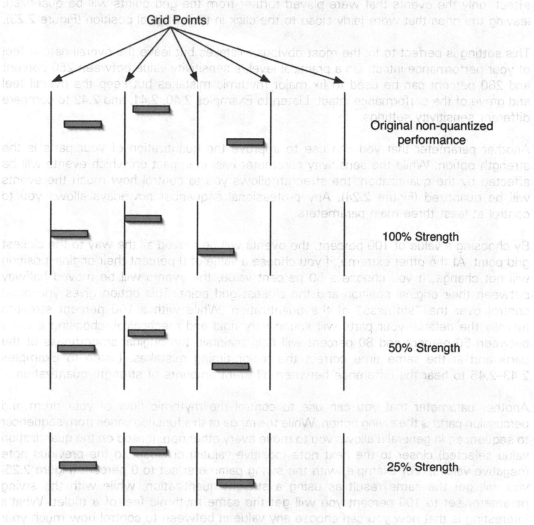

Figure 2.24 Quantization parameters: strength

individual use of the aforementioned quantization parameters (sensitivity, strength, and swing) can be extremely useful, you should experiment with different settings and values applied to different sections and tracks simultaneously. What makes a real drummer groove is the perfect combination of different rhythmic variations applied to different parts of the drum kit and to different sections of a piece. You can effectively re-create

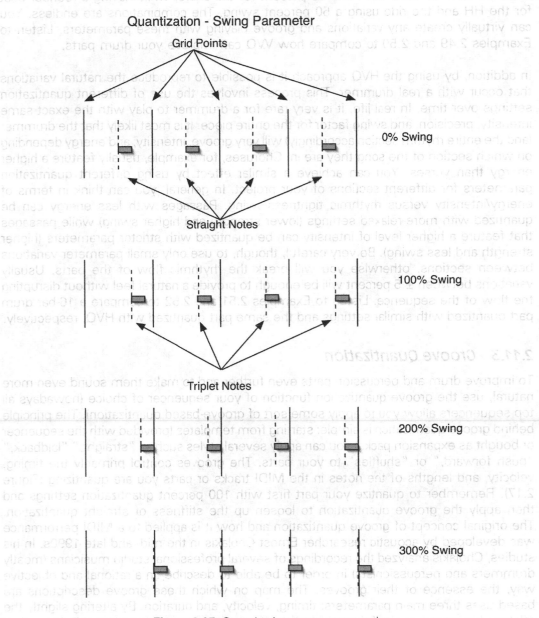

Figure 2.25 Quantization parameters: swing

such a combination by applying different quantization settings to different tracks (here referred to as vertical varied quantization or VVQ) and different sections of your drums and percussion parts (horizontal varied quantization or HVQ). For example, you can quantize the bass drum and the snare with a higher strength while leaving the HH part looser by setting its strength parameter around 70 percent. Or you can set the swing parameter to 20 percent for bass drum and 30 percent for the snare, while having a rounder beat for the HH and the ride using a 50 percent swing. The combinations are endless. You can virtually create any variations and groove playing with these parameters. Listen to Examples 2.49 and 2.50 to compare how VVQ can improve your drum parts.

In addition, by using the HVQ approach it is possible to reproduce the natural variations that occur with a real drummer. This process involves the use of different quantization settings over time. In real life, it is very rare for a drummer to play with the exact same intensity, precision, and swing factor for the entire piece. It is most likely that the drummer (and the entire rhythm section accordingly) will vary groove, intensity, and energy depending on which section of the song they are in. Choruses, for example, usually feature a higher energy than verses. You can achieve a similar effect by using different quantization parameters for different sections of your project. In general, you can think in terms of energy/intensity versus rhythmic tightness/swing. Passages with less energy can be quantized with more-relaxed settings (lower strength and higher swing) while passages that feature a higher level of intensity can be quantized with stricter parameters (higher strength and less swing). Be very careful, though, to use only small parameter variations between sections, otherwise you will break the rhythmic flow of the parts. Usually variations between ± 5 percent will be enough to provide a natural feel without disrupting the flow of the sequence. Listen to Examples 2.51 and 2.52 to compare a 16-bar drum part quantized with similar settings and the same part quantized with HVQ, respectively.

2.11.3 Groove Quantization

To improve drum and percussion parts even further and to make them sound even more natural, use the groove quantization function of your sequencer of choice (nowadays all top sequencers allow you to apply some sort of groove-based quantization). The principle behind groove quantization is simple: starting from templates (provided with the sequencer or bought as expansion packs) you can apply several styles such as "straight," "laidback," "push forward," or "shuffles" to your parts. The grooves control primarily the timing, velocity, and lengths of the notes in the MIDI tracks or parts you are quantizing (Figure 2.17). Remember to quantize your part first with 100 percent quantization settings and then apply the groove quantization to loosen up the stiffness of straight quantization. The original concept of groove quantization and how it is applied to a MIDI performance was developed by acoustic researcher Ernest Cholakis in the mid- and late 1990s. In his studies, Cholakis analyzed the recordings of several professional studio musicians (mostly drummers and percussionists) in order to be able to describe, in a rational and objective way, the essence of their grooves. The map on which these groove descriptions are based uses three main parameters: timing, velocity, and duration. By altering slightly the relationship among each of these parameters over time, a musician is capable of building a "groovy" performance that is peculiar to his or her own playing style. Take a look at

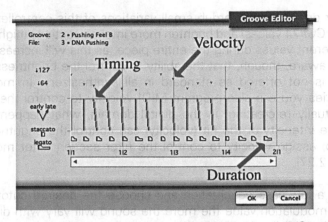

Figure 2.26 A groove quantization pattern in relationship to the click (Courtesy of MOTU)

Figure 2.26 to see what a groove looks like compared to the click in terms of velocity, duration, and timing.

In addition to the predefined templates that are either bundled with the sequencer or bought separately, you can create your own templates from audio files. This can open an incredible number of options. Imagine, for example, importing a CD track from your favorite album, featuring "Philly" Joe Jones or Dave Wackel, capturing their groove and applying it to your own parts. As in the case of the quantization parameters you learned in the previous paragraphs, groove quantization works best if it is varied and applied to different parts/sections with slightly different settings. Depending on the style of the production you can, for example, use a pushing feel for the bass drums and snare, and a laid-back style for the HH/ride. You can use a more pushy feel during more intense passages (such as chorus and transitions) and a more laid-back feel for the verses of the song. Experiment with different settings and solutions, keeping in mind that each style usually requires slightly different adjustments. For example, for a bossa nova pattern you could set the bass drum and the snare with a straight groove quantization and the HH with a laid-back feel. In a rock pattern, try to have the ride groove quantized with a push forward feel. Listen to Examples 2.53 and 2.54 to compare the same drum part first quantized straight and then groove quantized. Notice how the former sounds stiff while the latter comes alive.

2.11.4   *Performance Controllers for Drums and Percussion*

The use of advanced MIDI controllers can help you create subtle variations in the sequenced drum parts in order to improve further the realism of MIDI percussive instruments. In particular, CC#74 can be used to control the brightness of a sound, re-creating the sonic features of a drumhead hit with different intensities. In the acoustic realm, when you hit a drumhead or cymbal with less intensity the sound generated will have a darker and more opaque sonic print, whereas hitting it harder will generate a sharper and brighter result. You can use CC#74 to re-create such effect artificially. After sequencing your drum parts, overdub small variations of this controller to darken softer passages (lower CC#74 values) and brighten more intense sections (higher CC#74 values). Try to insert different values during the entire piece, as this will increase the spontaneity of the parts. Be aware, though, that the ability to control the brightness of drum sounds through CC#74 is not offered as standard in all synthesizers. In most sample-based percussion libraries you can also program MIDI velocity to control the cutoff frequency of the filter, virtually re-creating, in the digital domain, what happens in the acoustic world, where the intensity of the hit (velocity) can control the brightness of the timbre (filter). To do so, assign velocity to control the filter section (filter modulation) of your sampler (Figure 2.27).

Set the filter to a low-pass equalization and choose the desired cutoff frequency. The higher the filter modulation value the more the sound will vary with different velocities. Listen to Examples 2.55 and 2.56 to compare two sampled cymbals played without and with filter modulation applied. I also advise to program a similar control for the pitch parameter. Use the MIDI velocity to slightly alter the pitch of the snare and the toms (Figure 2.27). This will add a random acoustic touch to your drum parts.

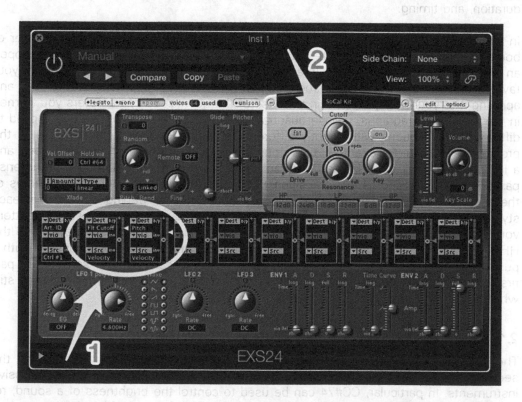

Figure 2.27 Example of velocity-controlled filter on the EXS-24 in Logic Pro, where MIDI velocity is set to control both (1) the cutoff frequency of a low-pass filter and (2) the pitch (Courtesy of Apple Inc.)

2.11.5 *Final Touches*

While MIDI and modern sample-based sound libraries offer an incredible resource to achieve realistic results for contemporary music productions, you can drastically improve your MIDI renditions by adding a few acoustic instruments (one or two usually will do the trick) to add a more human touch to the stiffness of sequenced parts. This is particularly true for drums and percussion parts. If your production features an "all-MIDI" drum part, the final result can be somewhat disappointing, no matter how much time you spend tweaking quantization parameters and MIDI CCs. By simply adding one or two parts played by real instruments, you will inject life back into your sequence. This is true also at the higher level of the entire rhythm section and even more at the full level of the entire production, as we will see in the next chapters. After investing hundreds or thousands of dollars in the latest and most sophisticated drum library, do yourself a big favor and go to the local music store to buy a couple of shakers and a tambourine. With a total of $15, you are now one step closer to dramatically improving your sequenced drum parts. With a little bit of practice (you will be surprised how hard it is to master the tambourine!) try to add a simple shaker part, after sequencing the MIDI tracks. The entire drums will come alive and the groove will be enhanced by the live instrument. Listen to Examples 2.57 and 2.58 to compare similar parts recorded, respectively, without and with the addition of acoustic instruments. If you want to go the extra mile, try also using a real HH and real cymbal. While this may require a bit more drumming skill than you ever wanted to experience, you will be very happy with the final results. The same can be said for percussion parts. It is worth investing in some bongos and other small percussion instruments just to add a bit of life to your MIDI percussion parts. Listen to Examples 2.59 and 2.60 to compare the same percussion part recorded without and with the additional real instruments.

2.12   Mixing the Virtual Drums and Percussion

While for the other instruments of the rhythm section at the mix stage we had to deal with a single element at the time, when it comes to drums and percussion things get a bit more complicated. In this case we have a multi-instrument entity that acts as one but is formed by several distinctive sonorities, all covering different areas of the audible spectrum. The first step when mixing drums and percussion is to pan the single pieces in a balanced way. There are two main options for panning: performer (drummer) view and audience view. The former has the HH and snare panned to the left, and the mid and low toms to the right, as the drummer would hear the sounds from his or her playing position. The latter features an exactly mirrored pan setting as a person in the audience would experience it (Figure 2.28).

Which one you choose is up to you and it doesn't affect the final result too much. It is important to keep in mind that through a cohesive panning you can achieve a much higher level of clarity and cleanness in your mix. Do not be afraid to pan cymbals and HH far from the center position. Usually high-frequency instruments can tolerate wider panning settings, opening nicely the stereo image of a drum set. The same can be said for high-pitched percussion instruments such as shakers and triangles. In general, try to avoid panning the bass drum anywhere else than in the center. The toms can be panned widely for a more dramatic effect, while more-conservative settings around the center will achieve a more mellow result. Other percussion instruments can pretty much be panned freely depending on the style or the arrangement. Once again, try to use the percussion as a way to open your mix up. In general, keep in mind that a wider pan setting creates a more dramatic result, giving the impression of a more direct and closer sound, whereas a more centered drum set will result in a more distant and mellow mix.

Because of its complex and wide frequency range, adding reverb to drums and percussion can be somehow intimidating and confusing. One of the biggest mistakes you can make

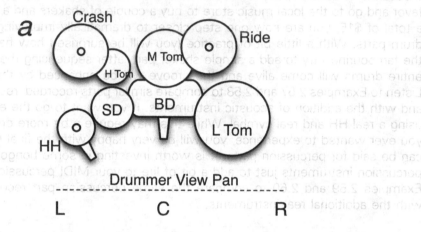

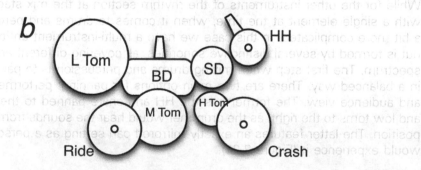

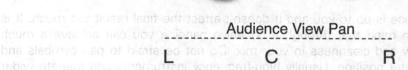

Figure 2.28 Pan settings for a drum set: (a) drummer view; (b) audience view

is to add too much reverb to the entire drum kit and therefore end up with a mix without focus and sharpness. In general, use a fairly short reverberation time (around 1.5–1.7 seconds) on toms, HH, and cymbals. On the snare you can afford to use a slightly longer reverb since its frequency range and its attack will cut through more easily than the

Table 2.5 Generic reverb parameters

Parameter	Description	Comments
Type	It describes the type and overall sonority of a reverb	Typical types include: hall, chamber, plate, gate, and reverse
Length (Decay)	It represents the decay or length of the actual reverb	It is usually expressed in seconds or milliseconds
		More-sophisticated reverbs allow for control of the decay parameter for two or more frequency ranges
Room Size	It controls the size of the room in which the reverb is created. Bigger rooms are usually associated with longer reverberation times	Sometimes, depending on the type of plug-in, in addition to Room Size you can find a control that allows you to control the Room Shape
Early Reflections (ER)	It controls the first bounces that follow the dry signal before the actual reverb	It can have several sub-parameters such as the ER size, gain, and pre-delay
Pre-delay	It controls the time between the dry signal and the early reflections	
Delay	It controls the time between the dry signal and the actual reverb	
Diffusion	It controls the distance between the reverb's reflections	A low diffusion features more scattered and distant reflections, while a high diffusion features closer and more frequent reflections.
		Low diffusion is usually more indicated for sustained pads and passages (i.e. string sections). High diffusion is more indicated for percussive instrument and fast transients audio material
High Shelf Filter	It cuts some of the high frequencies of the reverberated signal	This parameter is not found in all the reverbs.
		It is useful to avoid excess of high frequencies in the reverberated signal in order to re-create a more natural sound
Mix	It controls the balance between dry and wet signal	If the reverb plug-in is used as insert you should set the Mix parameter to 50 percent or lower. If you use the reverb as a send then you should set the Mix parameter around 100 percent

other pieces. Avoid adding reverb to the bass drum unless absolutely necessary. Low frequencies have a tendency to become muddy and washed out if used with reverb. To get the most punch out of your bass drum keep it as dry as possible and panned in the center. In order to tailor your reverb settings for drums, look at Table 2.5 to review the meaning and functions of the most common reverb parameters.

For drums and percussion, use a higher diffusion and lower slightly the early reflection level. Use a little bit of pre-delay to create a cleaner and clearer sonority. If you need to use reverb on a stereo mix of a drum/percussion set without being able to separate the individual tracks (a situation that you should avoid if possible), you can set your reverb parameters so that the low frequencies of the mix (mainly bass drums and low toms) either have a shorter reverberation time or feature a lower reverb level. You can achieve this effect by using the damping section of your reverb. This parameter allows you to

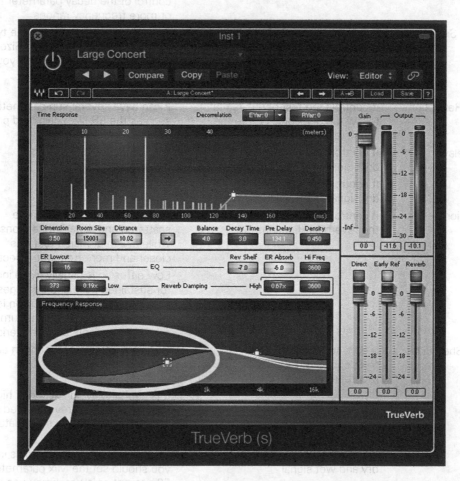

Figure 2.29 Use the damping parameter of your reverb to apply reverberation only to the higher frequencies of a drums/percussion mix (Courtesy of Waves)

control the amount of absorption that the reverb bounces will encounter. By damping the low frequencies of the reverb you create the illusion of not having reverb on low frequencies (Figure 2.29).

When it comes to equalizing drums and percussion, you should follow rules similar to those for the other instruments in the rhythm section. In general, try to carve out of each drum piece the frequencies that are specific to that particular instrument. You can most likely roll off, for example, the lower end of cymbals and HH since they usually do not contain any important sonic information below 500Hz. At the same time, use the mid-range section of a peak equalizer to shape the snare. Use Table 2.6 as a starting point for drum equalization.

After equalizing each single piece of the kit, spend some time listening to drums and percussion together as a whole. At this stage, pay particular attention to the balance between the single pieces and to their relative pan positions. The overall drum/percussion mix should be cohesive, clear, and balanced on its own, before adding the other elements of the rhythm section.

To improve further the realism of your MIDI drums/percussion parts, use the re-amping technique that we learned in the previous paragraphs of this chapter when discussing the guitar and the bass. In this case, use the best set of speakers you can get hold of

Table 2.6 Drums and percussion key equalization ranges and frequencies

Frequencies	Application	Comments
60–80Hz	Boost to add fullness to the bass drums	
100–200Hz	Boost to add fullness to snares	Do not overuse to control snare depth
200–300Hz	Cut to reduce low and unwanted resonances on cymbals	Be careful not to boost too much of this frequency range in order to avoid adding muddiness to the mix
400–600Hz	Cut to reduce unnatural "boxy" sound on drums	
5–6kHz	Boost for attack overall drums	A general mid-range frequency area to add presence and attack
7.5–9kHz	Boost to add attack on percussions	A mid- to high-range area that controls the clarity and the attack of the mid- to high-range instruments
10–11kHz	Boost to increase sharpness on cymbals	High-range section that affects clarity and sharpness
	Cut to darken drums and percussions	
14–15kHz	Cut to reduce sharpness on cymbals	

instead of a guitar amp (you can use also a couple of guitar amplifiers to achieve a more crunchy and cutting-edge sound). Send a mixed version of the drum/percussion tracks to the two speakers already panned as you like. Use two very good condenser microphones to capture the mix played through the speakers and record their output to a stereo track in your audio/MIDI sequencer. It is crucial that you use the best microphones and speaker you can afford. In addition, try to capture as much room ambience as possible in order to inject back into the stereo track a realistic environment response. This technique works especially well for studios that have good-sounding rooms. If you don't think you have a particularly good acoustic where you plan to re-amp your drums, don't bother with it. Listen to Examples 2.61 and 2.62 to compare a MIDI drum track mixed straight to HD and a re-amped version, respectively.

2.13 Final Considerations on Sequencing for the Rhythm Section: Tempo Changes

As you can see, there are several ways and techniques available to the modern composer and arranger to sequence convincing rhythm section parts without having to hire a rhythm section. Before moving on to the string orchestra in Chapter 3, some overall techniques that can be applied to the rhythm section as a whole, as well as the sequence in general, will be discussed, in particular the important issue of tempo and tempo variations. One of the main problems with a sequencer is that, by definition, it is extremely precise and reliable in terms of tempo. A sequencer is "stuck" with an extremely precise brain that does not allow instinctive reaction to musical events or external stimuli. If you have ever assumed that a sequencer sounds great because it is able to keep a perfectly steady tempo without any sweat, then you definitely need to read further because you are in for some surprises that can improve your sequencing skills considerably. As part of the goal to create as much variation as possible when sequencing for acoustic instruments, creating small tempo variations for the rhythm section can definitely improve the static nature of a MIDI production. To free up the sequencer the secret lies in the tempo track, sometimes also called the conductor track. In this track we don't find the usual MIDI or audio data that we are used to dealing with on a regular basis. Instead, we find meta-data that constitute the inner click of a sequencer. By adding, modifying, and varying the tempo track, we can teach the sequencer to be creatively and naturally "sloppy." This approach, if done right, will help the sequencer to loosen up and to match the natural groove of a real orchestra or ensemble. The insertion of small tempo changes data at strategic points of a sequence can greatly improve the overall groove and realism of the production, bringing your sequencing skills to a new level. If you analyze any acoustic band that has played in a studio session since the invention of the first recording device, you will notice that tempo played a crucial role in giving a distinguished signature to their sound. Several Miles Davis recordings of the 1950s and 1960s, for example, were characterized by a noticeable and steady increase in tempo from the beginning of the tune to the end. The same can be also said for rock and pop bands before the introduction and use of the click track. The fact that a band speeds up or slows down is not necessarily a bad thing if done smoothly and with the right intention.

The goal is to achieve similar results with the sequencer in order to regain that natural feel that is typical of an acoustic ensemble. The simplest way to improve the overall groove and feel of a sequence is to insert tiny tempo changes in crucial spots along the way. This technique is fairly simple and it doesn't require learning complicated and cumbersome procedures. The idea here is to teach your sequencer how to "humanly" make controlled mistakes when dealing with the tempo of your project. In general, it is better to increase the tempo of a sequence slightly, rather than decrease it. Most likely, a good band will get more excited during the performance of a piece. This will translate into an overall steady and almost imperceptible accelerando that will carry the audience through the piece, gently pushing the listener toward the end. Usually, these tiny tempo changes occur at crucial moments in the structure of a piece, such as during transitions between sections (such as the transition between a verse and a chorus) or during drum fills and rolls (where usually the drummer pushes the beat slightly to add excitement to the passage). Always apply very small changes to the tempo map of a sequence: anything between 0.5 and 1.5bpm will do it. Higher values will translate into obvious jumps in tempo that will make your virtual band sound sloppy and unsettled. You can also insert negative tempo changes, especially when going from an exciting and fully orchestrated section to a more calm and introspective one. The effect will be one of a rhythm section that is able to control the energy and release the tension created during a more exciting passage. With this technique, you are basically inserting small static tempo changes that will achieve the effect of having your sequencer making tiny human "mistakes." Table 2.7 describes where and when to insert tempo changes in order to free up your sequencer from a steady and stiff tempo map.

Listen to Examples 2.63 and 2.64 to experience how small tempo changes, inserted in crucial spots, can improve a sequence (Score 2.4). The first example features a piece without tempo changes, while the second example features the same piece with tiny tempo changes.

While static tempo changes are useful in re-creating how an acoustic ensemble would realistically play a piece, they are not very useful in creating more musical variations that are dictated by a human rendition of a particular composition. Accelerandos and rallentandos (Acc and Rall) represent two of the most important musical elements that set apart a MIDI sequence from a real performance. Fortunately, the sequencer can help us to re-create the tempo variations that would add naturally expressivity to a performance. Accelerando and rallentando can be easily re-created through the insertion of smooth tempo changes. The principle is easy: instead of using a single change to move statically from one tempo to another, in this case we set a beginning point and an end point in the sequence with a set start and end tempo. The sequencer will fill in the blanks, inserting for us a series of ascending (Acc) or descending (Rall) tempo changes between the two points. This is a great tool to quickly create tempo variations that are supposed to be performed by the sequencer according to a particular score. It gets even better, though! We can instruct the sequencer to use a series of different curves, ranging from linear to logarithmic, from exponential to polynomial. This feature allows us to control with extreme precision how a sequencer will react to an Acc or a Rall marking. When

Table 2.7 List of common critical points in a piece where small tempo changes can be inserted to re-create the feel of an acoustic rhythm section

Position	Variation (positive or negative)	Comments
Beginning of a piece	N/A	The tempo at the beginning of the sequence should reflect the intended original tempo of the piece. Sometimes, depending on the style, you can choose to set your initial tempo to a slightly lower value in order to average the future increases that you will insert
Transition from verse to the next verse	Between +0.5 and +1bpm	This is a good spot to insert a very tiny tempo increase since it is usually where the drummer would insert a fill or the ensemble would mark the beginning of a new section
Transition from verse to chorus	Between +0.5 and +1.5bpm	This is a transition where the rhythm section would definitely mark a change in excitement and therefore where the tempo would naturally increase
Transition from chorus to verse or to bridge	Between −0.5 and −1bpm	If you are looking for a place to let your virtual band "breath," this is the place
Transition from bridge to final chorus	Between +0.5 and +1bpm	This is where your ensemble really wants to shine and will push it to the limit

sequencing for orchestra, this feature is essential in order to achieve a more convincing MIDI rendition of a score.

Orchestras do not play (in most cases) with a metronome, but instead they follow a conductor. The tempo track (or conductor track) is your virtual conductor. You can create these dynamic tempo changes in several ways: by programming the changes selecting the range in bars in which the tempo change will occur and specifying the start tempo and end tempo that the computer will use to calculate the variation, or you can manually draw in the changes with the pencil tool; a more musical method involves the use of the tap tempo feature, where you can literally conduct your sequencer in real time and record the tempo changes as you would be the conductor of your virtual orchestra. This option allows you to conduct your sequencer by inputting MIDI messages from a controller, which acts as the baton of a conductor (Figure 2.30). You can have very natural tempo changes that your virtual orchestra will follow during the performance of your composition. While not all sequencers include this feature, some of the professional ones do. In general, you need to put the sequencer in "slave" mode (external sync), where the source of the sync is not SMPTE or MIDI time code (MTC) but is a metronome played live by you on your MIDI controller. Listen to Example 2.65 for a demonstration of the tap tempo feature applied to

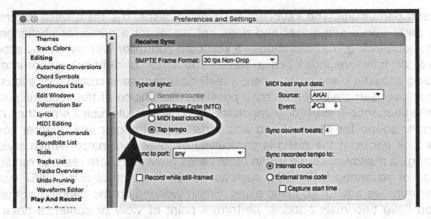

Figure 2.30 The "Tap Tempo" feature in Digital Performer (Courtesy of MOTU)

a rubato part. This technique can be applied not only to the rhythm section but also to any sequence style and orchestra section. When sequencing orchestral parts, using tempo changes and tap tempo can definitely increase the quality of your productions.

2.14 Summary

Although most rhythm players read music well, some (in the more popular music fields) may not; it is good to be aware of the capabilities of the specific players. The written parts for the rhythm section serve mainly as a guide, and professional rhythm players are often relied upon to help create a part. With master rhythm parts, only the essential parts are notated specifically. The chord symbols indicate the harmony to be heard and serve as a synopsis for the corresponding written notes. Directives in English or Italian guide the players otherwise. The drummer in this case must be most attentive as his or her part is practically non-existent. For the arranger or music producer, last-minute options are readily available as the players can view all of the written aspects of the music, enabling them to experiment with the creation of their individual parts.

Careful listening to rhythm sections on recordings is essential to one's development as a writer for the rhythm section. Listen to the section as a whole and listen laterally to the individual instruments. Compare inactivity versus activity within the instruments of the rhythm section and of the section as a whole with the greater ensemble; register the placement of the instruments in relation to each other and also of each instrument independently; tone color has a great effect on style and mood. Remember that many rhythm players specialize in certain styles of music and finding the right match for your music is very important to the success of the recording.

In this chapter, you learned advanced sequencing techniques to create convincing parts for the instruments of the modern rhythm section, including keyboards, guitars, bass, drums, and percussion. The keyboard is, in most cases, the main MIDI controller used to input parts into a sequencer and does not represent a particular challenge when it comes to sequencing. You can choose a keyboard controller with synth-key action, semi-weighted, or fully weighted like a piano. The latter is definitely recommended for sequencing acoustic piano parts where a realistic response is crucial. For acoustic piano, use a multilayer sample-based library if possible. The choice of the acoustic piano sound used in the sequence is extremely important. As a general rule, use a smoother, mellower, and warmer sound for classical and jazz parts, and an edgier and brighter sonority for pop and rock tracks. If the piano is particularly exposed (no matter which style you are sequencing), a mellower sound will in most cases provide a more realistic rendition. Take advantage of advanced MIDI CCs, such as CC#66 (sostenuto) and CC#67 (soft pedal), to further improve a MIDI acoustic piano. When working on the panning of the acoustic piano you have two main options: performer point of view or audience point of view. Use a light equalization to correct problematic frequencies embedded in the samples or annoying frequency interactions (such as masking) occurring with the other instruments of the mix. In terms of reverberation, use a convolution reverb instead of a synthesized one. For classical composition, you can use a longer reverberation time between 2.3 and 3.2 seconds, whereas for jazz productions a more intimate and smaller reverb with a length between 2 and 2.5 seconds may be preferable. Similar techniques and settings can be used for the electric piano and the organ.

The synthesizer (hardware or software) is an endless resource in terms of color, sonorities, and timbres. Of the many types of synthesis, the most commonly found in contemporary synthesizers include subtractive, additive, FM, wavetable, sampling, physical modeling, and granular synthesis. Synthesizers are available as either hardware or software. A software synthesizer is usually available in two formats: as a plug-in or as a standalone.

Another extremely versatile instrument of the rhythm section is the guitar. When it comes to guitar sounds and patches that are commonly used in contemporary productions, there are three main types: the nylon (sometimes also called classical), the steel string (folk), and the electric guitar, with all its variations, such as clean, saturated, and distorted. Sample-based multisample libraries are the best options for these sonorities, although a valuable alternative is represented by wavetable and PM synthesis. The best controller to sequence guitar part is the GMC. Using this device will allow you to input convincing guitar voicings (a task that is much harder to do from a keyboard controller). A GMC can be used also in a more creative way. Since each string has an individual pick-up, by assigning different strings to different MIDI channels and patches you can create a one-man band and generate interesting layers and ensemble combinations. Use key switches to insert in different spots slides, pull-offs, and other colors that are specific to the guitar vocabulary. Some libraries offer specific performance-related noises that are peculiar to the guitar's sonority. These usually include string noises, fingers sliding on the strings, frets buzz, etc. The most effective way to create convincing strummed guitar parts is to use a GMC. If you have to use a keyboard controller, use the function called "flam," which is available in most MIDI sequencers. The use of effects such as delays, chorus, and flanger at the mixing stage can really improve the overall sound of a MIDI-sequenced guitar patch. Re-amping is also an extremely valuable technique to render more-realistic guitar parts. When using equalization, try to be gentle and not to overdo it unless absolutely necessary. In terms of reverberation for background parts with a busy rhythmic signature, use slightly shorter reverberation times, between 1.5 and 1.7 seconds. For solo guitar parts or up-front melodies you can usually afford to apply slightly longer reverberations, up to 1.9 seconds.

The bass is less challenging than guitars when it comes to sequencing it and it can be rendered more easily through the use of a conventional keyboard MIDI controller. The electric bass (or bass guitar) sonority can be convincingly reproduced using both wavetable synthesis and sample-based techniques. As for the guitar, you can add a touch of realism by inserting performance noises here and there when sequencing for the bass. In terms of mixing, you should be conservative when panning any bass track since low frequencies usually feel more natural if panned in the center of the stereo image. In the case of MIDI bass parts, re-amping is a great technique to bring virtual tracks to life and remove the typical boxy sonority of the synthesizer. A touch of equalization on bass tracks is used to achieve a deep and full-bodied sonority and to avoid muddiness. The low register of the bass usually calls for very little reverb (or most likely no reverb at all). Instruments that cover mainly the low end of the frequency range, such as bass and bass drums, usually require much less (or no) reverberation.

Generic drums and percussion can be fairly accurately rendered with wavetable synthesis, but for acoustic sonorities you should definitely choose the route of a professional multi-sample library (or software sample playback/plug-in). For drums/percussion sequencing, try to avoid using your keyboard as a controller and use MIDI pads instead. Always use separate tracks for each piece of the kit or of the percussion set. The only exceptions are bass drums and snare, which will give you a much more natural feel if recorded at the same time. Try to avoid repetitions (copy and paste). Instead, play the part for longer sections (16 or 32 bars), or ideally for the entire length of the piece. Use quantization to fix parts rhythmically that sound sloppy, but instead of using straight quantization take advantage of the advanced quantization parameters (such as strength, sensitivity, and swing) and groove quantization. Try to add at least one real percussion instrument to your drum parts. Most of the time a simple shaker or tambourine audio track will do the trick. As in the case of the piano, there are two main options for panning: performer view and audience view. Take advantage of a wider panning setup to enhance intelligibility and presence. In terms of reverb settings, a fairly short reverberation time (around 1.5–1.7 seconds) may be used on toms, HH, and cymbals. On the snare you can use a slightly longer reverb. Avoid adding reverb to the bass drum unless absolutely necessary. Set the diffusion parameter of the reverb to higher values to reduce the edginess of the reverb tail. When using the equalizer on drums, carve out the frequencies that are specific to the particular drum piece on which you are working.

As part of the goal to create as much variation as possible when sequencing for acoustic instruments, creating small tempo variations for the rhythm section can definitely improve the static nature of a MIDI production. The use of accelerandos and rallentandos (through the insertion of smooth tempo changes) is another technique that will allow you to inject some new life into the virtual rhythm section.

2.15 Exercises

Exercise 2.1

(a) Write a contemporary "groove"-oriented piece for the rhythm section using instruments of your choice. Make at least one part carry an ostinato (a repeated rhythm or melodic pattern) that serves as a focal point. The remaining parts should work around the ostinato, incorporating lesser degrees of strictness. When you're finished, list the instruments in order from most strict to least strict. This will give you some perspective regarding the interaction and responsibilities of each of the players.

(b) Sequence the above groove using the following techniques:

   (i)   groove quantization for drums and percussion;
   (ii) sensitivity, strength, and swing for the bass part;
   (iii) add at least one real drum/percussion instruments (shaker, tambourine, etc.) to the drums/percussion part.

Exercise 2.2

Take a harmonic progression that uses chord symbols only; write the individual chord tones in the following ways and practice playing them.

(a) If you're a pianist, create guitar voicings at the keyboard by using only the index and pinky fingers of each hand to form a series of wide four-note voicings.

(b) If you're a guitarist, create piano voicings using only your left hand to form a series of close four-note voicings (within an octave range).

(c) If you don't play either of these instruments, do both techniques.

(This works best with progressions in jazz and Brazilian bossa nova styles.)

Exercise 2.3

Play a two-bar drum set pattern into your sequencer and transcribe what you've recorded, using conventional noteheads for the drums and x noteheads for the cymbals.

Exercise 2.4

Sequence Score 2.1 (from Appendix A) using the following techniques:

- groove quantization on drums
- performance noises for bass and guitar parts
- re-amp the guitar parts
- subtle tempo changes
- add a real shaker and/or tambourine part.

3 Writing and Sequencing for the String Orchestra

3.1 Introduction: General Characteristics

The full string orchestra is comprised of five sections. They are the first violin, second violin, viola, cello, and bass sections. The size of each section can vary, depending on the size of the orchestra. Economic considerations may ultimately determine how many string players can be hired. Whatever the size, there is usually a standard ratio that offers the best balance within the string orchestra. There are more first violins than second violins and the numbers in each remaining section continue to decrease. This is because the violin (being smaller and having shorter strings) is not always as loud as the larger string instruments and the higher notes in the first violin part need more players to establish a greater resonance. Each section, with the possible exception of the basses, is usually set up with an even number of players. There are two players per "desk," or music stand. The player on one side will usually play the higher note in the event of an indication to divide (divisi), while the other player is usually responsible for page turns. An average breakdown of players in a very large orchestra might be 16 first violins, 14 second violins, 12 violas, 10 cellos, and eight basses (abbreviated as 16–14–12–10–8). This sum total of 60 string players will provide an idea of how expensive a group of this size can be. This is why, in today's economy, along with the available technological advantages of sound production, string sections are usually much smaller.

Traditional orchestration books guide the reader within the context of a nineteenth-century live orchestra that could not use microphones. (Most symphony orchestras today still refrain from using microphones for amplification purposes when performing traditional orchestral repertoire.) Many string players were needed in order to establish a favorable balance with large brass sections, not to mention the additional woodwind and percussion sections. Today, it is certainly wise to respect and be aware of the traditional context, but it is possible and sometimes necessary to allow the technology of the twenty-first century (microphones in live performance and studio recording) to help achieve a proper balance with the rest of the ensemble.

One common technique that is applied in studio recording is "layering." Usually a smaller ensemble of strings is hired (as few as eight to 12 players) to record. After the initial pass is complete, the string ensemble will overdub, playing the same material as part of the second pass. Usually a third pass is required to create a lush, full sound.

The sound of a string quartet (violin 1, violin 2, viola, and cello) is quite different from the larger string orchestra. It offers an intimacy that is harder to capture within the string orchestra. The trade-off is that the tone quality is much more dry and exposed. It is more difficult to hide any problems with intonation and a homogeneous blend is inferior to the larger string orchestra. Both, however, have their unique artistic contributions and the arranger should consider carefully which ensemble is more appropriate. Arranging for a string quartet requires a bit more careful planning regarding harmonic voicing. If more than four notes within a harmonic structure are desired, the use of double-stops becomes mandatory. An excessive use of double-stops can hinder a performer's ability and in some cases may be impossible to perform. Although the string orchestra solves this problem, it may be advantageous for the music arranger to avoid the temptation of writing too many notes within a harmonic structure. Many large-scale orchestral works do show the strings playing a multitude of pitches divided within each section and, certainly, this can be quite effective when there are 60 string players on stage. Today, however, there are more times when the string section is much smaller and they are often playing with a rhythm section or a loud brass section. When working in this environment, it may be better to stay within four notes (five notes if basses are used), adding only what is necessary to keep the sound strong and full. It is also important to realize that a wider chord voicing will sound more full when the range of the voicing's spread is at or beyond two octaves. If extra notes are needed, it is better to split the bass, cello, and viola sections as they have a greater carrying power. Sometimes it can be best to let the strings (with the exception of the basses) play a melodic idea in octaves against voiced brass. These techniques provide expedience and enhancement of performance with regard to the following factors: it becomes unnecessary for string players to play an excessive amount of double-stops, string players do not have to decide "who will take what note?" in divisi passages, the "layering" technique is straightforward, and the sound is at its fullest. Of course, there are times when it is necessary or desirable to expand the harmony within a composition, especially if the strings are the featured section. The context ultimately determines the choices to be made.

In general, it is important to listen to many different contexts of string ensemble writing to understand the idiomatic aspects and technical capabilities of this group of instruments. One distinct advantage the string orchestra has, in contrast to the winds and brass, is that there are no concerns regarding the necessity to breathe in order to produce sound. Endurance factors that are critical to these other instrumental groups are inconsequential with strings. As a result, it is easy for string players to increase intensity—quite suddenly in fact—and stay there without any strain. Strings are great for creating lush pads, long, serene notes, or high scratchy notes that last indefinitely. The concerns regarding range are minimal since the string orchestra spans almost the entire length of the piano keyboard.

Unlike wind instruments, the execution of repeated notes is much easier. Combine the repeated notes with strategically placed accents and the strings can create quite a rhythmic impetus. What they lack in color (compared to the multitimbral woodwinds) they make up for in the creation of a beautiful transparent and/or velvety texture.

3.2 Creating Sound

String players create sound primarily by drawing the bow (made of horsehair) across the strings using the right hand. In the case of the "open" strings, this is all that is necessary. In most cases, the left hand works in conjunction with the right by pressing down (a "stopped" note) on the string being bowed. This helps the tone production with the addition of vibrato (and sometimes volume), but most importantly the fingers of the left hand create changes of pitch within the confines of the one particular string. Each finger (the thumb is not included) is placed consecutively, creating "positions" on the string. This of course is very similar to what guitarists and electric bassists do. The main difference is that the fingerboard does not contain frets, so the positions (or locations) of the notes are not locked in (or readily visible). This presents a challenge to the performer since the correct intonation of each note is ultimately determined by the sensitivity of the performer's ear in conjunction with the accurate physical placement of the fingers. With this in mind, it is important for the music writer to be respectful of these challenges in the performance of a string instrument. Unlike the piano, the notes do not automatically "pop out." It is important to make each individual part melodic in the sense that it is "singable."

3.2.1 "Open" String versus "Stopped" String

With the exception of the lowest string on these instruments, the three higher open strings may also be played on the adjacent lower string by stopping that string with the finger at the appropriate point. Where this opportunity arises, there are distinctions to be made concerning the choice of playing a note on the open string or with a stopped position on the lower string. Each approach has its advantages and disadvantages. Please refer to the outline below.

Open	Stopped
Easy to play	More technically demanding
No vibrato available	Vibrato and multiexpression available
Offers a natural fade of the note	The release of the note is more apparent
Notes are full and bright	Notes are slightly less resonant
Pizzicato technique is enhanced	Notes are shorter when plucked
Easier to play two notes	Double-stops are harder, but blend better

3.2.2 Bowings

The player can draw the bow in two directions: the "up" bow and the "down" bow, and each is indicated with its own specific marking (Figure 3.1).

The string player will play this using a bow stroke for each note. The markings illustrated above indicate to the player whether a down stroke or up stroke is desired. After the initial indications it is assumed that the pattern is to be continued.

Figure 3.1 Separate bow strokes

More importantly, though, it is essential that the orchestrator indicate the bowings (notes to each stroke of the bow) for the string player. This is done simply by writing a phrase marking (as you would indicate a slur for wind instruments) over or under the notes to be played within a bow stroke (Figure 3.2).

The bowings indicated here create a more flowing and interesting phrasing, particularly in measure 2. It is also not necessary to use any other bow markings (up or down) since this will occur naturally.

Figure 3.2 Indicating specific notes per bow stroke

If there is no such indication, the string player will bow each note separately as seen originally in Figure 3.1. At times, separate bow strokes can be a desirable effect (particularly for rhythmic momentum), but usually detached bowing may make a legato phrase sound choppy and stiff. Therefore, it is usually more common to indicate bowings (phrase markings). The "up" and "down" bow strokes have slightly different characteristics that are worthy of mention. The up bow starts at the point of the bow that is farthest from the hand and is pushed upwards across the string progressing toward the opposite end. It is effective with melodic pick-ups that lead to a strong downbeat. It also tends to start with a softer attack and increase in volume (Figure 3.3).

Figure 3.3 The "up" bow

The down bow works in the opposite fashion. Starting at the point of the bow that is closest to the hand, the player draws the bow downward across the string toward the opposite end. It creates a stronger, more percussive attack and tends to decrease in volume as the bow moves across the string (Figure 3.4).

The "down" bowstrokes reinforce the accents.

Figure 3.4 The "down" bow

Of course, the professional string player strives to smooth out the contrasts of the up and down strokes in favor of a balanced sound but, when a specific musical effect is called for, the string player will tend to favor one over the other.

There are other factors to consider as well. The velocity (speed) and weight (pressure) of the bow against the string affect the volume of tone considerably. A slower speed allows more notes to be played within a bow stroke. But, to create more volume, the bow must move across the strings with greater velocity, thereby decreasing the number of notes within the bow stroke.

It must be mentioned that, ultimately, bowings along with up and down indications are subject to interpretation. Try to suggest what you think will work most naturally in accordance with your musical intentions. It may not always be necessary to indicate an up bow or down bow, but always write bowings to indicate the expression, shape, and flow of the music. As a rule, defer to the concertmaster (the principal first violinist) if a change is preferred.

Helpful hint: Even though string players have spent long hours of practice trying to smooth out the necessary directional change of the bow, it is good to use this essential physical motion to your advantage. Since there is a natural accent that occurs as a result of the directional change, bowings can be written in a way that will bring out desired accents within a melodic phrase. This is particularly useful for an angular or syncopated rhythmic melody.

3.3 Performance Techniques for Greater Expression

There are many techniques that offer a variety of color and effects. Since the string orchestra is rather monochromatic in tone color, these effects become even more useful for contrast. Before examining the effects, it should be remembered that the normal position of the bow is between the bottom of the fingerboard and the bridge (the piece of wood with the notches that holds the strings in place on the body of the instrument). This position offers the fullest and most typical tone of each string instrument.

For a more edgy, glassy, or somewhat metallic effect, the string player will place the bow near the bridge. The musical effect can be cold, haunting, fragile, and tense. The Italian term for this technique is *sul ponticello*.

For a warm, dry, and intimate effect, the string player will place the bow over the bottom of the fingerboard. This dulls the tone somewhat, making it favorably darker or providing a floating quality. The Italian term for this technique is *sul tasto*.

For staccato passages (notes of short duration), string players can play with the bow "off the string" or "on the string." Off the string, known in Italian as *spiccato*, works well with softer and lighter passages. With this technique, the player bounces the bow off the string, creating a wonderfully buoyant quality. On the string, known in Italian as *martellato*, is heavier and better for loud, vigorous passages.

A variation on the previous technique is referred to, in French, as *louré* (loor-ay). This effect combines a sense of an interrupted legato texture. There are several notes, usually repeated pitches, that are played within one direction of the bow. These notes, played on the string, are articulated slightly by a gentle slowing of the wrist.

The bowed tremolo effect is created with a rapid movement of the bow within a fixed pitch. This effect also sounds good within the context of a voiced chord (Figure 3.5).

Its musical effect ranges from mild to aggressive forms of tension to exhilaration. It works at a very soft volume level (usually in conjunction with *ponticello*) and at a loud volume. At the loud volume level, the tremolo gives energy to the note and keeps its core sound thick and active.

Figure 3.5 Bowed tremolo technique

Figure 3.6 Fingered tremolo technique

The fingered tremolo is more delicate and floral. This effect is accomplished primarily with the left hand. The performer plays two different pitches on the same string (the open string does not work as well here), moving rapidly between the two by lifting one finger to expose the other position, thereby creating the effect of a trill (Figure 3.6).

Notice that specific strings are indicated for the various instruments in order to avoid the use of an open string. Remember also that the interval must not be too large owing to the physical confines of the hand. The most commonly used intervals are seconds (actually called trills), thirds, and fourths.

A less common bowing technique is the use of the wooden part of the bow. It is played against the string, creating a light tapping sound along with a less apparent harmonic of the string's fundamental pitch. This is referred to in Italian as *col legno* (col len-yo). It is good for creating an unusual percussive sound, but is not conducive to more commercial endeavors.

3.3.1 Portamento

Similar to the portamento on a synthesizer, this effect is created with the left hand sliding along one string from one pitch to the other and is indicated with a straight line drawn between the written pitches. The distance can be great or small as desired, but is certainly limited to the length of the string. The effect may be exaggerated or as subtle as a vocalist's sliding from one pitch to another when using an "ooh" or "ah" syllable.

3.3.2 Pizzicato

This Italian term indicates to the player that the notes should be played without the bow. The player instead uses the index finger of the right hand to pluck the string. (It is important to realize that the player is usually holding the bow with the other fingers in preparation for the next entrance requiring the use of the bow and, therefore, dexterity is somewhat limited.) In extended passages, it is helpful if the strings can abandon the bow to have more physical freedom. This requires more time to retrieve the bow when needed for bowed passages.

The pizzicato effect essentially creates notes of short duration that are rather light in weight. It is not wise to write overly complex parts using this technique. Precision may suffer greatly with so many players attempting to perform the part. If the part is simple, the effect can be quite interesting and a refreshing change from the more typical arco passages.

The effect takes on more depth in tone and rhythmic feel as it is employed within the lower strings. Many times the basses (playing pizzicato) support rhythmically the other string sections in the playing of their bowed passages. This is important to consider because it significantly lightens the feel of the music, making it more buoyant or gentle.

In particular, with the use of the basses, is the general concern of not being too heavy. Pizzicato, being lighter in texture, scoops out the heaviness of a bowed bass and, in turn, the robust quality of the bass enhances the rather thin sound of the pizzicato. The term arco indicates a return to the use of the bow. It is not necessary to indicate this unless the player is coming out of a pizzicato passage. However, pizzicato is always to be indicated for orchestral string players. The exception to this rule is within the context of a jazz band, where the bassist is almost always playing pizzicato. In this scenario, arco must be indicated to the player if the use of the bow is desired.

3.3.3 Harmonics

A harmonic is essentially a much higher overtone that is drawn into the foreground in lieu of the original fundamental note. With regard to the music in general, this effect offers an eerie quality. It is thinner and more fragile than the *ponticello* effect and is best used in slow and quiet drone-like passages.

There are two types of harmonics. Natural harmonics are created on the "open" strings. The player touches the string lightly in various places and, depending where the finger is placed, a different overtone from the series emerges. Normally, with natural harmonics, a small circle is written above the note to be heard in the harmonic series but, as a beginning student of orchestration, it is acceptable to use the following method, which also applies to creating artificial harmonics. Figure 3.7 shows a common harmonic on the violin's open G string. The diamond-shaped notehead on C, a perfect fourth above the open note G, indicates where the player touches the string without "stopping" it. The overtone that is created actually sounds two octaves higher than the written G.

(sounds 2 octaves higher)

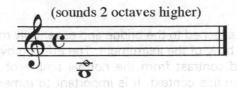

Figure 3.7 Natural harmonic on the "open" G string

Artificial harmonics are created with stopped notes. While holding one finger in the position of the stopped note, the player uses another finger, placing it very lightly a perfect fourth higher (Figure 3.8).

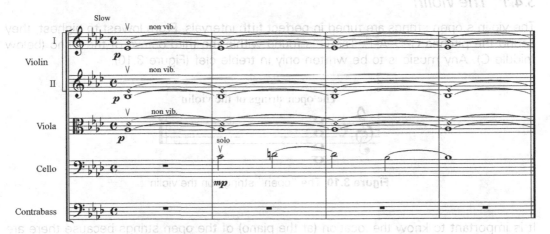

Figure 3.8 Artificial harmonics in the violins and viola

This also brings out the overtone, which is heard two octaves above the fundamental pitch. (The scientific principle behind the creation of these sounds goes into much more detail than can be afforded here and it is suggested that this topic be studied further using a traditional text on orchestration.)

If a small circle is written above the note, although incorrect for an artificial harmonic, the player will understand your intention and provide the sound of the harmonic (Figure 3.9).

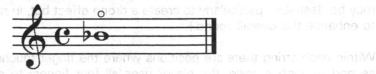

Figure 3.9 Alternate notation of a harmonic

3.3.4   Mutes

These small devices are attached to the bridge and essentially muffle, to a certain extent, the vibrations within the body of the instrument. The effect provides a veiled tone quality and can be quite a vivid contrast from the normal timbre of the string orchestra. All volumes work well within this context. It is important to remember that the player will need several seconds to put on and remove the mute. In Italian, *con sordino* means "with the mute" and *senza sordino* means "without the mute."

3.4   The Specific Instruments

3.4.1   The Violin

The violin's open strings are tuned in perfect fifth intervals. From lowest to highest, they sound the pitches G–D–A–E. The G is found within the third octave of the piano (below middle C). Any music is to be written only in treble clef (Figure 3.10).

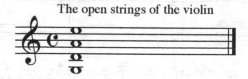

Figure 3.10 The "open" strings on the violin

It is important to know the location (at the piano) of the open strings because there are technical and tonal characteristics that must be considered:

- It is impossible to play lower than what the string is tuned to.
- Double-stops may be impossible or possible in accordance with string usage. To accomplish this, two strings must be employed and they must be adjacent strings.
- It is impossible for a violinist to play A and C (middle C) since both notes are found on the G string. However, these notes are possible one octave higher (Figure 3.11).

The most natural way to play this double-stop would be to play the A and C on the respective D and A strings. If the violinist were to use the open A string, the C would need to be played on the D string. This would, however, prevent the violinist from using any vibrato on the note A since it is being played on the open string. (At times this sound may be desirable, particularly to create a drone effect but, in most cases, vibrato is used to enhance the overall sound.)

Within each string there are positions where the finger touches down on the string. To ascend through a scale, the player uses all four fingers to execute this process. The upper range of the lowest string (G) overlaps into the next higher string (D) and this

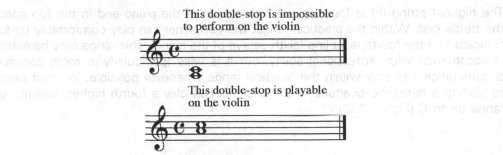

Figure 3.11 Double-stops on the violin

holds true for the remaining strings, excluding, of course, the highest string (E). This provides the player with options that are determined with regard to ease of performance and timbral expression. This overlapping provides a somewhat homogenized sound throughout the registers. The reverse aspect can be implemented as well when a phrase is dedicated to one string.

Each string has a distinct timbre. The G is the darkest and huskiest. The D is the weakest and most intimate. The A is stronger than the D and slightly darker in quality than the E. The E string is the brightest and projects well.

Sometimes the orchestrator may prefer that the violinist play a musical phrase on a particular string. This is indicated in the following way: sul G, sul D, sul A, or sul E.

A specific choice regarding a string can enhance the mood of the music. The musical example in Figure 3.12 shows that the entire phrase can be played on the D string.

In this scenario, the music will be a bit more mellow in the first half. If the phrase were played on the G string, the music would become more intense. The reason for this has to do with range. Since the G string is pitched a perfect fifth lower, the notes of the musical phrase sit higher in its register. The same notes heard on the D string are in that string's lower register. (A more vivid example might be represented with male and female voices singing in unison; the same notes will feel higher and be more intense with the male voice and the female voice will sound more relaxed.) For the second half of the phrase, the violin stays on the D string to avoid the open A string where, if played, the expression of mood will be lost.

Figure 3.12 Indicating string preference for tone and expression

The highest string (E) is found in the fifth octave of the piano and in the top space of the treble clef. Within the practical range, the performer can play comfortably up to the G located on the fourth leger line (sixth octave of the piano). This range may be extended in accordance with professional ability, but it is wise (particularly in more commercial circumstances) to stay within the practical range whenever possible. Without resorting to playing a harmonic overtone, a professional can play a fourth higher, extending the range up to C (Figure 3.13).

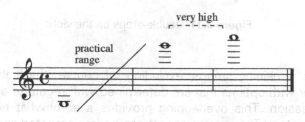

Figure 3.13 Violin range

Keep in mind that the extreme high register on a violin requires more violins to make the sound large enough to compete with heavier instrument groups (rhythm sections and brass).

3.4.2   Functions of the Violins in an Orchestration

- Playing a lead melodic line: usually both the first and second violins play in unison in the lower and middle registers and split into octaves when the first violins are in the high register.
- Playing a counterline: similar procedure as above. The first violins and second violins may also play in counterpoint to each other (one takes the melody; the other takes the countermelody).
- Creating a "pad" effect: the violins are voiced with the other sections of the string orchestra to create a homophonic texture.
- Generating rhythmic momentum: this can be done on repeated notes or in an arpeggiated fashion with a few or more notes.
- Solo violin: this allows for individual expression and creates a sense of intimacy.

3.4.3   The Viola

The viola's open strings are tuned in perfect fifth intervals. From lowest to highest, they sound the pitches C–G–D–A. The C is found at the beginning of the third octave of the piano and the three higher strings are at the same location as the violin's three lowest strings (Figure 3.14).

Figure 3.14 The "open" strings on the viola

The significant overlap between the viola and violin strings helps create the homogenized sound that is typical in the string orchestra. Music for the viola is to be written primarily in alto clef. In this clef, middle C (at the piano) is found on the third line of the staff. Since the viola's bottom range extends downward by a perfect fifth, it becomes necessary to use this clef in order to avoid an excessive amount of leger lines below the staff. When the viola part travels into the extreme high register, the treble clef may be used (Figure 3.15).

Figure 3.15 The extreme high register for viola

It is important to know the location (at the piano) of the open strings because there are technical and tonal characteristics that must be considered:

- It is impossible to play lower than what the string is tuned to.
- Double-stops may be impossible or possible in accordance with string usage. To accomplish this, two strings must be employed and they must be adjacent strings.
- It is impossible for a violist to play its lowest C and E since both notes are found on the C string. However, these notes are possible one octave higher. The most natural way to play this double-stop would be to play the C and E on the respective G and D strings.

As with the violin, the upper range of the lowest string (C) overlaps into the G string and this holds true for the remaining strings, excluding, of course, the highest string (A). This provides the player with options that are determined with regard to ease of performance and timbral expression. This overlapping provides a somewhat homogenized sound throughout the registers. The reverse aspect can be implemented as well when a phrase is dedicated to one string.

Each string has a distinct timbre. The C is the darkest and huskiest. The G is a bit thinner, but still has a dark, full quality. The D is warmer and progressively thinner in quality than the G. The A string is quite prominent and has the most intensity. On the higher strings, the overall sound is closer to the violin than the cello.

Within the practical range, the performer can play comfortably up to the G located on the space above the treble clef (fifth octave of the piano). This range may be extended in accordance with professional ability, but it is wise (particularly in more commercial circumstances) to stay within the practical range whenever possible. The range can also be extended with harmonics, but if volume is required it will be necessary to use a full and normal tone.

3.4.4 *Functions of the Violas in an Orchestration*

- Playing an accompaniment: usually play an ostinato in the form of a rhythmic pattern or a simple guide-tone line to suggest the overall harmony.
- Generating rhythmic momentum: this can be done on repeated notes or in an arpeggiated fashion with a few or more notes.
- Playing a melody: usually an octave or two below and sometimes in unison with the violins.
- Playing a counterline: the first violins and second violins may play in counterpoint to the violas, which are often doubled with the cellos (one takes the melody, the other takes the countermelody).
- Creating a pad effect: the violas are voiced with the other sections of the string orchestra to create a homophonic texture.
- Pizzicato: more resonant than the violin.
- Solo viola: this allows for individual expression and creates a sense of intimacy.
- In general, the viola is huskier and can be more mellow or colder than the violin. A bit larger than the violin, it tends to also respond a bit slower. Within the string section the violas are strategically located as a "bridge" between the upper strings (violins) and the lower strings (cellos and basses). They are the "glue" for the entire string orchestra, when the ensemble range becomes extended, and have the timbral diversity to thicken the violins or the cellos.

3.4.5 The Violoncello

Unlike the viola, which looks like a larger violin, the cello (as it is commonly known) is much larger and requires the use of a peg to suspend the instrument above the floor as it is placed between the legs of the performer. Its open strings are tuned in perfect fifth intervals. From lowest to highest, they sound the same pitches as the viola (C–G–D–A) but are heard an octave lower (Figure 3.16).

The open strings of the cello as written

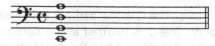

Figure 3.16 The "open" strings on the cello

The C is found at the beginning of the second octave on the piano. The C and G strings, located in the bass register of the piano, are used primarily for the purpose of playing a bass line. In traditional classical music, the cello section often plays the bass line with the bass section. The cello section's notes, being heard an octave higher, bring into focus for the listener the very low bass notes. The two higher strings (D and A) are heard within the third octave of the piano. It is here that the cello is used most regularly to play melodic material. The significant overlap of these two strings with the viola's bottom two strings (not with exact pitches, but in register), helps create the homogenized sound that is typical in the string orchestra. Music for the cello is to be written primarily in bass clef. In this clef, middle C (at the piano) is found on the first leger line above the staff. Since the cello's bottom range extends downward by an octave, in contrast to the viola, it becomes necessary to use the bass clef in order to avoid an excessive number of leger lines below the staff. When the cello part travels above middle C, the tenor clef is employed. This clef places middle C on the fourth line and reduces the amount of leger lines. If the cello extends into the extreme high register the treble clef may be used (Figure 3.17).

Figure 3.17 Clef options for cello

It is important to know the location (at the piano) of the open strings because there are technical and tonal characteristics that must be considered:

- It is impossible to play lower than what the string is tuned to.
- Double-stops may be impossible or possible in accordance with string usage. To accomplish this, two strings must be employed and they must be adjacent strings.
- The cello functions as a "bass" instrument (the bottom two strings) and a melody instrument (the top two strings).

As with the violin and viola, the upper range of the lowest string (C) overlaps into the G string and this holds true for the remaining strings excluding, of course, the highest string (A). This provides the player with options that are determined with regard to ease of performance and timbral expression. This overlapping provides a somewhat homogenized sound throughout the registers. The reverse aspect can be implemented as well when a phrase is dedicated to one string.

Each string has a distinct timbre. The C is the darkest and huskiest. The G is also quite full and favorably a bit lighter to the point that, in special circumstances, this string may be used for a melodic passage (particularly in the role of a solo cellist). The D is robust but mellow and the A is highly expressive. These top two strings are used most frequently in a melodic capacity.

Within the practical range, the performer can play comfortably up to C located on the third space within the treble clef (fifth octave of the piano). This range may be extended in accordance with professional ability, but it is wise (particularly in more commercial circumstances) to stay within the practical range whenever possible. The range can also be extended with harmonics, but if volume is required it will be necessary to use a full and normal tone.

3.4.6   Functions of the Cellos in an Orchestration

- Playing an accompaniment: usually play an ostinato in the form of a rhythmic pattern or a simple guide-tone line to suggest the overall harmony.
- Generating rhythmic momentum: this can be done on repeated notes or in an arpeggiated fashion with a few or more notes.
- Playing a melody: usually an octave or two below the violins, an octave below or in unison with the violas.
- Playing a counterline: the first violins and second violins may play in counterpoint to the violas, which are often doubled with the cellos (one takes the melody, the other takes the countermelody).
- Creating a pad effect: the cellos are voiced with the other sections of the string orchestra to create a homophonic texture.
- Pizzicato: more resonant than viola.
- Solo cello: this allows for individual expression and creates a sense of intimacy.
- In general, the cello is fuller and more distinctive than the viola. As a section, they are strategically located as a "bridge" between the violas and the basses. They have the timbral diversity to thicken and add "bottom" to the violas and add clarity to the basses. The cello section can function in the context of the string orchestra as one cello can function in the context of the string quartet, adding depth and weight to the entire upper string ensemble. At the same time, the cellos are light enough to work well with a rhythm section.

3.4.7 *The Double Bass*

Although it looks like a cello, the double bass is much larger. It also requires the use of a peg to suspend it above the floor, but the performer must stand in order to play it effectively. More commonly known as the bass, its open strings are tuned in perfect fourth intervals. From lowest to highest, they sound E–A–D–G. Some basses have an extension on the lowest string that can lower its range down a major third to the note C. The E is found at the beginning of the first octave of the piano and, with the extension, the C is located near the very bottom of the piano keyboard. The lower two strings and the upper two strings are located, respectively, in the contrabass and bass registers of the piano and are primarily used for the purpose of playing a bass line. The notes are written in the bass clef, but a register transposition of an octave is required (Figure 3.18).

open strings of the bass as written   their location and sound on the piano

Figure 3.18 The "open" strings on the bass

Since the bulk of the bass's range exists well below the bass clef staff, this would require an excessive amount of leger lines. To bypass this problem, it was decided that all bass parts should be written an octave higher. As a result, the reading of music becomes much simpler since most of the notes now lie within the bass clef staff.

In traditional classical music, the bass section is doubled with the cello section. The cello section's notes, being heard an octave higher, bring into focus for the listener the bass notes. The very low bass notes add bottom and weight. It is important, however, to realize that the basses should not be used too frequently, especially in the lower half of the instrument when playing arco. The sound can become very heavy and sinister. The bass section works well when playing pizzicato. The lightness of this technique is complemented by the low, rich, and robust tone, and can provide a wonderful sense of buoyancy and momentum to the orchestra. The highest string (G) is the only one that begins to sound a bit like a cello. As a result, the blend or overlapping found in the other string instruments is not as beneficial here. It becomes quite apparent when the basses enter, so the orchestrator should use them in a conscious way. When the bass part travels above middle C for an extended period, the tenor clef or even treble clef may be used. It is somewhat uncommon for the bass section of an orchestra to play in this register except for a special effect. However, it can be quite effective as an expanded cello section. (This scenario might emerge more out of necessity in the context of a small string ensemble, where the bass might be used as an auxiliary cello.) With regard to non-orchestral players, the bass clef may be used with leger lines instead of the less common tenor clef.

It is important to know the location (at the piano) of the open strings because there are technical and tonal characteristics that must be considered:

- The actual sound of the strings is heard an octave below what is written.
- It is impossible to play lower than what the string is tuned to.
- Double-stops are not always clear since the instrument is in such a low register. It is best to use them in the upper register of the instrument. Remember that two strings must be employed and they must be adjacent strings.

As with the other string instruments, the upper range of the lowest string (E) overlaps into the A string and this holds true for the remaining strings, excluding, of course, the highest string (G). This provides the player with options that are determined with regard to ease of performance and timbral expression. This overlapping provides a somewhat homogenized sound throughout the registers. The reverse aspect can be implemented as well when a phrase is dedicated to one string.

Each string has a distinct timbre. The E is the darkest but also the muddiest. The A becomes clearer but is still a bit rough. The D is robust but mellow and the G is the thinnest, enabling the greatest melodic expression and clarity. The highest string (G) is found in the second octave of the piano but, with the register transposition, is written in the top space of the bass clef.

Within the practical range, the performer can easily play up to the written D located on the space above the first leger line of the bass clef (but sounding at the D located in the third octave of the piano). This range may be extended quite considerably in accordance with professional ability, but it is wise (particularly in more commercial circumstances) to stay within the practical range whenever possible. The range can also be extended with harmonics (which work quite well on bass), but if volume is required it will be necessary to use a full and normal tone.

3.4.8 Functions of the Basses in an Orchestration

- Playing an accompaniment: usually play a bass line to support and define the overall harmony.
- Generating rhythmic momentum: this can be done with notes played pizzicato or played arco using short bow strokes.
- Playing a melody: usually an octave below or sometimes in unison with the cello section.
- Playing a countermelody: most commonly heard in the context of a fugue where each section has a melody to play.
- Creating a pad effect: the basses are voiced with the other sections of the string orchestra to create a homophonic texture.
- Similar to the dual (bass and melodic) functions of the cello, the double bass section can work quite effectively when split into two parts. Some may play a bass line using

pizzicato technique while the others, acting more as cellos, play a more legato passage with the bow.

- In general, the bass provides the bottom for the entire orchestra. Its low notes clarify the upper harmonic and melodic structures.
- The basses are too heavy to work well with a rhythm section. It is best to stay out of the way and let the rhythm section bass, along with the piano and guitar, do the work of laying down the harmonic and rhythmic foundation.

3.5 Orchestration Examples

For context and analysis, please refer to the separate scores. There are 10 scored examples for the string orchestra that will show some of the various possibilities that are commonly utilized. The scores are available at the end of the book in Appendix A.

Score 3.1 features the violins playing the melody in octaves. The violas play a harmonic guide-tone line. The cello section also fills out the harmony with arpeggios (notice that the low Ds, when played on the open string, facilitate the large melodic leaps; the open string sound works well here as the note D serves as a dominant pedal within the composition). The basses add rhythmic momentum and buoyancy using the pizzicato technique.

Score 3.2 (Audio Example 3.26) features the first and second violins in melodic counterpoint. The violas and cellos offer a homophonic texture along with rhythmic momentum via repeated notes. The basses cushion the bottom with a slow-moving bass line.

Score 3.3 features the violas and the cellos playing the melody in unison. The violins provide an active background figure against the slower moving melody; their separate bow strokes increase the percussiveness. The bass section divides, with some basses providing a smooth arco bass line while the other basses add momentum, playing a rhythmic figure using the pizzicato technique.

Score 3.4 features most of the string orchestra using the pizzicato technique. There are soloist features for violin, viola, and cello playing arco passages, in contrast to the pizzicato texture.

Score 3.5 (Audio Example 3.21) features an arrangement based on a Bach cello suite, with each section playing in melodic counterpoint. The cello section is split to support the viola counterline and double the bass line at the octave. The absence of bow markings should be noticed as it is customary, within the Baroque style, for all of the string instruments to use separate bows (one bow stroke for each pitch).

Score 3.6 (Audio Example 3.8) features the entire string section serving as an accompaniment by creating a lush pad. The harmony is thickened through the use of double-stops. A careful examination of the upper string parts will reveal an interlocking of the voices (rather than merely assigning the two highest notes to violin 1, the next two lower notes to violin 2 and so forth, the double-stops are paired to create the interval of the sixth, which is preferable to the fifth in accordance with the player's standard desire).

Score 3.7 features a solo viola. The other sections use harmonics that create an eerie backdrop to the viola's haunting and plaintive melody. Notice that the key of the composition allows the open strings to be used for the harmonics. This facilitates the process and enhances the resonance of the sound.

Score 3.8 (Audio Example 3.24) features the entire string section playing a melody in octaves. The effect is dramatic and bold at all dynamic levels.

Score 3.9 (Audio Example 3.11) features the cello and bass sections playing a melody in octaves, while the upper strings add tension using the bowed tremolo technique.

Score 3.10 features the string section creating a textural flourish using the fingered tremolo technique. The pizzicato basses add a light, buoyant rhythmic figure.

3.6 The Harp

Although the harp is not really considered a part of the string orchestra, for the purposes of consolidation it is included within this chapter.

The harp, in the grander sense, is a string instrument as is the piano. In comparison to woodwind and brass, one would think that these two instruments are essentially interchangeable but, in fact, they are more different than similar. These contrasts are outlined below:

Harp	Piano
47 strings	88 keys to corresponding strings
Strings plucked with fingers	Strings struck by hammers attached to keys
Harpist does not use the pinky fingers	Pianist can use all 10 fingers
Pedals are used to change string pitch	Each string (note) has its own key
Excessive chromaticism is difficult	Excessive chromaticism is easy
Sight-reading can be difficult	Sight-reading is much easier
More transparent, less powerful	Less transparent, more powerful
More piercing in very high register	More delicate in very high register
Controlled decay with hand muffle only	Controlled decay with hammers
Sustain is normal	Sustain is elective by using the sustain pedal

The mechanics of the harp are rather complex to understand when first becoming acquainted with the instrument, but it is essential that the arranger understand these principles in order to craft a part that is playable and, even better, sight-readable. The range of the 47 strings spans almost the entire length of the piano keyboard (from the lowest C♭ to the highest G#). You may be wondering how 47 strings can be almost equal in range (or span) as the piano keyboard's 88 keys. The reason is that the 47 strings, in effect, represent the white keys of the piano. (There are a total of 52 white keys.) Now you may be wondering how to acquire the notes that are so readily heard on the black keys. Well, there are seven pedals, each one representing the notes from the musical alphabet (A, B, C, D, E, F, and G). The harpist operates these pedals with the use of the feet. Three of the pedals are operated with the left foot and the remaining four are operated with the right. Each pedal also has three positions (high, middle, and low, similar to a gear shift in an automobile), which, in effect, alter the sound of the string to create chromaticism (i.e. to make A become A♭ or A♯). The highest position sounds the lowest pitch of the string (i.e. it is considered the "open" string). This position essentially puts the string in the "flat" position (i.e. A♭). When the pedal is moved downward to the middle position, a metal disk at the top of the string moves to "stop" the string, similar to what a cellist would do with the index finger, making the string sound one half-step higher (i.e. the string now sounds as A natural). When the pedal is moved down to the lowest position, a second disk moves onto the string, which forces the string to sound one half-step higher (i.e. the string now sounds as A♯). This double action, created by the disk mechanism in conjunction with the operation of the pedals, defines what is known as the double-action pedal harp that is used in professional orchestras today.

3.6.1 Spelling Matters

It is important to realize that, unlike the piano keys, each pedal has its own letter name (e.g. on the piano A♯ and B♭ share the same black key; on the harp A♯ is created by a completely different pedal than B♭). This is another reason why it is not always easy simply to give "the piano part" to the harpist and expect it to be read immediately. You may also notice some awkward enharmonic spellings, which are necessary for the harpist as these spellings relate to the pedals being used versus correct spelling within the laws of music theory. Another common technique is for repeated notes: If a constant repetition of the note F is played solely on the F string, the finger must interrupt the sustain by restriking the note. However, if the E string is tuned to E♯ (which sounds as F) both strings can now decay naturally and the repeated notes will flow into one another.

3.6.2 Mass Edit

Another important consideration is that, although there are almost seven full octaves comprising the range of the harp, each pedal is in control of all of the corresponding notes (i.e. the A pedal affects all of the A strings on the instrument; if the harpist moves the pedal down to the lowest position, all of the A strings will sound as A♯).

3.6.3 Pedal Charts

A pedal chart is used to help the harpist make the proper adjustments to sight-read a piece of music effectively. To create a proper pedal chart, the arranger must memorize the location of the pedals. Here is a simple way to help remember the order: the two pedals closest to either side of the harp column spell the word "be." The B pedal is located on the left side (operated with the left foot) and the E pedal is on the right (operated with the right foot). From there, simply go outward, following the musical alphabet in its normal sequential order (i.e. D–C–B | E–F–G–A).

The same piece of music is displayed in the keys of C major and E major. The pedal charts for each key can be viewed in Figure 3.19. The chart in Figure 3.20 prepares the harpist to play the notes of the G harmonic minor scale.

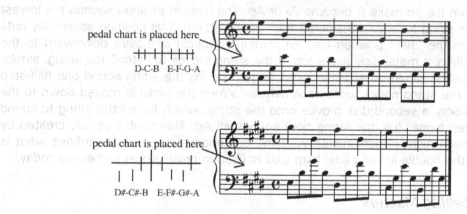

Figure 3.19 Two pedal charts for the harp

Figure 3.20 The pedal chart that establishes G harmonic minor

3.6.4 Pedal Changes

Most harp music requires pedal changes at some point within a given piece. If there are only one or two pedal changes to be made, then the arranger may simply indicate those changes without writing a new pedal chart. For example, if the G string needs to be changed to G#, simply write G# in the space between the grand staff. When pedal changes occur with both legs simultaneously, the pedals on the right will be placed higher in the space between the treble and bass staves. The left leg pedal changes will be written below the ones indicated with the right. When many pedal changes are required at once, it is better to write a new pedal chart.

The arranger must consider the physical capabilities of the harpist. It is not possible to make more than two pedal changes simultaneously as this can only be done when the pedals are operated by both feet. If two pedals are located on one side of the harp (i.e. one foot), it will understandably take a bit longer for the harpist to accomplish the task. This is another reason why the arranger should memorize the location of the pedals.

3.6.5 *The Harp Glissando*

One of the most recognizable sound qualities of the harp is the production of the glissando. This effect produces a dramatic sweeping effect when played loudly but can also, when played softly, produce an intimate and delicate sound that suggests the animation of an object being unveiled. Although these effects have their place, they should not be overused. When they are desired, however, some careful planning is involved to enable the harpist to produce the effect. If a harpist needs to play a glissando using all of the notes of the C major scale, a pedal chart for this scale would be all that is necessary. The harpist would then simply move the hand along the strings as a pianist would on the white keys to produce a similar effect. The difficulty occurs when, for whatever reason, the arranger decides to omit some of these notes. For example, let's suppose that the arranger needs the harpist to produce a glissando using the notes of a C major pentatonic scale (C–D–E–G–A). The harpist cannot simply bypass playing the F and B strings as this will become too choppy and the effect will be destroyed. In order to "omit" these notes, the F string must be tuned to F♭ (sounding as E) and the B string must be tuned to B♯ (sounding as C). See Figure 3.21 to view the pedal chart for these changes. Now the harpist can play all of the strings and only the notes of the C major pentatonic will be heard.

pedal chart is placed here

D-C-B♯ E-F♭-G-A

Figure 3.21 A harp glissando sounding as a C major pentatonic scale

3.6.6 *Ensemble Context*

The harp is not a loud instrument so it is best to use it in more intimate settings or expose it in certain places where the music texture thins out. It is highly effective in lighter passages, adding a buoyant, delicately percussive momentum to string sections

It has a lucid but floating quality at softer dynamic levels, as well as in the lower registers, and, when doubled at the unison with dryer instruments, can really enhance the ensemble sound.

3.7 Sequencing for the String Orchestra

By now, if you have carefully read this chapter, you should be comfortable with the terminology and principles involved when writing for strings. This is a crucial step in order to obtain a professional and realistic-sounding MIDI production of your orchestral arrangements. If you conceive of a MIDI string part as you would write it for a real acoustic instrument, you will achieve a much more realistic result at the sequencing stage. In the following pages, you will learn how to translate your orchestral score into a full, realistic and professional production using your MIDI studio and a few tricks that I am going to share with you. Now that you have your score ready on paper, it is time to move into your project studio.

3.7.1   Synthesis Techniques and Library Options

One of the first aspects to consider when sequencing for the string section is the sounds that you will be using to reproduce the parts. If you are familiar with this subject you probably have heard composers and MIDI programmers promote this or that MIDI synthesizer, sampler, or sound library. When it comes to orchestral sounds, each composer seems to have a sound source that she or he claims not to be able to live without. This is your first lesson: always be open to experimenting with new sonorities and new techniques. Simply because the synthesizer you used five years ago was capable, at that time, of giving you a decent orchestral sound palette, it doesn't mean that technology is still capable of providing the best options in terms of realism, flexibility, and overall quality. If you haven't updated, or at least checked out, some new options in terms of orchestral sounds, now it is the time to do it! Your choice of palette is crucial for a successful rendition of an orchestral score with your MIDI equipment. This point cannot be stressed enough. It is time to retire that sound module you have trusted for the last five years and check out some new options that will make your life easier and your sequences outstanding.

In your project studio, certain synthesis techniques are better than others to re-create string instruments. Chapter 2, analyzing the rhythm section, briefly discussed how the subtractive, additive, wavetable, and FM syntheses are able to offer an extremely effective and versatile palette of colors for the modern composer. These techniques, however, are not well suited to effectively re-creating and reproducing acoustic instruments. This is the reason why, when choosing a source for orchestral sounds, sample-based orchestral libraries are recommended. In recent years, the addition of physical modeling techniques to sample-based libraries has improved even further the capabilities of creating full-scale orchestral renditions within the realm of a project studio.

The advantage of using sample-based sounds instead of synthesized ones lies in the fact that, while a synthesizer tries to re-create complex acoustic waveforms artificially through the use of one or more oscillators and filters, a sampler stores the original recording (samples) of an acoustic set of waveforms that can be played (triggered) using a generic MIDI controller. This technique has the huge advantage that, if the original samples are recorded and programmed accurately, the results can be astonishing. But don't trash that old synthesizer just yet! Later in this chapter you will see how you can still use it to improve your sequencing techniques for strings. Listen to Examples 3.1 and 3.2 for a comparison between a part sequenced with a synthesized patch first and a sample-based one later.

While in the past hardware samplers constituted the majority of sample-based machines, nowadays computer-based samplers (also referred as software samplers) have the large majority (if not the totality) of the market. A sampler, either hardware or software, is basically an empty container that needs to be loaded with the right sounds in order to generate the required sonority. Samples (organized in programs and banks) are loaded into the RAM or streamed directly form the hard disk of the computer. A program or patch can be played through the MIDI controller. The sound sets that you buy to load into your sampler are called libraries. Libraries come in different formats (depending on the sampler you have), sizes, prices, and sound palettes. Some of the advanced libraries, such as VSL and EastWest, come bundled with their own instrument players, while others run on third-party software samplers (Native Instruments' Kontakt is one of the most popular software samplers) such as LA Strings, Spitfire Audio, or Cinematic Strings.

When it comes to string orchestra libraries the choice is pretty wide and, as mentioned earlier, choosing the right one for you can be a daunting process. The good news, though, is that there are string libraries available for pretty much any budget, from around $200 up to $10,000. The bad news is that, usually, with sampled orchestral libraries you get what you pay for and cheap libraries are usually, well, cheap. String samples fall mainly into two categories: libraries that are sampled completely dry and that need a good amount of tweaking at the mix stage with the reverberation (reverb) settings, and libraries that have been sampled in a live environment in order to capture, along with the string instruments, the acoustic response of the hall. The latter have the advantage of being ready to use. They are inspiring to play and they often require very little extra reverberation. Their drawback is that in some cases they tend to be too wet and therefore not as edgy as other libraries. They also can be difficult to place in the mix along with other instruments that originally were not recorded in the same location. The former have the advantage of giving edgy sonorities that work well for more classical and conventional MIDI renditions, but their drawback is that they sometimes are not very inspiring to play without the addition of the necessary reverberation. Table 3.1 summarizes some string libraries and their main sonic characteristics. One thing that it is very important to keep in mind is that, no matter how much money you will spend on your sampled library, your score will sound only as good as what you write. This means that to achieve the best results you need to think in terms of writing for a real string section first; then, and only then, you will be able to take full advantage of the extremely flexible power of a sophisticated sampled library.

Table 3.1 Different reverberation settings for the four zones of the orchestra's virtual stage

Library	Platform	Sonic environment	Comments
EastWest Hollywood Strings	VST, AU, RTAS, AAX	Wet	A versatile and complete orchestral library with very string sections. Very playable and easy to mix. Comprehensive sounds and articulations
IK Multimedia Philharmonik	VST, RTAS, DXi, AU	Medium-wet	Versatile and extremely playable. Good selection of both sections and solo instruments. A perfect "Swiss Army Knife" for any situation. Doesn't have key switches feature
Vienna Symphonic Library (VSL) Special Editions	VST, AU, RTAS, AAX	Dry	One of the best strings and orchestral libraries in the mid-price range. Flexible articulations and sonorities. Good entry point if you don't have any orchestral libraries
Cube and Super Package	VST, AU, RTAS	Dry	This is the ultimate orchestral library. It includes all possible instruments and articulations. Extremely powerful and comprehensive
LA Strings	Kontakt Sampler Format	Medium-wet	Fantastic string library. Very versatile, capable of producing textures from mellow and dreamy to rhythmic and powerful
Cinematic Strings 2	Kontakt Sampler Format	Medium-wet	Perfect for any movie, TV, or video game score. Versatile articulations and intuitive interface
8Dio Adagio Strings	Kontakt Sampler Format	Dry	Very realistic and edgy sounds. Nice addition to other libraries
Native Instruments—Session Strings Pro	Kontakt	Medium-wet	Good contemporary/pop string section. Good flexibility
Spitfire Audio	Kontakt Sampler Format	Wet	Huge selection of different ensembles and styles. They provide an excellent quality with libraries that can accommodate any situation from film to chamber orchestra, from large ensemble to string quartet
Orchestral Tools—Berlin Strings	Kontakt Sampler Format	Medium-wet	Excellent smooth and lush strings with built-in runs, adaptive legato, and bow noise

With the rapid evolution of technology, formats, and systems, it is hard (and somehow useless) to give a complete picture of the orchestral libraries that are available for the modern composer. New editions and upgrades are constantly developed and introduced on the market. What are worth outlining are the features and techniques that you need to master to get the most out of the libraries. Not all products are the same, of course, but the best ones share some common features that are essential nowadays to create realistic renditions of your scores.

3.7.2 Multilayer Patches

One of the main secrets in re-creating a convincing and realistic MIDI rendition of orchestral compositions is to use patches that are not static and repetitive, but instead are able to change over time, providing the highest possible level of variation. The sound of a violin, for example, changes constantly even when repeating the same note over and over. Every time you play a particular note it will have a slightly different attack, release, harmonic content, etc. Such variations mainly depend on the dynamic levels at which a note is played. Usually, softer notes have a tendency to have a slightly longer attack, while stronger notes feature a sharper attack. A simple "strings" patch on your old synthesizer is not going to be able to produce all the different colors and variations that make a string instrument so special and versatile. This is the reason why your string library (and more in general your orchestral library) should be a multilayer one. A multilayer patch is not only made up of several samples assigned to different notes, but in addition has several samples (in extremely sophisticated libraries up to 127) assigned to each key. The sample that will be triggered for a particular note can be selected depending on different parameters (MIDI velocity is the most common). Figure 3.22 shows how a multilayer string patch works. In this case, there are three layers that form a violins patch. Each layer is triggered by a different range of velocities. The sample "Violins pp" is triggered by velocities between 1 and 39, the sample "Violins mf" is triggered by velocities between 40 and 90, and the sample "Violins ff" is triggered by velocities between 91 and 127.

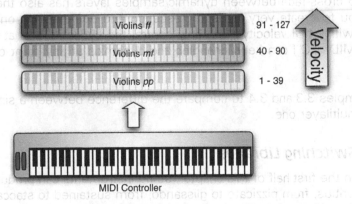

Figure 3.22 Three-layer violins patch with the samples triggered by velocity

This type of patch allows you to achieve a much more detailed and realistic sonority. Now you can easily follow the dynamics of your score and reproduce them with a much higher fidelity with your MIDI gear by simply using different levels of velocity when playing or editing the MIDI parts. Multilayer libraries are the essential starting point to achieve realistic renditions of strings and orchestral scores in general. The higher the number of layers, the more accurate these renditions will be. Keep also in mind that the drawback of using a library with a high number of layers per patch is that your RAM (and hard disk in the case of a direct-from-disk-stream library) needs to be large enough to hold all the samples and that your CPU has to be fast enough to handle the necessary calculation involved in the triggering of the samples. In some cases, instead of velocity, other MIDI controllers, such as CC#1, are used to cross-fade gently from one sample to another. This option gives a smoother transition between the different samples since it allows you to cross-fade seamlessly between the different layers forming the patch as shown in Figure 3.23.

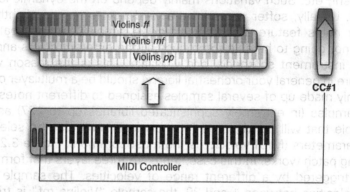

Figure 3.23 Three-layer violins patch with the samples triggered by the modulation wheel CC#1

Using a CC to cross-fade between dynamic/samples layers has also the big advantage of allowing you to create very realistic crescendo and decrescendo on long sustained notes. While with MIDI velocity I can only control the dynamic level at the start of the note, with a MIDI CC I can keep changing the dynamics at any point of the sustained note.

Listen to Examples 3.3 and 3.4 to compare the difference between a single-layer violins patch and a multilayer one.

3.7.3 Key-Switching Libraries

As described in the first half of this chapter, string instruments can produce an incredible variety of sonorities, from pizzicato to glissando, from sustained to staccato. One aspect that sets professional sample libraries apart from the rest of the orchestral sounds is the ability to provide flexible and musical ways quickly to integrate all the tonal colors of a string instrument within the realm of a MIDI studio. While the multilayer option gives you the ability to avoid a repetitive sonority, it doesn't provide the modern composer with as much flexibility as a real string section. The beauty of writing for acoustic strings is that you simply have to indicate on the score the type of sonority you have in mind in a specific passage—let's say, "pizzicato"—and the performer will switch to pizzicato in that precise spot. Wouldn't be nice if your virtual orchestra could do the same? Don't worry, this is another feature that you should look for in your sample library: key switches. In a traditional MIDI setup you would have different string sonorities and colors (pizzicato, *col legno*, sustained, staccato, détaché, etc.) assigned to different MIDI channels and/or devices. While this option is a valid one for small-scale productions, for larger and more complex compositions it can be tedious and can interfere with creativity. Every time you would need to switch the violin section from one sonority to another you would have to change MIDI track and record only the notes or passages that are meant to be played with that particular articulation. Key switches allow you to quickly instruct the virtual section or orchestra to switch to a different set of samples while remaining on the same MIDI channel and devices. When you load a key-switch-enabled patch in your sampler, you load not just one sonority (e.g. pizzicato), but several at the same time (e.g. sustained, tremolo, sforzando). Contrary to the multilayers technique, with key switches you don't change samples with velocity or control changes (CCs), but instead you use prepro-grammed MIDI notes, conveniently placed outside the playable range of the instrument you are sequencing, to switch quickly between all the available articulations of a particular patch. Refer to Figure 3.24 to see how a typical key-switch patch would be set up.

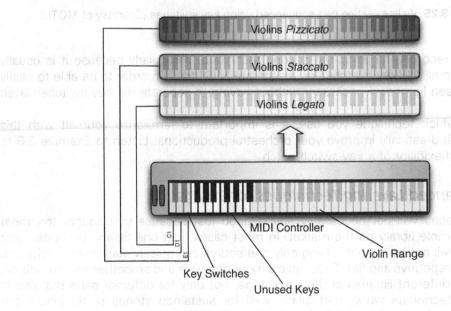

Figure 3.24 Typical key-switch violins patch

There are two main techniques in using key-switch patches. First, the switches can be recorded while you are actually recording the part. This approach has the advantage of saving time and allows you to perform the part as it was written. This translates into a more flowing and realistic performance. The main disadvantage of this technique is that it requires you to have good skills on the MIDI controller since the key-switch changes have to be included as part of your performance. If the part is particularly fast and articulated it could be challenging. The second approach is to sequence the entire part with a fairly generic sonority (e.g. "sustained") and on a second passage overdub the key-switch changes. This approach has the advantage of releasing the pressure of performing both tasks at the same time. Figure 3.25 shows how a MIDI part was sequenced using key switches.

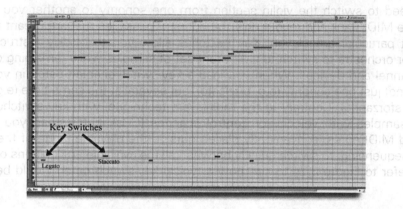

Figure 3.25 Violins section part sequenced using key switches (Courtesy of MOTU)

In general, I recommend using the second technique, particularly because it is usually better to keep all the key switches on a separate MIDI track in order to be able to easily switch between libraries if necessary without having to separate the key switches later.

No matter which technique you use, it is important to familiarize yourself with this feature. It will drastically improve your orchestral productions. Listen to Example 3.5 to evaluate the flexibility of a key-switch patch.

3.7.4 Advanced Layering Techniques for Strings

Your productions will not necessarily sound good just because you bought the most expensive sample library on the market. In most cases, just one library, or, better, just one source, will not be enough. Using only one source can quickly stamp your orchestral renditions as repetitive and flat. To achieve a more cohesive and smoother sonic rendition, you can use different libraries at the same time, not only for different parts but also in layers. This technique works particularly well for sustained strings parts, where the repetitiveness of a loop can be boring. By layering two different samples at the same time, you are virtually re-creating new harmonics and adding new life to your sections. Samples that have different sonic characteristics in terms of ambience can be mixed and matched. Use a layer of dry and wet patches. The former will provide the necessary attack and presence, while the latter will add a nice acoustic environment and will help smooth the harshness of the dry samples. The amount of dry and wet samples you use can vary depending on the library. As a general rule, 70 percent dry/30 percent wet might be used for fast and busy passages and 30 percent dry/70 percent wet for sustained and pad-like sections. Remember that these are general figures for you to start experimenting with and that results may vary depending on the sample library used. Most of the string sample libraries tend to be a bit bright and thin, sometimes giving a clean and aseptic sound that doesn't match the smoothness and resonance of a real string section. To avoid this problem you can also layer a small percentage (between 5 and 10 percent) of a good old-fashioned synthesized patch, like the one found in Roland (JV and XV series) or Korg (Trinity or Triton) synthesizers. This will not only enrich the sonority of the layered patch, but also smooth the combination of the two sample-based patches. Listen to Examples 3.6–3.8 for a comparison between the same part sequenced using different combinations of layered patches (Score 3.6). When layering different patches to form a cohesive master patch, individual MIDI volume tracks should be created in the sequencer's mix window, as shown in Figure 3.26.

After the part has been sequenced, you can quickly fine-tune the blend between the individual colors and sonorities. To add an even more realistic effect try to pan each patch with a slightly different value (Figure 3.26). This will create a natural separation between the patches that will increase the clarity of the section.

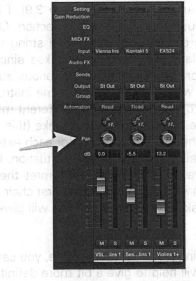

Figure 3.26 Having individual control over each component of a layered string patch gives you flexibility in terms of colors and texture (Courtesy of Apple Inc.)

While the layering of sampled and synthesized patches contributes to increasing the cohesion of the MIDI string orchestra, for the ultimate sound fidelity, layering a real solo string instrument for the violin, viola, and cello sections can be recommended. In most cases, you can avoid layering a solo bass unless the bass section is heavily featured and exposed in the mix. The use of a real string instrument for each section gives a natural edge to the MIDI instruments. Its natural harmonics will enrich the repetitiveness of the MIDI patches, while the natural acoustic imperfections of the instrument (such as detuning and noises) will give natural life to the sequenced part. The idea here is to be able to get as close as possible to a real large string section without hiring a full one! So don't go crazy and don't start overdubbing 14 real instruments. This will defeat the purpose of achieving realistic results with the use of MIDI. Instead, try to get the best out of what is affordable. Hiring a string player for this purpose is an ideal solution for small budget productions, giving results very close to those obtained with a full real string section. Always have an acoustic instrument doubling the highest line of each section. If you have a divisi passage, try to double each line. Each string player should record three good takes. You can use the alternate takes to blend better the acoustic instrument with the MIDI since in some situations it can be hard at the mixing stage to achieve a natural blend between sampled/synthesized and a single acoustic string instrument. A single string instrument, if left alone against the MIDI section, tends either to stand out too much (and therefore come across as a solo instrument) or to be completely buried (and therefore defeat its purpose). Make sure to check with your musicians that they are OK with you using more than one take in the final mix and whether they need to be paid extra for double- or triple-tracking. Listen to Examples 3.9–3.11 to compare a MIDI-only string orchestra, a MIDI string orchestra with one acoustic instrument layered per each section, and a MIDI string orchestra with one acoustic instrument layered per each section and three takes used in the mix (Score 3.9). I highly recommend triple-tracking real strings when sequencing for the string section. Often, even a so-so library can sound fantastic if used in conjunction with a real string instrument that has been tracked three times. I particularly like using three takes since two may cause some intonation problems and four or higher could create serious audio phasing problems. In order to avoid phasing issues when recording the same instrument several times, you should try to use different microphones and slightly different mic positions for each take. Try to use a closer microphone position for the first take (this would represent our first violinist take), and then move the mic further away for each extra take. This will re-create the effect of a larger section being recorded in a live situation. In order to minimize even more the risk of phasing, have the string player interpret the part in slightly different ways, a stronger attitude for the first take (such as a first chair string player would have) and less aggressive takes for takes two and three. This will give the impression of having tracked three different players.

In some cases, when a real string player is not available, you can add to the MIDI section a MIDI solo violin patch. This will help to give a bit more definition to the section. In this case we DO NOT overdub several solo violinist; one will do the trick.

3.8 MIDI Sequencing Techniques for Strings

The previous section looked at the options available to achieve the best sounds for a virtual string orchestra. Once you have found the layers combination that best suits your composition, it will be time to start sequencing the parts and to use all the MIDI tools available in your sequencer to improve the rendition. Keep in mind that while the patches you use are very important for a successful MIDI rendition of a string orchestra they are not enough to guarantee a good production. If this were the case, then you might as well go out and buy the most expensive orchestral library and sequence your parts! Instead, there are several other factors that come into play, aspects that can have a huge impact on the final result. First of all, when writing and sequencing string parts (this is also true for any other part, not only for strings) always conceive and play the parts as single lines. Stay away from the "piano" approach, where you sequence a four-part string section on one MIDI track. Sequencing and playing each line separately allows you to re-create the natural dynamics and interaction between the different sections of the string orchestra. In addition, it allows you to use a separate patch or layers for the different string sections, avoiding a generic "string" patch that would kill any serious intent of creating a realistic rendition. Avoid copying and pasting sections that are repeated throughout the piece. As shown for the rhythm section, the copy-and-paste technique will flatten the natural flow of the production. As much as possible, try to play the entire parts.

3.8.1 Attack and Release Control

Another common mistake when sequencing string parts is to use the sustain pedal to sustain long notes instead of using the regular Note On–Note Off message. It is crucial that each note of the string orchestra has its attack and release set individually. If you use the sustain pedal to sustain certain notes artificially, you will be limited in terms of MIDI editing. It is particularly important to have control over the attack and release of each note since string instruments are extremely variable when it comes to the attack and release of their vibrating strings. As discussed earlier in this chapter, what makes a string instrument so particular is the ability to achieve a wide range of sonorities through the use of different bowing techniques and strokes. By striking each note with a different intensity you can achieve slightly different sonorities ranging from silky and smooth to harsh and edgy.

In order to have the same control over the dynamic and the different colors of a MIDI patch, a series of controllers can be used that, when combined together, allows us to achieve almost the same flexibility available to the acoustic string player. The most basic technique involves the use of CC#7 (volume) to control the dynamic of a string line and its global flow. After sequencing one of the lines of a string orchestra with your MIDI controller, go back and overdub subtle volume changes (CC#7) using one of the sliders of your controller. Record the changes in real time. It is more musical and it allows you to control the flow of the line smoothly. This technique is particularly helpful when dealing with long, sustained notes, where dynamic changes can really bring your part to life. While overdubbing the volume changes, think like a string player and imagine that the slider you are controlling is your bow. Raising the slider is like adding more weight to the bow, which translates into more pressure and therefore more volume. Lowering the slider means a gentler touch and therefore a more delicate sonority. It takes some time to get used to musically controlling your virtual players with a slider, but once you get used to it, it will become an essential part of your productions. This technique can be particularly useful in smoothly controlling the attack and release of long, sustained notes. Most of the time, string patches have fairly harsh releases that, even when played at lower velocities, do not provide a subtle fade-out. Through the use of CC#7 you can quickly fade out notes in a natural way. Look at Figure 3.27 to see how the dynamic flow of a string line can be effectively controlled through the use of volume changes.

After overdubbing the volume changes, you should open the sequencer's graphic editor and fine-tune the slopes and curvatures of the crescendo and decrescendo in order to control precisely how the line plays out.

This technique allows you quickly to control the dynamics of a string line, but if you want to achieve more accurate results you should put in a bit more work at the editing stage of your production and use, in addition to volume changes, CC#11 expression. This

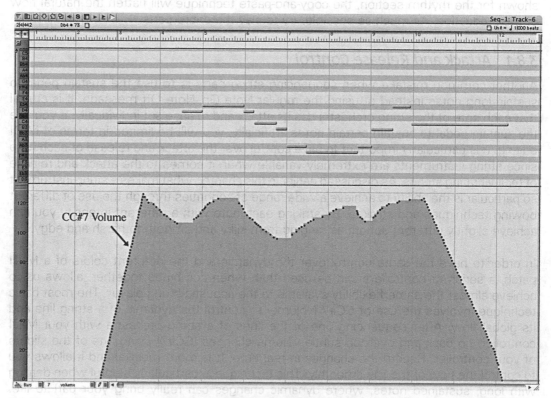

Figure 3.27 A string line where the dynamic was controlled with CC#7 (volume) (Courtesy of MOTU)

controller allows you to manage very accurately the attack and release of each note of a line. As mentioned earlier, what makes string instruments particularly flexible is the ability to have control over the attack and release of each note. CC#11 expression can be used to re-create the same effect on a MIDI patch. This controller works very much like CC#7 volume. It ranges from 0 to 127 and it can be assigned to a slider of the keyboard controller. The difference lies in that CC#11 works as a percentage of CC#7. Imagine CC#7 as the volume on the amplifier and CC#11 as the volume on your guitar. The former is used to change the overall loudness of the signal while the latter can be used to change the relative loudness of a certain section or passage (or, as in this case, single notes). Take a look at Figure 3.28 to see how the two controllers interact.

In Figure 3.28, CC#7 is set to 100 (out of 128). When CC#11 is set to 127, the overall output volume will be the same as the one set by CC#7. This means that a value of 127 for CC#11 is equal to 100 percent of the value of CC#7 (A). At the other end, a value of 64 for CC#11 halves (50 percent) the value of CC#7 (in this case 50, as shown in B). If CC#11 is set to 0, then the overall output of the MIDI channel will be 0 (0 percent of CC#7, as shown in C). The use of CC#11 for complex volume changes is extremely flexible. Imagine the case where, after programming several volume changes for each note of a complicated line, you decide that you want to raise the overall volume but without altering the overall shape of the programmed changes. You can easily do so by raising CC#7 and leaving CC#11 untouched. Now you can see how important this controller is.

CC#11 can be used to keep precise control over the attack of each note of a string line. This will allow you to re-create the natural variation occurring in an acoustic string instrument. To do so, first sequence the line as you would normally do. Then, select every other note from the track you just sequenced and cut and paste them into a brand new track. The idea here is to re-create two separate MIDI tracks, one for the up bows and the other for the down bows. Then assign each track to a different MIDI channel. This is useful for two reasons: first, you need the two tracks to be on different MIDI channels in order to be able to use CC#11 to control simultaneously the release of the previous note and the attack of the following one; second, you can use slightly different patches for the up and down bows to create even more variety in terms of sonority (in an acoustic instrument the up and down bows usually feature small sonic variations). Now it is time to program small variations in terms of attack and release for each note of the up- and down-bow tracks. To do so, use the MIDI graphic editor of your sequencer and with the pencil tool insert CC#11 in and out ramps for each note. Usually, depending on the part, it is enough to do it for any duration longer than one-eighth notes. For smaller values it is irrelevant. Try not to use the exact same curve for each note, but instead use some variations and, most importantly, make sure that the envelope of each note makes sense musically. Remember that at the end it is the musical effect that matters! Look at Figure 3.29 for a visual rendition of how this technique works.

In Figure 3.29, notice how, for each sustained note, CC#11 was used to control the attack and release. Listen to Examples 3.12 and 3.13 to compare a string part sequenced without the use of CC#11 and the same part sequenced with CC#11. While at first this

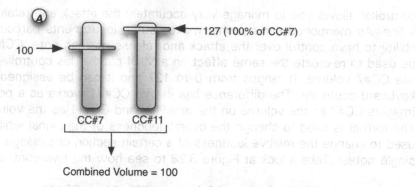

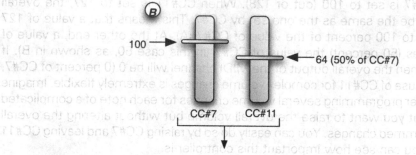

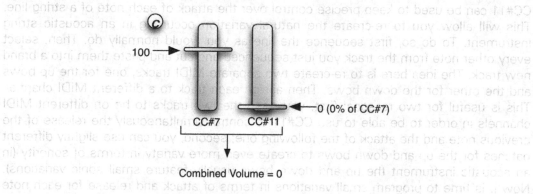

Figure 3.28 CC#11 works as a percentage of CC#7

technique might seem time-consuming, with a little bit of practice you will be able to use it quickly and efficiently. It is recommended that you use this technique for more-exposed string lines (usually violins and sometimes violas), while using it for background lines only if necessary and if time allows.

While CC#11 is very useful to control the loudness of a part, by itself it can't help reproducing the changes in tone that a string instrument is capable of when played at different dynamics. In order to realistically sequence any acoustic instrument, it is crucial

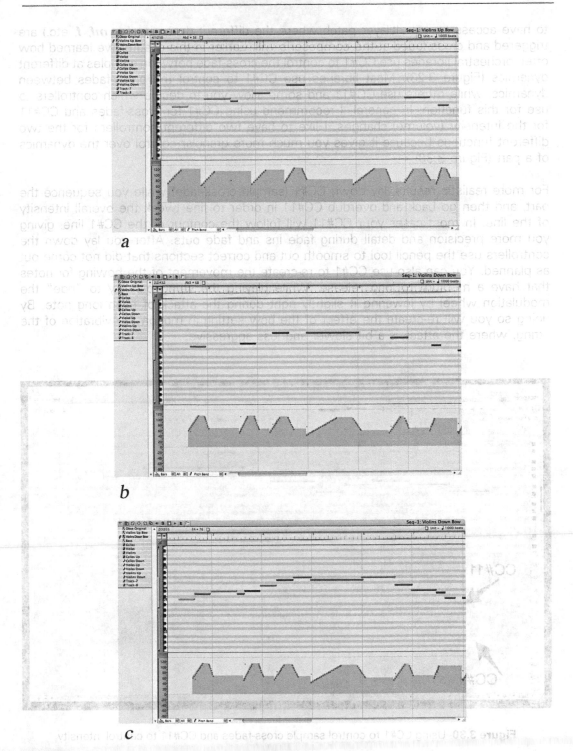

Figure 3.29 String line where the notes were spread over two different MIDI tracks and MIDI channels: (a) up bow; (b) down bow; (c) combined tracks (Courtesy of MOTU)

to have access to a multilayer patch where the different dynamics (*p*, *mf*, *f*, etc.) are triggered and cross-faded at the composer's will. Earlier in this chapter we learned how often orchestral libraries use CC#1 to control the cross-fade between samples at different dynamics (Figure 3.23). Most libraries use CC#1 to control the cross-fades between dynamics, while others use CC#11 and some allow you to decide which controllers to use for this function. In general, I recommend using CC#1 for cross-fades and CC#11 for the intensity (volume) changes. I like to have two different controllers for the two different functions because it gives you much more granular control over the dynamics of a part (Figure 3.30).

For more realistic results, lay down CC#1 (sample cross-fade) while you sequence the part, and then go back and overdub CC#11 in order to fine tweak the overall intensity of the line. In most cases your CC#11 will follow the contour of the CC#1 line, giving you more precision and detail during fade ins and fade outs. After you lay down the controllers use the pencil tool to smooth out and correct sections that did not come out as planned. You can also use CC#1 to re-create the movement of the bowing for notes that have a medium to long release. While playing the string part try to "ride" the modulation wheel by lowering it slightly right during the attack of each long note. By doing so you will re-create the effect of the bow setting in motion the vibration of the string, where the attack is a bit slower and less aggressive.

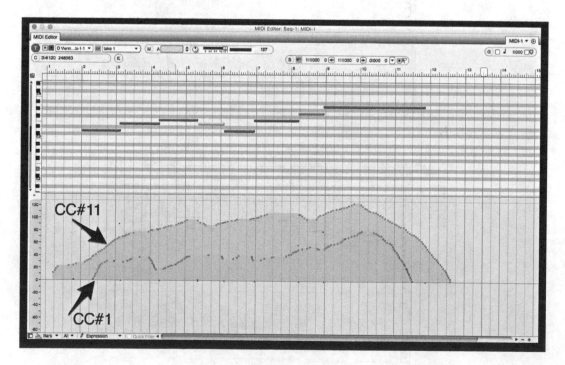

Figure 3.30 Using CC#1 to control sample cross-fades and CC#11 to control intensity

3.8.2 *Performance Controllers*

The three controllers used above (CC#7, CC#11, and CC#1) are ideal to control the dynamic of a MIDI string line. In the MIDI standard, there are a few other controllers that can be effectively used to shape a MIDI part even further. For a general review of these controllers, refer to Table 1.2 in Chapter 1. If you feel comfortable with the previous concepts, now it is time to dig a little deeper and use more-advanced tools to make your virtual orchestra shine. Among the extended controllers that the MIDI standard includes, you may find particularly useful, when sequencing for orchestra in general, and for strings in particular, those that allow you to shape the response of a patch. Remember that avoiding repetition is one of the first goals when working with MIDI renditions of acoustic instruments. Using advanced controllers and MIDI messages to avoid repetitions and to inject some new life to your parts is crucial. Among the messages and controllers that you may find particularly useful when sequencing for strings are aftertouch, breath controller (CC#2), portamento (CC#5, CC#65, CC#84), soft pedal (CC#67), and hold 2 (CC#69). They all control specific parameters and features of a MIDI channel, giving the flexibility of a real acoustic instrument. Let's take a close look at these options and at how we can creatively and effectively use them in our productions.

Aftertouch is a specific MIDI message that is sent after the Note On message. When you press a key of a controller, a Note On message is generated and sent to the MIDI OUT port; this is the message that triggers the sound on the receiving device. If you push a little bit harder on the key, after hitting it, an extra message, called aftertouch, is sent to the MIDI OUT of the controller. The aftertouch message is usually assigned to control the vibrato effect of a sound but, depending on the patch that is receiving it, it can also affect other parameters such as volume, pan, and more. In most samplers (hardware or software), this message can be assigned to control any parameter you want. In some devices it controls either the speed of the vibrato or the amount of vibrato in order to have a greater flexibility when playing long, sustained notes. In wavetable synthesizers it is usually used to control either the brightness of a string patch or the amount of vibrato.

The breath controller (CC#2) can be assigned to any parameter the programmer wants. Usually (as in the case of aftertouch), it is used to control the vibrato amount, the filter (brighter or darker sound), or the loudness of a patch. Most samplers allow you to program a patch to respond to any controller in the way you want. CC#2 can be used to control either the intensity or the amount of vibrato for string patches.

Portamento is the ability to slide two subsequent notes with a temporary bend of the pitch from the first note to the second. This function is used in synthesizers more than in samplers, even though it is available on both platforms. Portamento can be used to re-create a natural slide effect between two notes. If used on synthesized strings, try to be particularly careful since it can sound too fake. You can control the amount of portamento applied to a patch with CC#84. Its value sets the note number from which the portamento effect will start (0 = note C1 and 127 = note G8). If you want to turn off portamento quickly for the entire MIDI channel you are transmitting on, use CC#65 (where values from 0 to 63 are off, while values from 64 to 127 are on). CC#5 controls the rate or speed of the portamento: lower values will give a much slower portamento time, while higher values will translate into a faster portamento time. These controllers are particularly useful in re-creating the natural slide between two notes. While pitch bend can be used for situations in which slide notes occur rarely, portamento is ideal when slide notes are featured more constantly. Listen to Example 3.14 to hear how portamento can be used effectively to sequence string parts.

Soft pedal (CC#67), where on (values 64–127) lowers the volume of the notes being played, can be particularly useful in controlling the dynamic range of a string passage that needs to be suddenly quiet (like a subito piano passage). Be aware that this controller is not very common in patch programming and therefore you might need to assign it manually in your software synthesizer or sampler.

Hold 2 (CC#68), if set to the on position (values 64–127), lengthens the release of the notes that are being played, creating a legato effect between sustained notes. The length of the release is controlled by the release time of the voltage-controlled amplifier (VCA). Use this controller to achieve a smoother transition between notes in a string section. Be advised, though, that not all patches implement this controller and that sampled patches will most likely need to be manually programmed to respond to hold 2.

By combining all the aforementioned controllers, you can shape your virtual string section and have it following your own performance style. It takes time to master these techniques, so don't get frustrated at the beginning if you find it time-consuming. You will see that in a matter of days your strings and orchestral renditions will improve considerably.

3.8.3   *Extended Performance Controllers*

The MIDI standard includes not only CCs that are targeted to control the actual performance of a sequenced part, but also controllers that can affect directly the sonority of a patch. These CCs are mainly used by synthesizers, but they can also be applied to sample libraries, if programmed correctly. They fall into the category of sound controllers, as discussed in Chapter 1 (Table 1.2). Keep in mind that not all patches respond to sound controllers. It is up to the programmer to assign certain parameters to such CCs. In particular, sample libraries tend to avoid any preset CC assignments, favoring instead an open programming approach that is left to the final user for a more customizable experience. Therefore, if you are using synthesized string patches you can, in most cases, take advantage of these CCs right out of the box, whereas if you use sample libraries you will have to do a bit of patch programming of your own. No matter which sound-generation technique your orchestra palette is based on, you will find that these controllers can be extremely useful in rendering the virtual orchestra. Among the 10 sound controllers implemented in the MIDI standard (CC#70–CC#79), those that are particularly functional, in a string orchestra setting, are CC#72 (release control), CC#73 (attack control), and CC#74 (brightness). Let's take a look at each controller in more detail.

CC#72 (release control) allows you to change the release stage of the amplifier envelope of a patch.

The VCA of a synthesizer can be altered through the use of an envelope generator (EG), which is a multistage controller that allows the synthesizer to control over time the amplitude of a waveform. The release stage controls how long the sound sustains after the MIDI Off message is sent (after you release the key on a keyboard controller). It can be particularly useful to shape the release feature of certain patches in order to adapt it quickly to the requirements of the part. By lowering the release time (lower CC values), the patch will be more staccato, making it ideal for short and quick passages. A longer release time (higher CC values) gives a more legato and sustained effect that will fit pad-like passages. By assigning CC#72 to a MIDI fader control, you will get a better feel of the response of the patch while performing it.

CC#73 (attack control), like the previous controller, regulates the steepness of the attack stage of the amplifier envelope. This stage allows you to control how quickly the waveform will reach its maximum amplitude after the MIDI On message is sent (after you press the key on a keyboard controller). When sequencing a string part, this is particularly useful since it gives control over the edginess of the patch. A faster attack (lower CC values) is better suited for fast and staccato passages, while a longer attack (higher CC values) can provide the smoothness required for sustained and relaxed passages. As for CC#72, you could input this controller from a MIDI fader while performing the part, if possible.

CC#74 (brightness) is both easy to use and highly effective. As the name says, it controls the brightness of a patch. In more technical terms, it controls the filter cutoff frequency of the voltage-controlled filter (VCF) of the sound generator in a synthesizer or sampler. In general, lower values give a darker sonority while higher values give a brighter and edgier sound. By controlling the brightness of the patch, you give the illusion of changing the pressure and intensity of the bowing. For subtle and intimate passages, you can use lower CC values, which will translate into a darker and warmer effect. For edgy and up-front sections you can use higher CC values. The sensitivity of the controller changes from patch to patch and from synthesizer to synthesizer, therefore it is hard to give a set of values to use. Another effective way to use this controller is randomly to program small variations (around ± 5) in a string part to re-create a natural variation in the bowing strokes (Figure 3.31), variation that is missing in a synthesized or sample-based patch.

The most effective way to use the aforementioned controllers is to combine them together and take advantage of their features on a section-by-section basis. Remember that variation is the key to a more realistic and convincing virtual string orchestra and therefore try to use these techniques to create as much variation and surprise as possible. Work with these controllers in combinations. For example, usually, in a real acoustic environment, a brighter sound corresponds to a sharper attack, a sonority that can be easily re-created through lower values of CC#73 and higher values of CC#74. In contrast, a darker and mellower sonority is usually paired to a slower attack and release, an effect that can be re-created with higher values of CC#73 and lower values of CC#74.

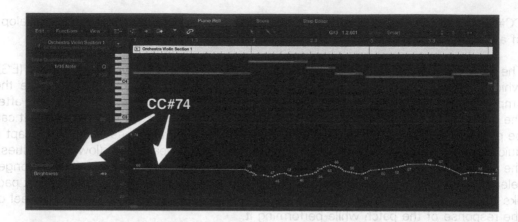

Figure 3.31 The insertion of small changes in CC#74 can re-create the natural variations in bowing intensity that would happen with an acoustic instrument (Courtesy of MOTU)

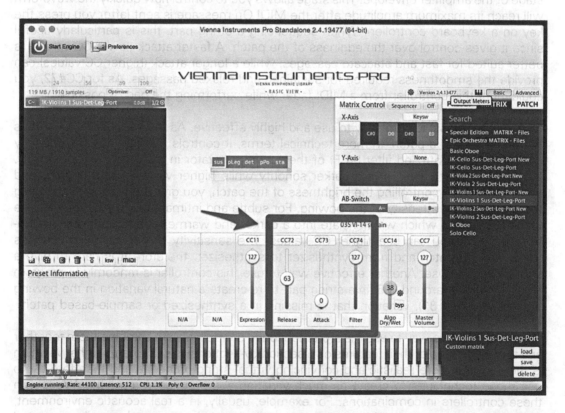

Figure 3.32 Most sample-based libraries allow you to assign CC to individual parameters of their sound engine (Courtesy of VSL)

Synthesized patches have the advantage of responding, in most cases, to the majority of these controllers automatically since the original programmer took the time to implement them in the patches. Sample-based libraries usually do not come with preprogrammed performance and sound controllers automatically assigned. This is both bad and good news. It is bad in the sense that you won't be able to take advantage of these controllers right out of the box. The good news is that a sample-based library allows you to program pretty much any parameter of its sound engine to most of the 128 controllers available in the MIDI standard. The number of controllers you can assign to parameters of a library depends on the sophistication of the product, but in general most mid-priced orchestral libraries have several options for this purpose. In Figure 3.32, for example, CC#72, 73, and 74 were assigned, respectively, to release, attack, and filter in VSL Instrument.

3.9 Hardware MIDI Controllers for String Sequencing

There are several options when it comes to MIDI controllers for sequencing string parts. One of the most used is the keyboard controller. If you have done orchestral sequencing before, you will already know that the mechanics of a MIDI keyboard and a string acoustic instrument have very little in common. This is the reason why it is fairly hard to render the flexibility and multicolor character of a string instrument from a keyboard. Don't worry, though; don't throw away your MIDI keyboard controller just yet. This section analyzes some techniques and alternate MIDI controllers that will improve your string orchestra sequencing experience.

If you want to use a MIDI keyboard controller, the first thing to do is to make sure that you have a device with the capability of sending CC messages through the use of onboard knobs and/or faders. Most controllers these days feature at least one or two assignable faders. Usually they are marked "Data Entry" or "Controllers." When sequencing for orchestra, it is highly recommended that you use a controller that has more than two assignable data entry faders, since you will end up using several CCs at the same time to shape the MIDI parts. For this reason it is useful to have a separate MIDI controller that is dedicated to CCs only. You can use a smaller MIDI keyboard controller featuring fewer keys but more data entry faders and knobs to program the CCs and the main keyboard controller to sequence the parts. A preferable option is to use a tablet running software such as Touch OSC or Lemur. The advantage of this option is that you can create and configure your own soft controllers (knobs, faders, etc.) and assign them to send any MIDI CC you need. The tables can connect to your DAW via a regular Wi-Fi network, reducing the need for extra cables and wires. See Figure 3.33 for an overview on how to use this configuration.

Each knob and fader is completely programmable. In most devices, you can store different configurations or sets of knob/CC assignments, giving you the flexibility to have sets for strings, brass, woodwind, percussion, etc. Each set can be instantly recalled with the push of a button. This approach is not too expensive and it is extremely practical. While using a knob or fader to send, for example, volume changes is a pretty good option, there are more musical and natural ways to take control of your virtual orchestra

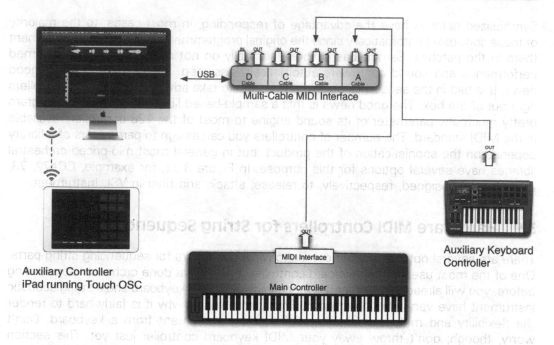

Figure 3.33 Multicontroller setup where the main controller is used to input the notes and the smaller controller or a tablet is used to input the CCs either in real time or as overdub (Courtesy of Apple Inc. and Novation)

One device that can be particularly effective is a breath controller (BC). This MIDI device allows you to send any CC data by blowing air into its mouthpiece. You can program the type of CC sent through the BC with the bundled software application. While the advantage of using such a device is particularly obvious when sequencing for wind instruments (specific BC techniques for brass and woodwind are discussed later in this book), it is also very useful when sequencing string parts. You can assign the BC to CC#11 (expression) and smoothly control the volume of your parts while playing. Assign it to CC#1 (modulation) and you will be able to add a touch of vibrato. The applications are endless and you will find yourself wondering how you could have sequenced without it. You can use it in a multicontroller environment as learned before, in addition to the main keyboard controllers. Look at Figure 3.34 for a diagram of such a setup.

The use of MIDI controllers such as a guitar-to-MIDI converter (GMC) is another valid technique for improving the inputting stage of string parts. Such devices are useful when sequencing pizzicato passages. The natural response of the string of the guitar (or bass) allows you to reproduce the feel of a real string instrument, giving you the right dynamic and velocity ranges. Because of the nature of the instrument, a GMC is particularly indicated to sequence string parts since, compared with a keyboard, the feel and the response of the guitar are closer to those of string instruments. The phrasing will be more natural and, because you have to sequence single lines separately, it will greatly improve

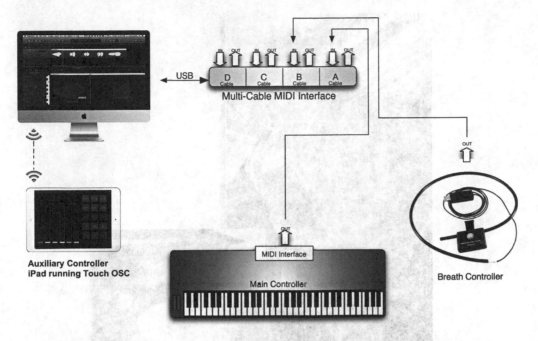

Figure 3.34 Multicontroller setup where the main controller is used to input the notes, and the breath controller and the smaller controller (or tablet) are used to input the CC (Courtesy of Apple Inc., TEControl, and Novation)

the final result. The combination of a GMC and a BC is a great way musically to combine vibrato and pitch bend (using the GMC) and volume changes (using the BC). The use of the GMC is particularly effective with solo string lines, where the expressivity of the acoustic instrument is more exposed. On such occasions the flexibility of this controller shines. Bends, slurs, and vibrato passages are easier to re-create musically on the guitar than on the keyboard. Another alternative to the MIDI keyboard controller is the wind controller (WC). This device allows you to input MIDI notes through a clarinet/saxophone-like interface that you play exactly like an acoustic wind instrument. While a WC can be most effectively used to sequence wind instruments, it can also be used to sequence string parts. With this device you can send several controllers at the same time by using the controller wheels located on the rear end of the instrument (Figure 3.35).

A WC is ideal to sequence solo lines and single lines that feature a lot of volume and vibrato changes. The use of this device for sequencing wind instruments is discussed later in the book.

3.9.1 Special Sequencing Techniques for Strings

In most cases, your sample library will have all the necessary articulation you plan to use in your score. All major sample libraries carry articulations such as legato, staccato, tremolo,

Figure 3.35 The Synthophone, one of the most advanced wind controllers on the market, played by Berklee College of Music professor and Synthophone expert Randy Felts

détaché, and pizzicato. The most advanced ones include also more-specific effects such as trills (usually mainly major and minor thirds), *sul tasto*, mute, and *sul ponticello*. Some libraries go as far as giving you several options for each articulation where you will find, for example, different lengths of staccato, legato, no vibrato, and more. If your library does not include such advanced options, don't worry. There are a few techniques to re-create some of these sonorities quickly in a pretty convincing way through the use of CCs and some MIDI programming. Let's take a look at the most common techniques.

3.9.2 *Legato, Staccato, and Détaché*

In most cases even the simplest library comes with these articulations. You can adjust the edginess of the patch through the use of the attack, decay, sustain, release (ADSR) curve. A shorter attack time will give a more dramatic sonority, while a longer attack time will give a mellower and more melodic effect and is more indicated for pad-like passages. The same can be said for the patch's release time. Program a longer release time to achieve a more legato effect between notes. Shorter release times are particularly indicated for staccato passages and for fast runs. You can adjust the attack and release parameters directly from the editor of the sampler instruments. These editors can vary, but generally they are accessible from the library editor of the sampler. In some cases the parameters are available directly from the main panel of a plug-in. Another option, to have quicker and more flexible control over these parameters, is to use CC#73 and CC#72 to control, respectively, the attack and release time of a patch, as shown earlier in this chapter. To create the détaché effect, start from a regular legato patch and program a slightly shorter attach time and a slightly shorter release time. The shorter attack time will give the effect of a more defined bow stroke.

3.9.3 **Sul Tasto, Sul Ponticello, *and Mute***

These sonorities are typical of the string instrument vocabulary. What makes them particularly useful in a string section is the ability to re-create a darker and more intimate sound (*sul tasto*) or an edgier and colder sound (*sul ponticello* and mute). If your library does not provide such sonorities, you can reproduce them fairly accurately by adapting a regular string patch. Use the filter equalizer of the sample instrument to shape a full body string section sound. The darker sound of *sul tasto* can be re-created by using a high shelving filter to cut some of the high frequencies of the original patch. The exact cutoff point can vary, but in general start from around 10kHz with a 25dB attenuation. This will give a mellow sound. If you need a darker sound, lower the frequencies and use a 26dB attenuation. Be careful not to overdo it since a very dark sound will start sounding too muddy and it will be lost in the mix very easily. To reproduce the *sul ponticello* sonority once again, you should start from a regular sustained patch and use the equalization section to shape the sound. For this sonority, it is recommended that you use two filters. The first is a high-pass filter set around 750Hz and a peak equalizer set to boost 16dB around 6kHz with a Q point of 0.8. These settings will give a good starting point to bring up front the edginess of your string patch. CC can also be used to automate the equalization section of the synthesizer or sample library. Remember that you can use CC#74 to control the cutoff frequency of the built-in equalizer of the synthesizer in order to move between two different sonorities without having to use separate MIDI channels. Re-creating the mute effect on a string instrument is slightly harder since the addition of the mute changes considerably the vibration pattern of the string. You can get close, though, using a high shelving equalizer set between 5 and 6kHz with a gain reduction of 28dB and a peak equalizer set between 1 and 2kHz with a gain of 16dB. While the shelving equalizer rolls off some of the edginess, the latter helps add some of the nasal sonority that is typical of a muted string.

3.9.4   Trills and Tremolos

The best way to render trills and tremolos realistically is to use a library that has these samples already included. It is fairly common, however, that even mid-sized libraries do not include such articulations. In some cases, if trills are included they may not cover all the intervals you need. Fortunately, a couple of techniques can be used to re-create such effects without having to buy a brand new orchestral library. The problem with programming trills for a virtual MIDI string section is that just triggering two notes very rapidly at a given interval will not produce very realistic results. If you try to do so, you will get the so-called "machine gun" effect, where you end up having an artificial repetition of two notes that has absolutely nothing to do with an acoustic trill. The main reason for this unwanted result is the fact that, in a sound generator (synthesized or sample-based), every time you trigger a Note On you also activate the full envelope generator for the amplifier, including the attack section, generating an unrealistic trill effect. When an acoustic instrument plays a trill, the bow moves across the fingerboard without "retriggering" each note of the trill; instead, the fingers on the fingerboard are used to alternate between the two pitches of the trill. A virtual translation of such an effect would be to be able to bypass the attack section of the envelope for each repetition after the

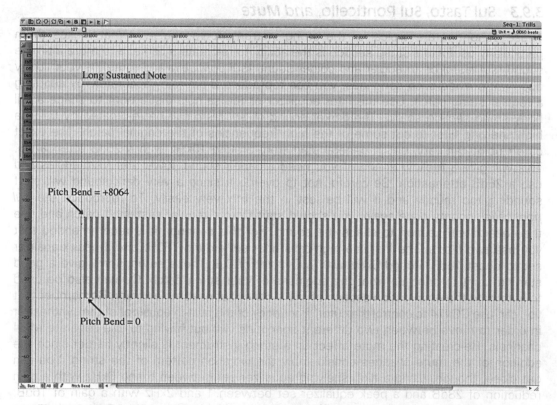

Figure 3.36 Trill effect created using a long sustained note and a fast pitch-bend alternation between values of 0 and 18,064 (Courtesy of MOTU)

first triggered note, leaving a smooth transition between the two notes that form the trill. This can be achieved using the pitch-bend MIDI message of the synthesizer. First of all, load a string patch that has mid- to fast attack and mid- to short release and a long sustain. Next, program the pitch bend parameter to give, at full positive value, the interval of the trill that you need (let's say minor second). Now sequence one single note for the entire length of the trill passage (Figure 3.36).

Insert pitch-bend messages alternating between values 0 and 18,064 at regular intervals (e.g. thirty-second or sixty-fourth notes). In most sequencers, this can be done by enabling the grid and setting it at the desired value. Use the pencil tool and a square wave line to quickly insert the messages for the entire length of the trill. Keep in mind that this technique is particularly effective for trills that span from minor second to minor third. For bigger intervals, depending on the patch and sound engine, the quality and realism of the final result will vary because of the speed at which the pitch-bend parameter reacts. Listen to Examples 3.15 and 3.16 to hear this technique used on a sample-based and a synthesized patch, respectively.

When trying to program a tremolo effect for a string section, we run into problems similar to those for the trill. Just retriggering the same note with a Note On/Off message will cause the dreadful "machine gun" effect mentioned earlier. A realistic tremolo can be re-created using the up/down-bow technique, as explained earlier, where each other note is assigned to a different MIDI channel. Then, on each MIDI channel, assign a slightly

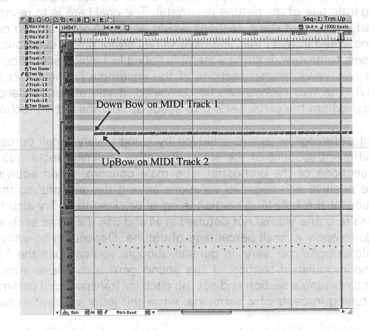

Figure 3.37 Tremolo effect: two different patches on two different MIDI channels alternating using the up/down-bow technique (Courtesy of MOTU).

different patch. For this particular effect you could use a patch with a fast attack and short release for the first MIDI track (down bow) and a patch with a slightly slower attack and longer release for the second MIDI track (up bow), as shown in Figure 3.37.

For a more realistic effect, try to detune the up-bow patch slightly, with values no higher than 5 cents. The detuning will add realism to the tremolo. In addition, avoid using a perfect 100 percent quantization setting. Use instead the randomization option of the sequencer in order to have each note slightly off grid. Use very small percentages of randomization; normally values between 5 percent and 10 percent will add a natural feel without making the part sound sloppy. Listen to Examples 3.17 and 3.18 to compare a tremolo effect created with only note repetition (can you hear the machine gun effect?) and a tremolo effect created using the technique above. For an even more realistic effect, try to split the tremolo part into more than two tracks/MIDI channels, assigning each track to a slightly different patch.

3.9.5 Detuning Techniques

One of the biggest problems with a virtual MIDI orchestra is the fact that it is always perfectly in tune, especially if you are using synthesized patches. While real acoustic string sections tend to be musically out of tune, creating a rich sound, virtual ones have the bad habit of being perfectly and constantly in tune with the rest of the ensemble with which they are playing. As mentioned earlier, lack of variation is your worst enemy when sequencing in general and when working with orchestral sonorities in particular. This is the reason why it is recommended that you make your virtual ensemble sound a bit "worse" than it could through a gentle use of the detune parameter. The goal here is to re-create some of the small imperfections that string sections would display in order to give natural life to their performance. Keep in mind that you have to be careful not to overdo it; you don't want, in most cases, to have your string ensemble sound like a high school band! Detuning can be used in two ways: across sections of the ensemble (vertically) or across the piece (horizontally). The two options can be combined, if you prefer, depending on the time available to program the parts. Let's take a look at how these two techniques work.

The vertical detuning system is achieved by using a very small percentage of detune between different sections of the string ensemble (a similar technique can be used for the other sections of the orchestra). The main concern, when applying detune, is to balance the positive and negative variations among the sections so that the combined pitch variation of the full string section is equal or close to 0. Try also to have the main melodic lines (e.g. the violins) not detuned at all and use the other sections (violas, cellos, and basses) to apply a small percentage of detune. Depending on which library is used, the detuning options may vary. In general, though, you can use the "fine" adjustment found in the modulation section of the sound generator. Make sure to use the fine adjustment to detune a section and not the pitch (or transposition) parameter. The former varies the tuning in cents of a semitone, while the latter varies it in semitones.

In a typical string section of violins, violas, cellos, and basses, the fine-tuning parameter of the top violins should be left unchanged. Use subtle variations for the other sections.

Usually a ± 5 cent variation is enough to add a realistic effect without making the virtual orchestra sound out of tune. Try to balance the spread between the sections. For example, use 25 cents for the violas, 13 for the cellos, and 12 for the basses. The center pitch is still set on 0, with the violas at one end and the cellos and basses at the other, balancing each other. Listen to Examples 3.19–3.21 to compare a string orchestra sequenced without detune, with a mild detune, and with a medium detune, respectively (Score 3.5). A similar technique can be used with a horizontal approach, where the detuning is applied not among different sections but among different passages of the piece. This technique aims to emulate the natural detuning that would occur with human players in different moments of a performance. The best way to apply this technique to sequences is to take advantage of the advanced performance parameters of your library's engine. Some advanced orchestra libraries allow you to vary the tuning of each note either randomly or via an artificial intelligent algorithm. One of my favorite tools for the creation of realistic detuning effects is the "humanize" sections in VSL Instrument Pro (Figure 3.38), where you can select different tuning preset curves that are selected automatically after each Note ON message, in a round-robin fashion. While you can program your own curves, the instrument comes bundled with several presets that cover most of the situations.

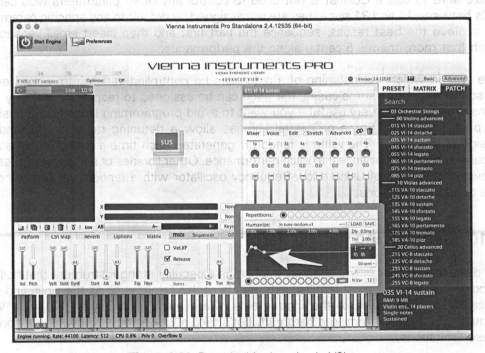

Figure 3.38 Round-robin detuning in VSL

If your library does not include such automatic algorithms, you can assign the fine-tuning parameter of a patch to a CC and then use it to program subtle random variations in pitch during the sequence (Figure 3.39).

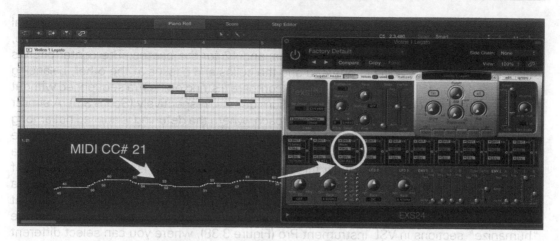

Figure 3.39 Fine tuning parameter assigned to CC#21 in Logic ESX24

Make sure to use a CC that is not used to control any other parameters (you can use CCs between 14 and 31 since they are not assigned by default to any specific parameter). To achieve the best results, sequence the part first and then insert small variations in pitch (not more than ± 5 cents) along the performance.

The way in which the detuning of a patch can be controlled varies from sound engine to sound engine. In some cases, the pitch can be assigned to respond to velocity variations. This option is very useful if you want to avoid programming later the CC assigned to pitch variations. Some orchestral libraries allow a detuning range to be randomly assigned to a patch. The value is randomly generated each time a Note On message is received, giving a human feel to each performance. Other libraries or software synthesizers provide the option of using a low-frequency oscillator with a random pattern to control the detuning variations.

3.10   Sequencing for the Harp

As discussed earlier in this chapter, the harp is a peculiar string instrument that has little in common with the most familiar instruments of this family such as the violin and viola. Because of the nature of the instruments, there are a few issues that arise when sequencing or programming a harp part. Let's take a look at how to overcome some of these limitations.

In terms of sonority, both synthesized and sampled patches can usually do a pretty good job in rendering this instrument. In general, synthesized patches are a bit warmer and "fatter," while sampled patches tend to be thinner and cut through the mix a bit more. A good sonority can be reached by layering a 30 percent synthesized patch with a 70 percent sampled one. Some of the biggest challenges in sequencing for harp are the glissandos, especially when sequencing from a keyboard controller. This is because, if you use the white keys of the keyboard to do a glissando, you will be able to use them only for keys related to the tonal center of C and its relative keys. If you need to sequence a glissando in the key of D major, it will be very hard to achieve a smooth and realistic effect since you can't just slide your finger on the keyboard. A solution to this problem can be easily found in some of the advanced features of the sequencer. In most applications there are MIDI filters that can be inserted either on a MIDI channel strip or as a filter object in the MIDI data chain between the MIDI controller and the sequencer input. The purpose of these filters is to reassign the pitch data of some of the Note On messages according to a specific scale or key chosen by you. For example, you can set the filter to transpose the original key of C (the one in which you are playing) to the key of D (the key in which you want your glissando to sound), as shown in Figure 3.40(a).

Although you can always play your glissando in the key of C and transpose it later, it may be preferable to play the part in the right key along with the rest of the tracks, giving a natural flow to the performance. This technique can be used not only to re-create glissandos in different keys but also in different scales. For example, to sequence a glissando based on the Eb dominant scale, you can reassign a C major to a Eb Mixolidian scale (Figure 3.40b). When sequencing for harp, remember also that it is better not to quantize the glissando parts unless absolutely necessary.

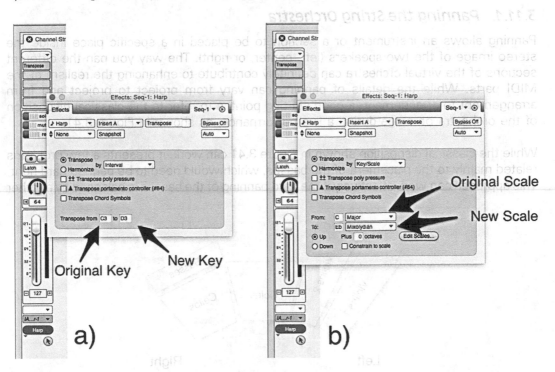

Figure 3.40 Real-time transposition MIDI filter allows you to create harp glissandos in any key (a) and scale (b) required (Courtesy of MOTU)

3.11 Mixing the String Section

As we learned so far, the actual sequencing of the parts of a string orchestra plays an important role in the rendition process. The mixing process is equally important. When writing using a MIDI sequencer you will have to wear several hats at the same time: composer, arranger, performer, and mixing engineer. This final step is often overlooked by seasoned composers for acoustic ensembles. It is seen as something that is not pertinent to their area of expertise and taken care of by someone else later in the production process. While this approach would have worked a few years ago, things have changed considerably nowadays. More and more, you, the composer, will find yourself having to deliver a full demo or a full production without leaving your project studio. This will put you in charge of not only writing a great piece of music but also taking care of all the other production aspects. This is why the mixing aspect of a virtual orchestra is as important as any other step. As the mixing engineer, your main tasks are to place each instrument and section effectively in the right environment and to let each part or line come across as clearly as possible. These very important goals involve paying particular attention to three aspects of the mixing process: panning, equalization, and reverberation. Let's analyze each step separately in order to gain a deeper knowledge of the mixing process for the virtual string orchestra.

3.11.1 Panning the String Orchestra

Panning allows an instrument or a section to be placed in a specific place inside the stereo image of the two speakers (left, center, or right). The way you pan the different sections of the virtual orchestra can definitely contribute to enhancing the realism of the MIDI parts. While the details of panning can vary from project to project and from arrangement to arrangement, a good starting point is to refer to the classical disposition of the orchestra on stage during a live performance, as shown in Figure 3.41.

While the classical disposition shown in Figure 3.41 can work, it presents a few problems related mainly to the placement of the basses, which would need to be panned hard right. This approach can work effectively if the hard panning of the basses is balanced by another

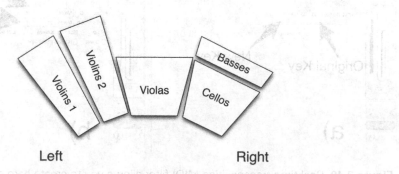

Left   Right

Figure 3.41 Disposition of the classical orchestra

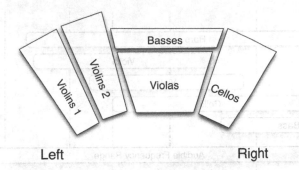

Figure 3.42 Alternative strings panning setup

instrument (such as the piano) or another low-range section (such as trombones and tubas on the left). In a more practical setting, you can set the basses in the middle (or slightly panned right) and spread the violins hard left, the cellos hard right, and the violas either centered or slightly right to balance a violins divisi panned slightly left (Figure 3.42).

This alternative pan setting has the advantage of providing a more balanced frequency distribution. The traditional panning (Figure 3.41) may be preferable when sequencing for classical ensembles, while the second one (Figure 3.42) may be more appropriate for contemporary and pop productions. Some of the orchestral libraries feature pan settings for each section of the orchestra that correspond to those found in a real acoustic orchestra. This approach has the advantage of saving time since you won't have to adjust the panning of each instrument/section manually. However, the preprogrammed settings can always be changed in order to fine-tune the panning according to your needs.

3.11.2 *Equalization for the String Orchestra*

The equalization settings in a virtual string orchestra environment can vary extremely depending on the patch and library used. In general, the goal is to achieve clarity and smoothness at the same time. Clarity is usually an issue and it is harder to obtain when using synthesized patches. These sonorities have the tendency to be muddy and fat, creating some problems at the final mix stage, especially for violins and viola sonorities. To mitigate this issue, a good starting point is to cut some of the low end of the tracks. Start by lowering by 3dB in the 200–300Hz range. Use a more drastic reduction if the patch is still too bass heavy. Add shininess to the synth patch by boosting 3dB (or more) between 5 and 6kHz. This frequency range can add a neat edge to the synth patch. Do not overboost it, because it could make the strings sound too harsh. Sample-based libraries have the opposite problem: sometimes they tend to sound a bit too thin and edgy. To obtain a more robust sonority, apply a 3dB (or more) boost in the frequency range between 100 and 200Hz. To reduce the harshness of a sample-based patch, cut no more than 3dB between 9 and 10kHz. Be particularly careful not to cut too much in this range since a drastic reduction would result in a lack of realism. When combining and mixing two or more string sections, try to carve for each its own space through the

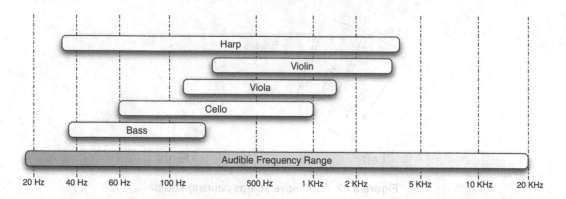

Figure 3.43 Fundamental frequency ranges of the five most important string instruments

use of peak equalizers. Based on the frequency range covered and on the range featured by each section, try to cut some of the frequencies that are not specifically highlighted. The goal here is to center the attention of the listener to those frequencies that are important for each section and to reduce the amount of masking created by the other frequencies that tend to overlap with neighbors' instruments. The fundamental frequencies of the string family are represented in Figure 3.43.

The graph in Figure 3.43 shows how the fundamental frequencies of each instrument overlap in the low and low–middle range. These are the ranges where more attention is needed when mixing and equalizing the string orchestra. Keep in mind that each instrument extends its frequency range through the generation of harmonics that go higher than its fundamental frequencies. Usually the harmonics have a lower intensity and therefore need a milder intervention (or none at all) in equalization compared with the fundamentals. In general, strong and sustained overlaps between frequencies of adjacent sections can be avoided by cutting some of the frequencies that fall in the overlapping areas. This will add clarity and smoothness to string orchestra mixes.

3.11.3 Reverberation

Another key aspect to be considered at the mix stage is the type and amount of reverberation that you will use for the virtual orchestra. If so far you have simply added a generic "hall" reverb indiscriminately to the entire string orchestra without worrying too much about the impact that the type and amount of reverb can have on the final result, now is the time to refine your mixing skills and to pay more attention to all the different subtleties that choosing the right reverberation implies. As anticipated in Chapter 2, convolution effects provide a much more detailed and effective solution when it comes to reverberation. This is particularly true when it comes to mixing the acoustic strings. The importance of placing a sample-based sonority in a natural and realistic environment is significant. Compare Examples 3.22, 3.23, and 3.24 and listen to a virtual string ensemble mixed without any reverb, mixed with a generic synthesized reverb, and mixed with a convolution reverb, respectively (Score 3.8). You can notice how the re-creation of a virtual environment plays a crucial role in rendering the virtual orchestra. Convolution reverbs give you the clarity and accuracy of real acoustic responses. Depending on the complexity and sophistication of your convolution reverb, you can choose among several presampled acoustic spaces featuring not only different reverberation times but also different acoustic responses. When choosing a reverb for the string orchestra, you have to keep in mind that you are dealing with a wide range of frequencies. The right balance is needed between acoustic ambience and clarity. Too much reverb (especially on the bass and cello section) will spoil the mix, watering down the definition of the instruments. Too little reverb and the virtual orchestra will sound fake and without life. As a starting point, reverberation times between 2.3 and 4 seconds can be used for a medium to large string ensemble. Usually, the larger the ensemble the slightly longer the reverberation time since, we assume, a larger orchestra would perform in a larger space. The exact reverberation time used depends on several factors. For faster pieces that require a more aggressive sonority and where the attack of the instruments plays an important role, use a shorter reverb (between 2.3 and 2.8 seconds). For more sustained and pad-like parts, set a longer reverb time (2.9 up to 4 seconds). You have to be particularly careful, when adding reverb to a string orchestra, to balance the reverberation mix for the low-frequency instruments such as basses and cellos. Low frequencies tend to travel longer in space and therefore in a real acoustic environment usually carry the reverberation longer than high frequencies. For this reason, use the built-in filter section of the reverb to damp, if necessary, some of the reverb's low end. Usually a 23dB cut around 200Hz is a good starting point, as shown in Figure 3.44.

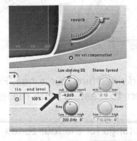

Figure 3.44 Filter section of a convolution reverb. Notice how cutting some of the low frequencies can help reduce muddiness (Courtesy of Apple Inc.)

Another technique that can be particularly effective is the use of slightly different reverberation settings for each individual section of the string orchestra. This approach will give you a much more realistic rendition of an acoustic environment. Because of the different positioning in space of each section in a real concert hall, not every section will be affected by the same reverberation. Using the same reverberation time and color for the entire orchestra will tend to flatten out the bidimensional space of the stage. Sometimes just having four different reverbs with slightly different settings can improve dramatically how the virtual orchestra will sound. One thing to avoid is having too many different reverbs with settings that are far apart. This will create a chaotic ensemble in which the listener will not be able to identify the specific environment of each section. Instead, divide the virtual stage where you would place the orchestra into four different zones, as shown in Figure 3.45.

Each zone has a different reverb parameter. Table 3.2 gives examples of different reverberation settings for each zone.

If you feel that setting up all the different reverberation zones and parameters is too time-consuming, don't despair. Several convolution reverbs (e.g. AltiVerb by AudioEase and MIR for VSL) provide a built-in engine programmed automatically to give the acoustic response of different zones of a hall with the parameters set up for the different sections of the orchestra. In MIR, for example, we can precisely position each instrument or section in any virtual space/room/hall we choose. In addition to accurate positioning, we can change the player's direction and the spread of the section (Figure 3.46).

Listen to Examples 3.25 and 3.26 to compare a completely dry orchestra and one treated with a multizone reverb (Score 3.2).

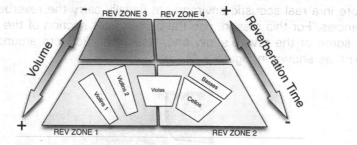

Figure 3.45 The virtual stage is divided into four zones. For each zone, use slightly different reverberation parameters to create a more realistic instrument placement

Table 3.2 Different reverberation settings for the four zones of the orchestra's virtual stage

Zone #	Reverberation time	Filter setting	Comments
1	2.6 seconds	Flat	
2	2. 3 seconds	Damp—2db around 100Hz	The slightly shorter reverb time and the filter on the low end of the spectrum give a bit more clarity to basses and cellos
3	3.1 seconds	Damp—3db around 150Hz	
4	2.9 seconds	Damp—3db around 200Hz	Zones 3 and 4 are normally occupied by other sections of the orchestra such as brass, woodwinds, and percussion

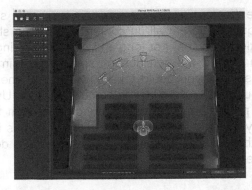

Figure 3.46 A full string section placed on a virtual stage using MIR

Thus, the reverberation aspect plays a crucial role in creating the final virtual orchestra rendition. Earlier in this chapter, two categories of sampled libraries—dry and wet—were discussed. Now it should be more clear how working with a dry library, even if harder and more time-consuming, will give you more options at the mix stage. These types of libraries should be your first option for more accurate and realistic results, while a wet library is recommended for projects where production time is limited.

3.12 The Final Touches

Well, your virtual string orchestra is almost ready. We learned several techniques to improve its programming, the performance, the sounds, and the mix. There is one more aspect that can improve the final result and can increase the level of "humanization" of the virtual players. One thing missing so far is the natural noise present in every recording of a large ensemble. It is true that in an acoustic studio recording the engineers would do anything to avoid capturing such noises on tape, but, since in a sequencer we are dealing with an extremely sterile environment where there is no real human interaction, we can arbitrarily (and extremely subtly) inject some "human errors" into the orchestra by adding some performance noises. Such noises include creaking floors, squeaky chairs, page turning, and breathing (especially for wind instruments). Be extremely careful not to overdo it. You don't want to transform the orchestra into a bunch of undisciplined and noisy amateur musicians! Some sample-based libraries include performance noises as part of their presets. Here you will find most of the common performance noise, including string pulling, squeaks, and bow rustles. In most libraries, these noises are already mapped to notes. If this is the case, simply create a new MIDI track and assign it to an orchestral noise patch. After you have sequenced all the other parts, record some noises as part of the performance. Listen to Examples 3.27 and 3.28 to compare a sequenced string orchestra without and with extra noises, respectively.

Another important aspect to be considered when sequencing for strings (and the orchestra in general) is quantization. The right balance between a solid and a natural performance

is often hard to achieve. If you quantize the part too much, the string section will sound too stiff, whereas no quantization at all (in conjunction with a sloppy performance) will translate into an unsettled rendition. For long and sustained string parts it may be better not to use quantization at all. If the part doesn't sound right, simply replay it. For faster and more rhythmic passages, you could take advantage of the advanced quantization parameters of your sequencer, such as strength and sensitivity. Use a negative sensitivity (between −60 and −70 percent) and strength around 80 percent to start with. If the part still sounds rhythmically unsettled, raise the strength. Since this issue is typical of every sequenced section of the orchestra, it will be analyzed in more detail in Chapter 5 when discussing the brass section.

3.13   Summary and Final Considerations on Writing and Sequencing for the String Orchestra

When writing for strings, be careful not to have the strings playing all the time. Even though they can do it, as there is no concern for breathing and endurance, the sound can eventually wear on the human ear. Listen to recordings with strings in various musical settings and become aware of their activity and inactivity within an orchestration. Since the color of strings in general is monochromatic, it is helpful to consider the use of contrasting registers, various textures (monophonic, polyphonic, and homophonic), and the multitude of available performance techniques to provide contrast and interest. When working with large string ensembles, remember that it is always possible to use soloists or any combination of smaller ensembles within the larger one.

Strings, when used with much more powerful instruments such as brass or a modern rhythm section, must be respected with regard to their limited carrying power. They must be scored in a way that will enable them to be heard. One way is to have them "poke through" the "windows" of an arrangement. Another way is to place them in registers where they are most intense when an arrangement blossoms into a grand forte. You must remember that a studio recording creates an artificial balance, suggesting that the strings are as strong as the drums and electric guitars. In a live setting, the strings must be written in a way that will help them be audible to the listener. If respect is given following the guidelines of conventional orchestration, the strings will sound even better when amplified with microphones.

Some commercial arrangers are careless about writing bowings, thinking that the professional string player will "make it happen." Even though this approach can work through professional competence on the part of the string player, it is better to write bowings for the strings. The up-bow and down-bow indications are not as important as the phrase marking that indicates how many notes are played with one bow stroke.

Time is of the essence in today's commercial marketplace and a recording session or rehearsal will be more expedient when the arranger has indicated bowings that provide an expressive approach and a clear intention.

With regard to the mechanics of performance, the arranger's awareness of instrumental technique is even more important for a harpist. Hopefully you have a new appreciation for the complexity of the harp and the dilemma that many harpists endure when sight-reading a part that has been poorly prepared by the arranger.

Concerning the generally constrained budget for many of today's music recordings, remember that the layering process is a great way to help bolster the strings when desirous of a lush string sound. The balance of the instruments should always be heaviest in the violins as the lower strings have greater carrying power. A good minimum number to work with in this capacity is 11 players: six violins, two violas, two cellos, and one bass. You will probably need three passes to create the illusion of a large string orchestra. Another suggestion is to have the players switch parts so that the timbral mix becomes even more blended. If your budget still needs to be tighter, try using a string quartet mixed in with MIDI tracks and always reserve any solo passages for real players if you have them.

When sequencing for orchestra, always remember that variety is the key word for a realistic rendition. Use light detuning, equalization, CCs, automation, reverberation, and performance noises to inject some life into the ensemble. If possible, use one, two, or three live players (each one triple-tracked) to double the highest part of violin, viola, and cello sections. Sections are much easier to sequence than solo instruments, especially when dealing with string instruments. Even though it is possible to render solo strings (especially cello) fairly accurately, the fact that a part is so exposed and up front can easily give away the sample-based origin of the instrument. Never try to use a synthesized string patch to sequence a solo part. If you cannot use a live player, at least use an advanced sample-based library. You can find very convincing solo libraries either bundled with the main sections or sold separately.

When sequencing string parts, always remember to use the right ranges of the acoustic instruments. Synthesized patches very often do not restrict you in terms of range, while professional sampled ones allow you only to use the real range of the instrument you are sequencing. Try to use professional sample-based libraries as much as possible. Remember that these libraries can usually be divided into two broad categories: dry and wet. The one you choose largely depends on the type of sequencing you are doing. Dry patches are a bit more flexible but require more time at the mix stage, while wet patches are easier to blend in with the rest of the orchestration. Always try to have one of each type available in your project studio. Take advantage of multisample patches. Not only are they made of several samples assigned to different notes, but in addition they have several samples assigned to the same key, with the ability of triggering a particular sample through velocity or other CCs, such as modulation (CC#1). Another technique targeted to create variations in a virtual string orchestra in sample-based libraries is key-switching (KS), which allows you quickly to instruct the virtual section or orchestra to switch to a different set of samples (all preloaded in RAM or streamed directly from the HD) while remaining on the same MIDI channel and devices. You will use preprogrammed MIDI notes, conveniently placed outside the playable range of the instrument you are sequencing, to switch quickly between all the available articulation of a particular patch. To activate the different sonorities during a performance, you can either record the switches while you are actually recording the part itself or, if you prefer, sequence the entire part with a generic sonority and on a second passage overdub the key-switch changes.

In most situations, just one type of sonority will not be enough to render realistically a string part. By layering different patches together you can reach a smoother sound, virtually re-creating new harmonics and adding new life to your parts. When working with sample-based libraries only, you can start by using 70 percent dry/30 percent wet for fast and rhythmically busy passages and 30 percent dry/70 percent wet for sustained and pad-like sections. You can also layer a small percentage (between 5 and 10 percent) of a good old-fashioned wavetable synthesized patch. This will not only enrich the sonority of the layered patch but also smooth the combination of the two sampled patches. The final touch is provided by the addition of a real string instrument (one for each section if possible, starting with the high register first in terms of importance) layered on top of the virtual string orchestra. Its natural harmonics will enrich the repetitiveness of the MIDI patches, while the natural acoustic imperfections of the instrument (detuning, noises, etc.) will give natural life to the sequenced part.

Having chosen the best sonorities and sequenced the actual parts, it is time to use some MIDI programming techniques to smooth out some imperfections and add some natural feel to the sequence. Use CC#7 (volume) and CC#11 (expression) to control the overall dynamic of a part. In particular, remember to use the up/down-bow technique to control the attack and release of each note. Use advanced MIDI controllers and messages such as aftertouch, breath controller (CC#2), portamento (CC#5, CC#65, CC#84), soft pedal (CC#67), and hold 2 (CC#69) to avoid repetitions and to inject some new life into your parts. Aftertouch in most samplers (hardware or software) can be assigned to control any parameter you want, but mainly you will find it as controller for vibrato. The breath controller (CC#2) can be assigned to any parameter, although it is often used to control the vibrato amount, the filter (brighter or darker sound), or the loudness of a patch. Portamento can be used to re-create a natural slide effect between two notes. Soft Pedal lowers the volume of the notes being played, like in a subito piano passage. Hold 2 lengthens the release of the notes that are being played, creating a legato effect between sustained notes. Performance extended controllers can target directly the sound parameters of a patch. When sequencing for strings, CC#72 (release control), CC#73 (attack control), and CC#74 (brightness) are particularly useful. If possible, take advantage of a multicontroller MIDI setup. This will allow you to program in advance and assign all the CCs to different faders and knobs, speeding up the sequencing process. Experiment with different types of controller, such as breath controllers, wind controllers, and guitar-to-MIDI controllers.

The mix stage of the production process is crucial for a successful rendition of the virtual string orchestra. This stage requires particular attention to be paid to three aspects of the mixing process: panning, equalization, and reverberation. Pan settings can vary from project to project and from arrangement to arrangement, but a good starting point is to refer to the classical disposition of the orchestra on stage during a live performance. In a more practical setting, you can set the basses in the middle and spread the violins hard left, cellos hard right, and violas either centered or slightly right to balance a violins divisi panned slightly left. Equalization settings can vary widely depending on the library and patches used. The goal is to achieve clarity and a cohesive and natural sound. Usually synthesized strings do not cut through as much as sample-based ones, whereas sampled patches tend to be a bit thin. Through the use of equalization try to compensate for any unbalance in the sonic spectrum of the string orchestra. Add shininess to the synth patch by boosting 3dB (or more) between 5 and 6kHz. Apply a 3dB (or more) boost in the frequency range between 100 and 200Hz to fatten up a sample-based patch. Based on the frequency range covered and on the range featured by each section, try to cut some of the frequencies that are not specifically highlighted by a particular section. Center the attention of the listener to those frequencies that are important for each section and to reduce the amount of masking created by the other frequencies that tend to overlap with neighbors' instruments. The importance of placing a sample-based sonority in a natural and realistic environment is significant. When choosing a reverb for the string orchestra, strike the right balance between acoustic ambience and clarity. The reverb should be felt but not heard. If you actually hear the reverb it is probably too much. For more aggressive sonorities, try using a shorter reverb (between 2.3 and 2.8 seconds). For more sustained and pad-like parts, set a longer reverb time (2.9 up to 4 seconds). If time allows, use slightly different reverberation settings for each individual section of the string orchestra. There are several convolution reverbs that provide a built-in engine programmed automatically to give you the acoustic response of different zones of a hall with the parameters set up for the different sections of the orchestra. To improve orchestral renditions even further, add some real noise such as squeaking floors and chairs, page turning, breathing (especially for wind instruments), or light coughs to remove some of the sterility of the MIDI environment from the orchestral performance.

3.14 Exercises

Exercise 3.1

Take a phrase from a popular melody and write it in treble, alto, tenor, and bass clefs. Keep the phrase located around middle C on the piano if possible.

Exercise 3.2

(a) Take a rhythmically active melody and write in the bowings along with "up" and "down" bow markings. Your choices should enhance the rhythmic flow and create an interesting syncopation.
(b) Sequence the above melody for a violin section using the "attack/release" techniques explained in Section 3.8.1, alternating the up and down bow on two separate MIDI tracks.

Exercise 3.3

Write several short musical phrases that feature the violin using one open string simultaneously with a moving melodic line on the adjacent string. Feel free to elaborate using the other open strings as well.

Exercise 3.4

(a) Write a mournful piece for viola that features an occasional pizzicato note on the A string. Reserving the lowest three strings for the bowed melody, use notes that will create an ebb and flow of tension with the pizzicato note.

(b) Sequence the above piece for a viola section using the key-switching (KS) technique to alternate between sustained and pizzicato. If you don't have a KS patch, use two different tracks/MIDI channels to alternate between the articulations.

Exercise 3.5

(a) Write a lyrical piece for solo cello in the key of G major. Using the open strings of the instrument, suggest the I, IV, and V chords by playing the root and 5th of each chord simultaneously. Place these chords on a strong downbeat at the beginning of a new phrase and follow each of them with an interesting melodic response. Since the open strings are being used, a cellist can easily jump from one function to the other.

(b) Sequence the above piece for a cello section using the following techniques:

    (i)   attack/release for each note (up- and down-bow technique);
    (ii)   layer a dry library and a wet library (if possible);
    (iii) key-switching;
    (iv) layer a real cello (triple-tracked) to sweeten the section.

Exercise 3.6

Create a pedal chart (or diagram) for the following scales: D major, F natural minor, B harmonic minor, and G melodic minor.

Exercise 3.7

Create a pedal chart that will enable a harpist to play a glissando that only sounds as the notes of an F major pentatonic (F–G–A–C–D). Hint: you must indicate pedal changes for the two remaining strings (B and E), spelling them enharmonically to sound in unison with the desired pitches.

Exercise 3.8

Write a pedal chart that will enable a harpist to play a glissando that sounds as a G9 dominant chord (G–B–D–F–A).

Exercise 3.9

Create a sequence choosing one or all of the following scores: 3.1, 3.3, 3.7, and 3.10. Experiment with all the sequencing techniques learned in this chapter, paying particular attention to layering, attack/release, use of CCs, and the final mix.

4 Writing and Sequencing for the Woodwind Section

4.1 General Characteristics

The woodwind instruments, as a section, offer the greatest diversity with regard to tone color. This is advantageous regarding expression and variety, but is challenging when a homogenized sound is desired. The challenge is greatest when working within a harmonic framework but is easiest when the harmony is assigned to one color group, for example three flutes, each playing one note of a triad. The woodwind instruments are at their best when featured playing independent lines. It is wonderful to hear, for example, a melody initially carried by a solo flute, picked up by a solo clarinet, and then transferred to a bassoon. Other times there may be a counterline used where, for example, the double reeds sounding in octaves might be heard against the flute and clarinet in octaves. There are many possibilities but, in general, monophonic and polyphonic textures of writing will present less difficulty than homophonic texture. Technically, the woodwind instruments are quite agile as they are able to execute pitches with greater accuracy than the brass and certainly the strings. Their articulation is very precise and their ability to jump registers is also very good. Sonically, they project well when heard against strings but are certainly weaker than loud brass. The trade-off, in comparison to the strings, is that the performers need air to create sound, so endurance should be respected when writing for the woodwinds.

4.2 The Woodwind Section

The woodwind section of a full orchestra is comprised of four distinct color groups. They are, in score order, the flutes, oboes, clarinets, and bassoons. (Collectively, the wind section can also be bisected into two large groups: the double reeds, consisting of the oboe, English horn (cor anglais), and bassoon, comprise one group, offering a nasal and dry sound; the other group, consisting of the flutes and clarinets, is more mellow, wet, and transparent.) The size of each section can vary depending on the size of the orchestra.

The largest section consists of three players within each color group, for a total of 12 musicians:

Woodwinds in threes:

- First flute: the principal chair, or section leader, usually plays any solo passages.
- Second flute: supports the principal, trading melodic phrases and adding harmony, as well as creating unison or an octave.
- Third flute: requires a "doubling" on piccolo or alto flute; this player also enhances the section in unison, octave, and harmonic textures.
- First oboe: the principal chair usually plays any solo passages.
- Second oboe: supports the principal, trading melodic phrases, adding harmony, and, less commonly, creating unison or an octave.
- Third oboe: requires a "doubling" on English horn; this player also enhances the section, thickening the harmonic texture and, less commonly, creating unison or an octave.
- First clarinet: the principal chair usually plays any solo passages.
- Second clarinet: supports the principal, trading melodic phrases and adding harmony, as well as creating unison or an octave.
- Third clarinet: requires a "doubling" on bass clarinet; this player also enhances the section in unison, octave, and harmonic textures.
- First bassoon: the principal chair usually plays any solo passages.
- Second bassoon: supports the principal, trading melodic phrases and adding harmony, as well as creating unison or an octave.
- Third bassoon: requires a "doubling" on contrabassoon; this player also enhances the section in unison, octave, and harmonic textures.

Smaller orchestras may use only two players (woodwinds in pairs) for each group; in this case, the second chair is responsible for "doubling." Traditionally, the principal chair has always been reserved as the solo chair, so it is better for that player to stay on the main instrument of the color group.

4.2.1 "Doubling"

In studio recording and pit orchestras, there may be only one player from each group. Economic considerations may ultimately determine how many woodwind players can be hired. The concept of "doubling" has been used abundantly in these situations, where a player might be hired to play instruments that cross one color group. Cited below are some possibilities.

Scenario 1

- First chair: flute and clarinet;
- Second chair: flute and oboe;
- Third chair: clarinet and bass clarinet;
- Fourth chair: bassoon and clarinet.

With the exception of the third chair, the doubling requirements in this scenario significantly narrow the number of musicians who are professionally capable of this task.

Scenario 2

A wiser choice, where possible, would be to keep the double reeds confined to a particular chair, as in the following scenario:

- First chair: flute and piccolo;
- Second chair: flute and clarinet;
- Third chair: oboe and English horn (double reed chair);
- Fourth chair: clarinet and bass clarinet.

It is more common for primary flautists and clarinetists to double on each other's instruments and less likely that a double-reed player would cross over to this color group. That said, woodwind players who specialize in studio work and theater work demonstrate their abilities on a wide array of woodwind instruments and are usually paid handsomely for it.

Scenario 3

The third scenario includes the saxophone as it is used abundantly in more pop and jazz styles:

- First chair: flute and soprano saxophone;
- Second chair: flute, clarinet, and alto saxophone;
- Third chair: oboe, English horn, and tenor saxophone;
- Fourth chair: clarinet, bass clarinet, and baritone saxophone.

Regarding instrumentation, there is no particular rule for doubling, except that the player must be able to play the instrument that is requested. More importantly, it is the responsibility of the contractor (or anyone else doing the hiring) to find a musician who can play the necessary instruments on a professional level. To anticipate this, the orchestrator should consider how to assign the instruments to the chairs. It is quite easy to write without any restrictions (especially within the MIDI domain), so it is important to consider these factors if the end result will be a live performance. One final word about doubling: it is essential for the orchestrator to allow the proper time necessary for the player to make a switch from one instrument to another. There are some factors to consider that will affect the transition accordingly:

- The size of the instrument (a flute is easier to handle than a bassoon or bass clarinet).
- The mouthpiece (in contrast to the flute's metal mouthpiece, a reed needs to be moistened to help produce a controlled sound).
- Intonation becomes precarious when a player does not have time to warm up the new instrument.

4.3　Creating Sound

Woodwind players primarily create sound by blowing air through a pipe or cylinder. The longer the pipe, the lower the basic pitch. Along the length of this pipe is a series of holes that help, in effect, to "shorten" the pipe to create a series of higher pitches. Keys cover the holes ("lengthening" the pipe) and the player's fingers work the keys in various combinations to create a multitude of pitches. The mechanics of these instruments make it easier for the performer to create various pitches quite accurately, in contrast to the more precarious string instruments. But, with the winds, there is more tension within the body due to the production of a constant airstream in conjunction with the wind resistance of the instrument. In addition, there is also a slight fluctuation in embouchure (the mouth's position on the mouthpiece) to accommodate different registers. Consequently, notes in the extreme upper register will require more physical exertion from the winds than from the strings.

4.3.1　*Tonguing (Articulation)*

The tongue is essential in the creation of sound. It helps to add expression with regard to the initial attack of a note. As consonants do in a spoken language, the tongue adds clarity and intensity. This is most apparent with short (staccato) notes. Woodwind instruments provide a clean, crisp attack in this regard. The double reeds are the best at this, offering a dense, precise, pinpoint staccato. The clarinet's staccato is a bit more blunt and the flute's staccato lies somewhere in between. A series of symbols notifies the player regarding the strength, subtlety, or duration of an attack. Figure 4.1 shows a variety of expressions using these symbols.

Figure 4.1 Common articulation markings

4.3.2　*The Slur*

A slur is indicated using a curved line over two or more notes. The first of these notes is to be articulated with the tongue while the remaining notes are heard via airflow and fingerings (the various depression of the keys), creating a smooth-flowing (legato) sound (Figure 4.2).

The notes within a slur are to be played in one breath. A slur that extends for a bar or more is commonly referred to as a phrase (Figure 4.3). The player may certainly play more than one phrase within one breath, but should never breathe during a phrase marking.

Figure 4.2 Slur markings

Figure 4.3 Phrase markings

As with strings, it is important that the orchestrator indicate phrase markings. If there is no such indication, the player might tongue each note separately. This works fine for more percussive articulations, but desiring any smooth lines will require the presence of a slur or a phrase marking.

4.3.3 *The Trill*

Trills are very effective on woodwind instruments. The speed of the trill is usually not measured, sounding more like a flutter that defies strict tempo. However, trills can still be timed within the greater framework of beats or bars. Figure 4.4 shows some trills as they should be written, indicating the starting note, the note to progress to, and the duration of this exercise. It looks as if there are too many beats in the bar but, using bar 1 for reference, the note C is the starting note and the D♭ is the note to progress to. The trill occurs back and forth between these two designated pitches within the confines of a whole note (or four-beat) period.

Figure 4.4 Trills

Trills work well in the majority of a woodwind instrument's range, but can be problematic in the extreme high or low registers. The most commonly used intervals are seconds (notes adjacent to each other). The tremolo is essentially a trill expanding to an interval of a third or sometimes a fourth. Wider intervals are not practical at fast speeds.

4.3.4 Grace Notes

Although this type of note is not exclusive to the woodwind, it is probably most effective when played by these instruments. As its name suggests, the grace note offers a delicate and playful character. As with the trill, the most common interval is the second. Figure 4.5 shows how the grace note is to be written.

Figure 4.5 Grace notes

4.4 The Specific Instruments

There are many instruments within the woodwind family so, owing to the confines of space, only the most common will be discussed here with regard to tone, technique, range (including registers and transposition), and musical function within an ensemble. The instruments are presented in their score order (how they would appear on a conductor's score from top to bottom).

4.4.1 The Flute (in C)

Tone: The modern flute is unique to the orchestral winds as it is made of metal. There are wooden flutes as well, but these instruments are more commonly associated with non-Western music or music from the Renaissance period (recorders). The modern flute also uses a metal mouthpiece instead of a reed. Its metallic sound is distinct, making it somewhat "colder" than the wooden instruments, yet also more transparent.

Technique: The flute is quite agile throughout most of the instrument, but it does become slightly more sluggish in its lowest notes. Register jumps are fairly easy within or slightly above an octave but, in contrast to the strings, the woodwind player's task is a bit more difficult, so excessively wide leaps at quick speeds should be avoided.

Figure 4.6 Range and registers of the flute

Range: There is no transposition involved and music is to be written only in treble clef. The instrument's range is from middle C on the piano (some flutes have a low B) extending upwards for three octaves (Figure 4.6).

The characteristics of its registers are as follows:

- Low register: dark, cold, thick, rather soft;
- Middle register: slightly thinner but sweet and warm;
- High register: exceedingly bright, penetrating, loudest projection.

Musical function: The flute offers the lightest, wistful quality, but is weaker in general than the other woodwind instruments, so care must be taken not to overpower this instrument when it plays in its weaker registers. The flute can lead the woodwind section, but must be in its strongest register (notes above the staff) in loud passages.

4.4.2 The Piccolo Flute (in C)

Tone: The piccolo, like the flute, also uses a metal mouthpiece instead of a reed, but its overall length is much shorter. Consequently, its sound is much smaller and potentially shrill. It has more of a whistling quality than the flute and can be heard easily over the entire orchestra when in its high register.

Technique: the piccolo is also quite agile throughout most of the instrument.

Range: There is no pitch transposition, but there is a register transposition. Because the piccolo's notes are so high, it would be necessary to use an excessive amount of leger lines to write them. Consequently, the notes are written an octave lower than the actual sound. The instrument's written range is from its low D to high C, almost three octaves higher (Figure 4.7).

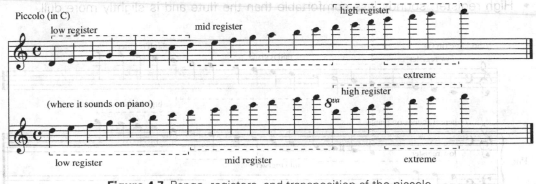

Figure 4.7 Range, registers, and transposition of the piccolo

The characteristics of its registers are as follows:

- Low register: dark, cold, haunting, very soft;
- Middle register: sweet and full;
- High register: exceedingly shrill, penetrating, loudest projection.

Musical function: The piccolo can be very effective in loud passages, but should be used selectively. Although it is used abundantly in marching band music, lighter and more subtle styles of music have less tolerance for its sound. One of its best uses is portraying light, comedic statements (in particular when paired with the tuba). It also works well when in unison or octaves with one or more flutes as it can strengthen the flute ensemble sound.

4.4.3 The Alto Flute (in G)

Tone: The alto flute, like the flute, also uses a metal mouthpiece instead of a reed, but its overall length is longer. Consequently, its sound is a bit deeper and more mellow and is generally a bit weaker in projection.

Technique: the alto flute is not quite as agile as the flute in C and also requires more breath.

Range: There is a pitch transposition involved. When its low C is played, the G, a perfect fourth below middle C at the piano, is heard. Consequently, the arranger must write the part a perfect fourth higher. The instrument's written range is from its low C up approximately two and a half octaves higher to G (Figure 4.8).

The characteristics of its registers are as follows:

- Low register: dark, cool, lush, soft;
- Middle register: lighter and thinner but still more mellow than the C flute;
- High register: sounds less comfortable than the flute and is slightly more dull.

Figure 4.8 Range, registers, and transposition of the alto flute

Figure 4.9 Flute and alto flute in unison

Musical function: The alto flute is used primarily to extend the register below the flute. It also works well when in unison with one or more flutes, but not so much for volume. The alto flute, when playing in unison with a C flute's lowest notes, can add more body and expression as these notes sit in the alto flute's higher register (Figure 4.9).

The best musical context is usually soft and lush in mood and the instrument is very effective in recording sessions. Henry Mancini and Claus Ogerman are two composers who feature a low- to mid-register flute section sound in much of their work.

4.4.4 The Oboe (in C)

Tone: The oboe is the soprano voice of the double-reed instruments. The double reeds, in general, offer a nasal, pungent tone that is quite distinct from the softer, wispier flute sound. The bamboo reed (which is folded over to make two sides, hence the term "double reed") is primarily responsible for this. The wood barrel provides warmth, in contrast to the metal barrel of the flute.

Technique: the oboe is quite agile throughout most of the instrument, but does become sluggish with less dynamic control in its lowest notes.

Range: There is no pitch transposition involved and music is to be written only in treble clef. The instrument's range is from B♭ below middle C on the piano, extending upwards about two and a half octaves (Figure 4.10). Its best range is from its low E♭ up to G directly above the staff, where the instrument offers the most flexibility technically and dynamically in conjunction with the sweetest tone.

Figure 4.10 Range and registers of the oboe

The characteristics of its registers are as follows:

- Low register: dark, coarse, thicker, less dynamic control;
- Middle register: sweet and warm (the best register overall);
- High register: exceedingly pinched, intense.

Musical function: The oboe should not necessarily be considered subordinate to the flute. Although accomplished players can "stay under" a flute (depending on the registers of both instruments), the oboe's strong, penetrating sound makes it potentially a lead voice. It can be thought of as the "trumpet" of the woodwind section (evident from music of the Baroque period). For most musical settings, a pair of oboes, or an oboe and an English horn, does not work well in unison (the effect is somewhat similar to bagpipes). It is better, when playing simultaneously, for these instruments to work in harmony or in octaves if space permits.

4.4.5 *The English Horn (in F)*

Tone: The English horn is the alto voice of the double-reed instruments. Similar to the oboe, its tone in the higher register is more difficult to distinguish, but its lower registers offer a noticeably deeper, darker, and more mellow tone, due to the longer barrel. The instrument has fewer higher frequencies and is therefore best used in solo or soft ensemble passages.

Technique: the English horn is less agile than the oboe, but has more control in its lowest register than the oboe when playing its lowest notes.

Range: There is a pitch transposition of a perfect fifth involved. When the English horn plays C, it sounds as F on the piano, a perfect fifth lower. Therefore, the arranger must write the part a perfect fifth higher using only the treble clef. The instrument's range is from its low B extending upwards slightly over two octaves (Figure 4.11).

Figure 4.11 Range, registers, and transposition of the English horn

The characteristics of its registers are as follows:

- Low register: dark, full, good dynamic control (most characteristic);
- Middle register: brighter but with more tension as it ascends;
- High register: exceedingly pinched, intense.

Musical function: The English horn's practical function is to extend the range of the double-reed sound below the oboe. It does, however, have its own unique tone quality and music writers, when confronted with an equal choice, may deliberate between using the English horn and the oboe. Within a double-reed ensemble sound, the English horn is analogous to the viola within a string section.

4.4.6 The Clarinet (in B♭)

Tone: Within the woodwind section, the clarinet has the warmest wood tone. It uses a single reed made of bamboo (distinguishing itself from the double reeds) in conjunction with a hard rubber mouthpiece. The general tone, in contrast to the double reeds, is more transparent and "wet." Stylistically, there is a wide expression of tone within the instrument. The Dixieland-style clarinet sound differs greatly from the orchestral clarinet sound. It is important for the music producer to know what is needed.

Technique: The clarinet is extremely agile throughout most of the instrument, but can become tricky regarding fingerings in its middle, "throat tone," register. Register jumps are easy within or slightly above an octave, but excessively wide and rapid leaps should be avoided.

Range: There is a pitch transposition involved and all music is to be written in treble clef. When the clarinetist plays C it sounds as B♭ on the piano. Its lowest written note is E below the staff and extends upwards by more than three and a half octaves. The uppermost notes are available depending on the skill of the performer (Figure 4.12).

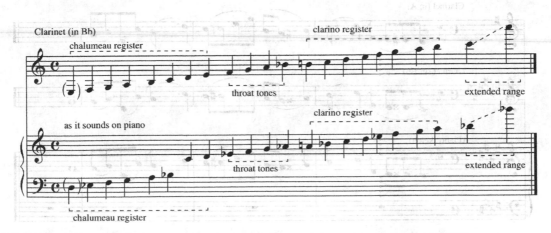

Figure 4.12 Range, registers, and transposition of the clarinet in B♭

The characteristics of its registers are as follows:

- Low register (*chalumeau*): dark, lush, thick, rather soft but full;
- Middle register (throat tones): slightly thinner and less vibrant;
- High register (*clarino*): bright, clear, loudest projection;
- Extended range: progressively intense and hard to control (only for professionals).

Musical function: The clarinet family is extensive, encompassing the widest range of all the woodwind instruments. The B♭ clarinet is amazingly versatile with regard to range, registers, and expression. Its weakest area, the throat tone register, should be respected since exceedingly difficult technical passages can challenge the player. Performance style is different from the flutes and double reeds as it is customary for clarinetists to avoid the use of vibrato in orchestral playing.

4.4.7 *The Clarinet (in A)*

Tone: this clarinet is very similar to the one in B♭, if only slightly darker and longer in size.

Technique: similar to the clarinet in B♭.

Range: There is a pitch transposition involved and all music is to be written in treble clef. When the clarinetist plays C it sounds a minor third lower as A on the piano. Therefore, the part must be written a minor third higher. The instrument's range and registers are the same as the clarinet in B♭.

Musical function: The main reason for using the clarinet in A pertains to the key of the music. Figure 4.13 shows a piece of music written in the concert key of E major. If a

Figure 4.13 Comparison of the transpositions for the clarinets in A and B♭

Bb clarinet were to play this, its key would be F♯ major. The clarinet in A is a better choice as its key would be G major.

4.4.8 The Bass Clarinet (in Bb)

Tone: The bass clarinet is much longer, requiring a metal rod that extends from the bottom of the instrument, touching the floor. Its neck (connecting the mouthpiece to the main body of the instrument) is made of metal and is curved to make it more comfortable for the performer to play. At the other end is a metal, flared bell. As a result, the instrument can obtain a brighter, raspy tone quality when needed. When played softly, it offers the same warmth as the other clarinets. Its sound is very deep in its lowest register, but becomes progressively thinner as it ascends through its range of notes, sounding more like a vocal falsetto quality.

Technique: the bass clarinet is less agile than the other clarinets and wide leaps, particularly downward into the low register, can be difficult to control.

Range: There is a pitch and register transposition, as well as a clef transposition. Its lowest written note (Eb) actually sounds a major ninth lower on piano as the note Db in the second octave (Figure 4.14).

Although it is acceptable to write music in bass clef, it is better and more prevalent to write in treble clef since this is customary for clarinet players. The instrument's written range is from its low Eb up approximately three octaves higher. The characteristics of its registers are as follows:

- Low register (*chalumeau*): deep, dark, cool, lush, full;
- Middle register (throat tones): though not as full as the lower register, the bass clarinet's sound fares better than the other clarinets;

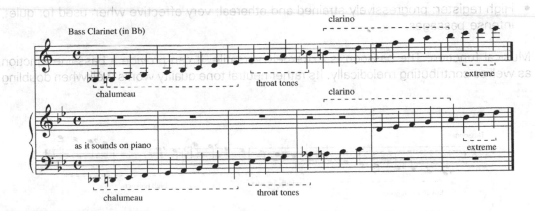

Figure 4.14 Range, registers, and transposition of the bass clarinet

- High register (*clarino*): the bottom half of this register is full and clear, but as it ascends the sound becomes more strained and ethereal;
- Extreme register: generally not used except for effect.

Musical function: The bass clarinet is used primarily to extend the clarinet sound well into the bass register, where it offers lush bass tones. It is wonderfully agile and lighter in tone than heavier instruments in this register, such as the string bass (arco), bass trombone, or baritone saxophone. Its agility also allows it to assume melodic functions in its higher register; the tradeoff in this regard is that the instrument tends to lose some of its tonal beauty, so it is important to consider the musical mood.

4.4.9   *The Bassoon (in C)*

Tone: The bassoon is a double-reed instrument. It is used most commonly as the bass instrument of the double reeds. (The contrabassoon is technically the lowest of this group, but is not commonly used.) It is more robust in sound than the higher double reeds and possesses three distinct sound qualities in accordance with its three registers.

Technique: although it is quite agile, the extreme ends of its range will curtail its flexibility.

Range: No pitch transposition is necessary. The bassoonist primarily reads bass clef and also reads tenor clef (for review, the use of the tenor clef for cello parts is applicable to the bassoon as well). Its lowest note is the B♭ below the bass clef staff and its range can extend slightly higher than three octaves, depending on the accomplishments of the performer (Figure 4.15).

The characteristics of its registers are as follows:

- Low register: deep and full but slightly sluggish; used primarily for bass lines;
- Middle register: lighter and thinner yet still robust; great for melodies and accompaniment figures;
- High register: progressively strained and ethereal; very effective when used for quiet, intense passages.

Musical function: The bassoon is extremely versatile. It can provide a bass-line function as well as contributing melodically. Its rather neutral tone quality works well when doubling

Figure 4.15 Range and registers of the bassoon

with other instruments: its staccato adds punch to pizzicato basses; it can thicken a cello line or pair with a French horn when playing a rhythmic, harmonic accompaniment; and it adds power and projection to the higher double reeds when playing a melody an octave lower. Unlike the bass clarinet, whose middle register tends to become more pale, the bassoon's middle register remains robust and much more useful in orchestral settings.

4.5   The Saxophones

This group of woodwind instruments is not really considered part of the orchestra because of its late arrival on the Western music scene. With the jazz bands of the swing era, the saxophone came into significant prominence, continuing strongly to the present day. The history of great individual sax players of the twentieth century is rich with diversity and individuality. The tone and expression of the instrument can vary greatly in accordance with the style of the individual. Many styles have been codified, but any further in-depth study cannot be afforded here. Suffice to say, the arranger should realize that the style of music should determine the proper sound and henceforth the hiring of the appropriate player.

4.5.1   Saxophone Tone Quality

Even though the saxophone is considered a woodwind instrument, it is essentially made of brass. This enables the saxophone to compete more equitably with brass sections (typically large jazz bands) and rock bands (or other electric instrument environments). It is categorized as a woodwind instrument because the player uses a mouthpiece with a single reed, similar to the clarinet. The mouthpiece can be made of hard rubber, plastic, or metal, having a significant effect on tone quality. (The rubber mouthpiece generally provides a warmer tone, while the metal mouthpiece is more edgy, which makes it suitable for rock and funk.) Today, many players have a mouthpiece setup that can accommodate the demands of a multistylistic music market.

4.5.2   Saxophone Range

All of the saxophones play from a low B♭ below the treble clef staff upwards to a high F above the staff (Figure 4.16). The fingerings for the saxophones are the same, making it easy for the performer to play any saxophone (be aware that not all saxophonists play or own all of the instruments). Today, many pop and jazz saxophonists have extended

Figure 4.16 Range of the saxophones

their personal range by playing notes that are considered to be "off the horn." Usually these notes are used in an improvisatory way by a soloist, so the arranger should not write these notes unless there is a specific player in mind. The bottom end of the range can be expanded on the baritone saxophone from Bb to A if the particular horn has that note available. Unless the arranger is aware of this option, it is best to stay within the conventional range. Ultimately, when writing for saxophones, the challenge is for the orchestrator to understand the various transpositions involved with each saxophone.

4.5.3 *Saxophone Transposition*

Because the various sizes of the individual saxophones (from small to large are the sopranino, soprano, alto, tenor, baritone, and bass saxophones), each is transposed differently.

Listed below are the saxophones, along with their corresponding transposition:

- Sopranino (in Eb): sounds a minor third higher; written a minor third lower;
- Soprano (in Bb): sounds a major second lower; written a major second higher;
- Alto (in Eb): sounds a major sixth lower; written a major sixth higher;
- Tenor (in Bb): sounds a major ninth lower; written a major ninth higher;
- Baritone (in Eb): sounds a major sixth plus one octave lower; written a major sixth plus one octave higher;
- Bass (in Bb): sounds a major ninth plus one octave lower; written a major ninth plus one octave higher.

The sopranino and bass saxophones are rarely used. Score 4.1 in Appendix A shows a melody transposed for the most popular saxophones.

Musical function: Unlike the orchestral woodwind instruments, the blend of the saxophones is similar to the uniform tone quality of the strings. They work well to create a homogenized, balanced harmony. This technique is most evident in jazz band writing, where closely voiced harmony is typical (see Score 4.2 in Appendix A and listen to Example 4.8).

Three types of voicing techniques are displayed: the "closed voicing," with the top note doubled an octave lower; the "drop 2" voicing, where the second note from the top is dropped an octave lower; and the five-way (or independent) voicing. The standard instrumentation for a saxophone section within a jazz band is:

- Alto 1 (lead player, sometimes doubles on soprano sax, flute, or clarinet);
- Alto 2 (sometimes doubles on flute or clarinet);
- Tenor 1 (sometimes doubles on flute or clarinet);
- Tenor 2 (sometimes doubles on flute or clarinet);
- Baritone (sometimes doubles on bass clarinet or clarinet).

Saxophones also sound great in unison or octaves and blend rather well with trumpet and trombone. Many bands (e.g. Blood Sweat and Tears, Chicago, James Brown, The Jazz Messengers, Tower of Power) use smaller horn sections in various combinations.

4.6 Concerns Regarding Orchestration

The woodwinds' diversity of color, weight, and balance presents a significant set of challenges to the arranger. Unlike strings and brass, the woodwinds do not blend, so the direct transfer of ideas from a piano to the instruments can be very misleading. Listed below are some considerations that underscore this problem:

- Register placement;
- Voice-leading;
- Balance within the orchestra.

4.6.1 Register Placement

Some woodwind instruments are strongest in their upper register; some are strongest in their middle register; others are strongest in their lowest register (Figure 4.17).

The tendency when sitting at a keyboard, especially in a MIDI environment, is to play harmonic voicings that fall within the shape of the hand. A close proximity of notes can be problematic, especially when orchestrated for a diverse group of winds. For the music written in Figure 4.17, the rather low register almost precludes the flutes from participating at all. The melody (top notes of the chord sequence) is certainly within the range of the flute, but there would be an imbalance if the harmony parts were played by stronger woodwind instruments. To rectify this problem, a stronger woodwind should double the flute in unison to provide a better balance. Another successful way to orchestrate this example as written is to use a uniform color group (all flutes, all clarinets, etc.). A flute ensemble sound in this register would require the use of the alto flute on the middle line and the bass flute on the bottom line. The sound would blend, but would be very soft. Three clarinets would work well here, offering a stronger, fuller tone quality, although still not loud since the lead voice is in the throat tone register. For a double-reed ensemble sound, the three parts could be orchestrated for oboe, English horn, and bassoon. This

Figure 4.17 An initial keyboard sketch

would be a more hardy but dryer sound and the oboe would need to be careful, when playing its low D, to avoid any coarseness. Using two English horns and a bassoon would address this concern and the result would be slightly more mellow, although still robust. Most orchestrators will rearrange a passage like this within the context of the instrumentation. In other words, the available combination of instruments and the necessary dynamic level will ultimately determine how this music will be voiced. When working with a full woodwind section there are more possibilities than with smaller chamber groups. Score 4.3 in Appendix A shows some solutions (a non-transposed score is used for ease of analysis) with regard to the given instrumentation.

There are many possibilities, but essentially there are some concepts that remain consistent within the choices: the flute is respected as being more delicate when in the staff and is doubled at the unison by an oboe or a clarinet; the oboe's natural projection is respected and is given either the melody or a line that will establish good counterpoint with the melody; and each color group is written with a sense of balance unto itself. As the ensemble decreases in size, more attention is given to the individual instruments, as they will become more apparent. With the quartet version, the original arrangement of the inner parts has been altered slightly to provide each instrument's line with a melodic sensibility. Notice in bar 2 that the oboe and clarinet parts cross to provide a better melodic line and a better resolution into the cadence.

4.6.2 Voice-Leading

The separate and distinct color of various woodwinds providing harmonic support will be noticed more than a group of instruments with a uniform timbre (see Scores 4.4–4.6 in Appendix A).

In Score 4.4 (Example 4.2 on the website), the music is mildly chromatic inside the key of D minor. All chromatic notes resolve to a neighbor scale tone. Notice also that within the concert sketch the voicings are impossible to play on piano. The music is conceived truly with the specific instruments in mind. More careful attention to the oboe part will show a slight difference from its part within the sketch. The repeated notes have been tied together (as a whole note) for a more sustained and more melodic sense. On piano, it is more natural for all the notes to be struck since they fade naturally, but an oboe's notes are very noticeable so repeated notes will make the music sound stagnant. The tied notes now make the oboe line and the music in general more graceful and interesting.

The second example (Score 4.5, Example 4.4 on the website) is in G major and is in a slightly more contemporary style that allows for some striking dissonance. Here again, notice that the oboe's notes have been tied. In this case, the oboe part acts as a pedal point on the dominant (fifth) of the key moving down to the tonic (root) of the key. This creates a pleasant stillness and clearly defines the tonal center against the more interesting dissonant clashes. The phrase ends in the deceptive key of E major and all notes in G major resolve stepwise into the new key.

The final example (Score 4.6, Example 4.13 on the website) is much more linear and ideally suited to the woodwind. It would also work with strings, but the four distinct woodwind colors really bring forth to the listener the individuality of each line. The piece begins in G major and contains some chromatic chord motion within the key and then modulates suddenly, in bar 3, to A♭ major. The melody is carried in the flute for the first two bars and then the oboe takes it for the statement in the new key. Notice that all chromatic notes resolve to an adjacent scale tone and that each part is a melody in itself. Finally, notice the concert sketch. It does not look like a piano part and cannot be played at the piano.

4.6.3 Balance within the Orchestra

Although the winds have no problem projecting when heard with strings, they are challenged significantly when having to compete with brass at a loud volume. It is important for the woodwind instruments to be placed in their strongest registers. Even so, their sound will still become absorbed somewhat into the brass sound, but the higher frequencies will emerge and the lower woodwinds will help fatten the overall texture (Figure 4.18). In this example, a C major chord is voiced with the bassoons in their fattest register. The English horn and oboes are placed in their middle register, where some higher frequencies can emerge, but the sound is still robust. The clarinets are placed above the oboes in their bright *clarino* register and the flutes are placed in their highest octave, where their carrying power is greatest. Careful observation will point out that there is an empty octave between the English horn and the first bassoon. This gap would most likely be filled by the French horns and/or trombones.

One way to score the winds with the brass is to have them "poke through" or find "a window" in between the brass figures. This antiphonal exchange between the two instrument groups enables the listener to hear and follow the musical ideas and instrumental colors more clearly (see Score 4.7 in Appendix A).

Figure 4.18 The full orchestral woodwind section scored at the forte level

In this example, the music is very loud and hostile. The winds answer the brass chords with a melodic fragment that is scored in octaves for clarity and power. The piccolo is placed in the highest octave of the music, but is in its middle register, where it will project but not be too shrill. The other flutes are heard an octave lower and, although they are not in their most powerful register, the two in unison will add strength and support the piccolo. The two oboes are scored in octaves; oboe 1 brings out the higher frequencies while oboe 2 adds body. A similar approach is used with the clarinets, but the third clarinet doubles in unison with the second to strengthen this register. Two bassoons play in unison in their middle register, where mobility is good and the tone is robust. The lowest octave is avoided, as this register will be rather muddy. In bar 3, a dynamic contrast is desired where the listener first hears the woodwind trill with the brass answering. The texture and volume are deliberately thinner here as some woodwind drop out. The trill is scored initially in a one-octave spread and, in bar 4, widens to two octaves with the entrance of the bassoons. This is intensified further with the first oboe and first clarinet joining the flutes in the top octave.

4.7   Sequencing for the Woodwind Section: an Introduction

The woodwind section represents a rare good example of sonorities and articulations that, in most cases, can be rendered fairly accurately by a virtual MIDI ensemble. The monophonic characteristic of the instruments that belong to this section and, for some of the instruments, their fairly limited number of sonorities, make it an ideal option when it comes to MIDI sequencing. This is particularly true for the oboe and the bassoon. As we saw for the string orchestra, there are several techniques and MIDI control changes (CCs) that can be effectively used to bring your orchestral rendition to a higher level of realism. The next sections will guide you through a series of advanced options and techniques to get the most out of your virtual woodwind section. Let's begin!

4.8   Synthesis Techniques and Library Options

Sample-based and physical modeling libraries are the two best options for woodwind libraries. At the moment there are no other suitable alternatives that are convincing enough to cover these sonorities. Basic subtractive synthesis has, for many years, fulfilled the needs of many MIDI orchestrators looking for semi-realistic clarinet and flute sounds. Through wavetable synthesis it is only partially possible to achieve a realistic rendition of woodwind instruments since the limitation of the size of the samples can drastically diminish the versatility of a patch. For this reason, you can get away with wavetable woodwind patches when the parts being sequenced are not very exposed. Solo passages that feature solo oboe or bassoon call for the use of sample-based patches, while the clarinet can be slightly more permissive (due to its similarity to a square wave and to its relatively low harmonic content). The saxophones are among the hardest sonorities to re-create successfully in a virtual environment. Their rich, reedy tones and their complex articulations are extremely difficult to render. Nowadays, only through sampling and "sample modeling" techniques can we accurately reproduce the sonorities of oboe, clarinet, flute, and bassoon and their variations. As for the string orchestra, a multilayered sample library should be the first option to consider when shopping for an acoustic series of patches. Chapter 3 described how such technology allows you not only to assign different sampled notes to different keys through the mapping process, but also to trigger different dynamically related samples through the use of either MIDI velocity or MIDI CCs. This is crucial in order to render effectively and realistically the different colors, dynamics, and timbres of the woodwind instruments. When looking for an orchestral library capable of providing the most versatile and convincing sonorities, a few things should be kept in mind. First, look for libraries that provide not only the basic woodwind instruments (flute, oboe, clarinet, bassoon, and soprano, alto, and tenor saxophones), but also their closely related siblings, such as piccolo flute, alto flute, English horn, bass clarinet, contrabassoon, and baritone saxophone. The addition of these variations will ensure that your sound palette is complete and that you will not be limited in the future. Look also for libraries that provide extra chairs for each instrument. I would definitely recommend at least a second chair for flute, oboe, clarinet, and bassoon. Try to avoid using the same sample for the second chair since it will create unpleasant phasing effects. As you learned in the first part of this chapter, the subtle differences in color of each instrument of the woodwind family and their chameleonic palette of combinations can make a big difference in creating a variety of colors and combinations. The goal here is to be able to reproduce the same combinations with your virtual ensemble. Another important aspect to consider when shopping for a woodwind library is the number of different articulations that such a library offers. Although woodwind instruments in general are capable of producing a fairly restricted number of alternative sonorities, you should make sure that you can count on more than just a basic sustained articulation for each instrument of the woodwind section. Table 4.1 lists the most essential articulations that should be available when sequencing for the woodwind section.

This selection of articulations provides a flexible palette that covers the majority of situations. Pay particular attention to the fact that, usually, you will need at least two different staccato patches. Staccato articulations may be set around 0.3 seconds (for very quick passages and transitions), and 0.5 or 0.7 seconds (for passages that fall in between real staccato and sustained). When discussing the options in terms of sample libraries for the string orchestra, the point was stressed of avoiding all-in-one patches of "strings" that provide generic, and often poorly sounding sonorities, and to use instead patches that cover single sections such as violins, violas, etc. When it comes to the woodwind section, it is sometimes useful to take advantage of sampled patches that use precombined ensembles. Especially when sequencing unison passages, these patches can provide a quick and valid alternative to a single instrument sequencing approach. Therefore, you should also check for libraries that provide some woodwind ensemble options in addition to the solo samples. While some libraries provide excellent solo patches and weak ensembles, others are the opposite. Therefore, combining two or more woodwind libraries will provide your virtual studio with the most cohesive and flexible sound palette possible. When it comes to sequencing woodwind, libraries that support key-switching (KS) are particularly useful. This feature was covered in Chapter 3 when discussing the

Table 4.1 Recommended woodwind instrument articulations for a sample library

Instrument	Articulations
Flutes	Sustained
	Legato
	Staccato 1 and 2
	Fortepiano
	Sforzato
	Flutter tonguing
Oboe and English horn	Sustained
	Legato
	Staccato 1 and 2
	Fortepiano
	Sforzato
	Marcato
Clarinet and Bass Clarinet	Sustained
	Legato
	Staccato 1 and 2
	Fortepiano
	Sforzato
	Marcato
	Glissandos
	Flutter tonguing
Bassoon	Sustained
	Legato
	Staccato 1 and 2
	Marcato
Saxophones	Sustained
	Legato
	Staccato 1 and 2
	Fortepiano
	Sforzato
	Marcato
	Glissandos
	Vibrato

string orchestra. In the case of woodwind, KS allows you to perform a passage, or a part, flawlessly without the need for creating individual tracks for each articulation. Using KS will translate into a smoother and more musical performance.

The acoustic environment in which the single samples were recorded has an impact on the realism of the patches you will use in your productions. As in the case of the string orchestra, most libraries can be divided into dry or wet depending on the amount of natural ambience and reverberation that was captured during the recording of the samples. For woodwind a slightly wetter library may be preferable. The natural reverberation is essential in order to inspire a more fluent and flowing performance when sequencing your parts. Dry woodwind libraries tend to be a bit too harsh for my taste, leading sometimes even the more legato passages to sound disconnected. This is a very personal point of view and therefore I strongly suggest listening to and comparing several woodwind libraries to choose the one that best fits your personal taste and compositional style.

Table 4.2 shows a summary of the most common multisample libraries for woodwind instruments and their key features.

Table 4.2 Some of the most commonly used multisample woodwind libraries available on the market and their main features

Library	Platform	Sonic environment	Comments
Wallander Woodwinds & Saxophones	VST, AU, RTAS	Dry	Excellent library with very natural sounds and great articulations. It is particularly indicated for MIDI wind controllers
Sample Modeling	VST2, AU, AAX, RTAS	Dry/wet	Excellent flutes, clarinets, and saxophones with some interesting oboe and flute, based on a combination of sample-based and physical modeling solution. Excellent for MIDI wind controllers
Vienna Special Edition	VST, AU, RTAS, AAX	Dry	This library provides versatile woodwind sonorities with an excellent selection of articulations and dynamics
EastWest/ Hollywood Orchestral Woodwinds	VST, AU, AAX	Wet	Extremely playable and well-rounded collection of woodwind samples including legato, staccato at different dynamics, sforzando, and some trills
Vienna Super Package	VST, AU, RTAS	Dry	This is the most comprehensive option for the majority of woodwind solo and section productions. It features an incredible collection of articulation and sonorities

As in the case of the string orchestra, the addition of one or two real instruments to the sampled ones can highly improve the quality of orchestral renditions. In general, it is easier to mix and match MIDI and real tracks when it comes to woodwinds. By simply assigning the top and bottom lines of a woodwind section to real players and leaving the rest to MIDI tracks, you can achieve amazing results. If you are on a tight budget, you could use a real player only for the top line (usually flute or clarinet). Sustained solo passages that greatly expose different articulations and fast passages should be left to a real instrument. The combination of MIDI and acoustic tracks for the woodwind section will be discussed in more detail in the following pages. For now, after choosing the best library that fits your style and taste, it is time to start learning some specific techniques to improve your sequencing skills for the woodwind section.

4.9 Sequencing Techniques for the Woodwind Section: Input Controllers

The first aspect of sequencing for the woodwind section to be considered is the controller you will use to input the parts. As you probably know by now, this is a very delicate topic for any sequencing project and often a very controversial one. Keyboard controllers in general, and especially those equipped with synth or semi-weighted action keys, can provide a valuable solution. For woodwind sequencing, a light-action keyboard may be preferred because it allows you to play in a more gentle and relaxed way than a full-weighted keyboard. With a more relaxed action you can focus better on the expressivity of woodwind parts, and feel more in control of the gentle attack of which woodwind instruments are capable. If your keyboard controller allows you to change the curve of the action, try to use a gentler slope (exponential curve). One of the biggest mistakes when sequencing and writing for wind instruments is to forget that the virtual players need to breathe, exactly like their real counterparts. Very often, sequenced wind instruments parts can be heard that keep going and going forever without any breath pauses. If you want to achieve realistic renditions of orchestral parts, you need to try to be 100 percent authentic in every small aspect of the performance. There are a couple of tricks you can use to avoid such a mistake. The first can be simply achieved by blowing air out while playing the passage you are sequencing, like a real wind player would do. Take a deep breath at the beginning of the phrase, start blowing the air out while you are playing or sequencing the part on your keyboard controller, and when you run out of air it is time to give the virtual wind player a short rest (break the phrase up) and to take another breath of air. You can use this technique at different stages of the production. While composing, it gives you a better idea of how long a phrase should be and where you should give the player a rest. At the sequencing stage it allows you to insert strategically very short breaks into a phrase to allow the virtual musician to breathe. Look at Figure 4.19 to compare the same phrase with and without breaks for breathing. If you want to bring this technique up a notch, then try to blow a bit harder when sequencing for the high register of the instrument and a bit softer for its low register. This small trick will guarantee an even more realistic placement of breathing pauses.

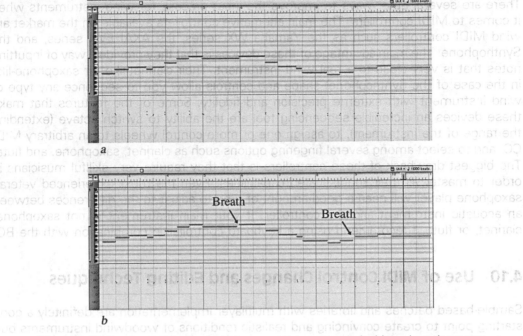

a

Breath

Breath

b

Figure 4.19 Compare the same flute passage (a) sequenced without breath pauses and (b) with two breath pauses inserted (Courtesy of MOTU)

Another way to create a realistic phrasing for wind instruments from a keyboard MIDI controller is the use of a MIDI breath controller (BC). This device was introduced in Chapter 3 when discussing alternative controllers to sequence the string orchestra. A BC is even more useful when used to sequence wind instruments parts. The most basic use of the BC is to control the length and intensity of the phrasing of a wind instrument. It allows you to assign any MIDI CC to its MIDI OUT and control the value of that CC through a mouthpiece into which you blow. The harder you blow, the higher the value of the specified controller; the softer the stream of air, the lower the CC value. You can see how this device can increase the realism of parts sequenced from a MIDI keyboard controller. While the BC can be used for several other advanced techniques, it can be very effectively used to control the flow and dynamics of the phrasing. If you assign the BC to send CC#7 volume (or CC#11 expression), you will be able to sequence extremely accurate wind instrument passages in terms of both phrasing and dynamics. Since the BC controls the volume of the performance, when you blow harder you will get a louder sound and, when you blow more softly, a softer sound. In addition, when you run out of breath (or if you are not blowing into the mouthpiece at all) the CC value will be equal to 0 and therefore (even if you play a note on the keyboard controller) the instrument will be muted. From now on, your woodwind parts will sound natural and playable without any unrealistic long phrases.

There are several other ways to improve the input process for wind instruments when it comes to MIDI controllers. The main alternative controllers available on the market are wind MIDI controllers such as the Yamaha WX series, the AKAI EWI series, and the Synthophone. The main advantage of these devices is that they provide a way of inputting notes that is very close to a real wind instrument. Their clarinet-like (or saxophone-like in the case of the Synthophone) shape and controls allow you to sequence any type of wind instrument with extreme precision and fidelity. Some of the features that make these devices an incredible sequencing tool are the ability to switch octave (extending the range of the instrument), to assign one or more control wheels to an arbitrary MIDI CC, and to select among several fingering options such as clarinet, saxophone, and flute. The biggest drawback of these controllers is that they require very skillful musicians in order to master all their sophisticated capabilities. Even the most experienced veteran saxophone player will need a good amount of time to adjust to the differences between an acoustic instrument and the controller. If your main instrument is not saxophone, clarinet, or flute, I recommend using a keyboard controller in combination with the BC.

4.10   Use of MIDI Control Changes and Editing Techniques

Sample-based patches and libraries with multilayer implementation are definitely a good starting point to create convincing and realistic renditions of woodwind instruments but, as shown when discussing the string orchestra, there are additional sequencing techniques that you can take advantage of to improve your productions. Once again, the goal here is to create enough variation in the woodwind parts to overcome the limitations of the repetitive samples. Just some small variation in dynamics and volume can go a long way toward adding some realistic touches to your parts. Let's analyze some of these techniques. Use CC#7 in conjunction with CC#11 to reshape the dynamic contour of a part. In order to avoid monotonic and lifeless passages that stay at the same volume for several bars, try to use these two controllers to add a touch of realism. More exciting passages tend to be louder, while calmer and introversive ones tend to be softer. For woodwind instruments, try setting one of the assignable faders/knobs of your keyboard controller to CC#11. This will control the dynamic shape of your part. Then set the value of CC#7 through a snapshot automation or by inserting the data manually with the pencil tool of your sequencer. This will function as the master volume of the track. At this point, you have two options: either sequence the part and in real time use CC#11 to shape its dynamic contour or record the part first and then, in a second pass, overdub the CC#11 automation. The latter choice gives you the possibility of focusing more on the performance first and on the dynamic afterward. I sometimes even close my eyes while overdubbing the controller and immerse myself totally into the part, almost conducting the instrument through the fader, creating a seamless connection between my hand and the sound of the instrument. You can achieve even better results with the aforementioned MIDI BC. If it is assigned to CC#11, you will be able to control the dynamic of a woodwind part smoothly by blowing into it. Either technique will guarantee that the dynamic of your wind parts is constantly moving. By just adding a little bit of dynamic, these parts will sound better right away. The combination of CC#7 and CC#11 allows you to readjust the overall volume of a track quickly, without having to redo the entire automation. Since CC#11 works basically as a percentage of CC#7, if you need to lower or raise the entire volume of a track without redoing its dynamic automation/variation simply change the value of CC#7 at the beginning of a part. Your overall volume will change, but the contour of the dynamic curves programmed will maintain the same shape (Figure 4.20).

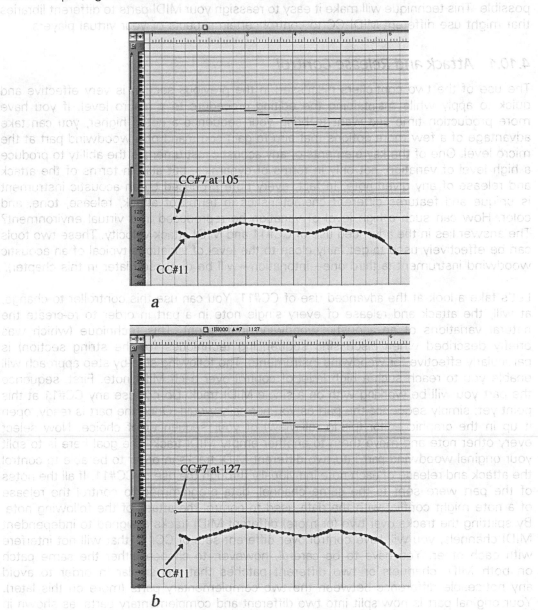

Figure 4.20 Combined use of CC#7 and CC#11 gives much greater flexibility in terms of dynamic changes (Courtesy of MOTU)

In Figure 4.20(a), the main volume (CC#7) is set to 105, while the overall dynamic curve is controlled by CC#11. In Figure 4.32(b), the part is made louder by changing the value of CC#7 to 127 without altering the curve of CC#11.

I recommend also recording the MIDI CC messages on separate MIDI tracks when possible. This technique will make it easy to reassign your MIDI parts to different libraries that might use different MIDI CC to control certain aspects of your virtual players.

4.10.1 Attack and Release Control

The use of the two controllers discussed in the previous section is very effective and quick to apply while maintaining the editing procedure to a macro level. If you have more production time and want to bring your sequence a notch higher, you can take advantage of a few more options that aim to gain command of a woodwind part at the micro level. One of the key elements of any acoustic instrument is the ability to produce a high level of variation, not only in terms of dynamics, but also in terms of the attack and release of any given note. In fact, every note produced by an acoustic instrument is unique and features different characteristics in terms of attack, release, tone, and color. How can such a high level of variation be reproduced in a virtual environment? The answer lies in the advanced use of CC#11 and MIDI attack velocity. These two tools can be effectively used to get fairly close to the level of variations typical of an acoustic woodwind instrument (a third one—intonation—will be discussed later in this chapter).

Let's take a look at the advanced use of CC#11. You can use this controller to change, at will, the attack and release of every single note in a part in order to re-create the natural variations of an acoustic woodwind instrument. This technique (which was briefly described when discussing sequencing techniques for the string section) is particularly effective for woodwind instruments. The following step-by-step approach will enable you to reach such a high level of control over each MIDI note. First, sequence the part you will be working with on a single MIDI track. Do not use any CC#11 at this point yet; simply sequence the part as you normally would. Once the part is ready, open it up in the graphic editor (piano roll view) of your sequencer of choice. Now select every other note and move them to another empty MIDI track. The goal here is to split your original woodwind part into two different MIDI tracks in order to be able to control the attack and release of each note individually through the use of CC#11. If all the notes of the part were sent to the same channel, data programmed to control the release of a note might conflict with the data used to control the attack of the following note. By splitting the tracks over two (or more) different MIDI tracks assigned to independent MIDI channels, you will have control over different sets of CC#11 that will not interfere with each other. You have to be careful, however, to choose either the same patch on both MIDI channels or two different patches that are similar in order to avoid any noticeable difference between the two complementary parts (more on this later). Your original part is now split into two different and complementary parts, as shown in Figure 4.21.

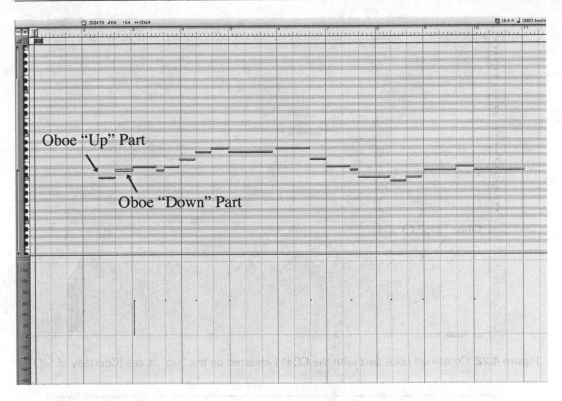

Figure 4.21 Split complementary parts on two different MIDI tracks assigned to two different MIDI channels (Courtesy of MOTU)

These two parts may be called "up" and "down," in reference to the up and down bow directions on a string instrument. Of course, in the case of wind instruments, this way of naming the two tracks is purely arbitrary, but it serves the purpose of quickly and easily identifying the two tracks. With the aid of the pencil tool, now insert small in and out fades using CC#11 for each note of the two tracks. The amount and speed of these fades can vary drastically depending on the type of material sequenced. Longer and gentler fades may be applied for long, sustained notes, and steeper curves for shorter notes. If you are looking for a more staccato effect, you may not need to apply any fades. The final result should look like Figures 4.22 and 4.23.

The final result will be a smooth and more musical effect. The addition of single attacks and releases for each individual note creates a much higher level of variation, allowing you to shape each note in a way that would not be possible otherwise. As mentioned before, you should experiment with the curve and the values that best fit the part you are working on. Make sure that you don't overdo it; the effect of the individual fades should be felt but not heard. Listen to Examples 4.1 and 4.2 to compare a woodwind ensemble part sequenced, respectively, without and with the addition of individual CC#11 fades (Score 4.4).

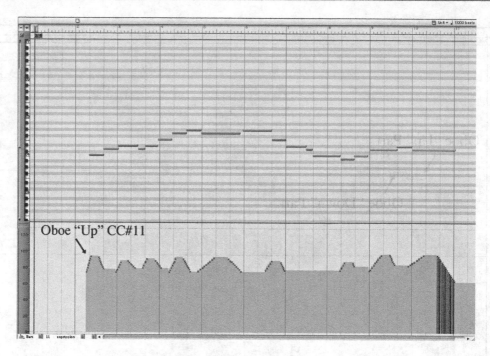

Figure 4.22 Combined oboe part with the CC#11 inserted on the "up" notes (Courtesy of MOTU)

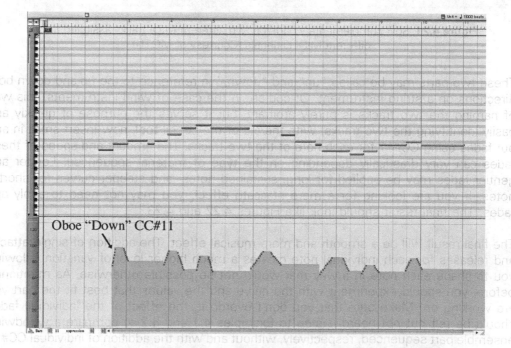

Figure 4.23 Combined oboe part with the CC#11 inserted on the "down" notes
(Courtesy of MOTU)

As mentioned briefly earlier, you can implement this technique with the same patch assigned to the two different MIDI channels on which each split is sent. This will create a more coherent sonority, but it can sound a bit repetitive over time. If you want to bring it a notch higher, you could use two slightly different patches on each track. Most orchestral libraries come with at least two variations of the same instrument (e.g. Oboe 1 and Oboe 2). By using different patches, you will create a brand new instrument that will alternate between the two patches every other note, creating an interesting and refreshing effect, especially suited for section passages. If your woodwind library does not have two similar variations of the same instrument, don't despair! While using the same patch for both tracks, try altering one of the two through the application of a filter and very small detuning. Make, for example, the second patch (the "down" part) slightly darker by using a very light low-pass filter. In addition, by detuning it slightly (only a few cents) you can give the impression of a more realistic and human performance. As you have hopefully learned by now, variation really is the key to a successful orchestra rendition.

If you think that creating single fades for each note of an entire orchestral score is too tedious and time-consuming (although it is totally worth it!), there is another solution

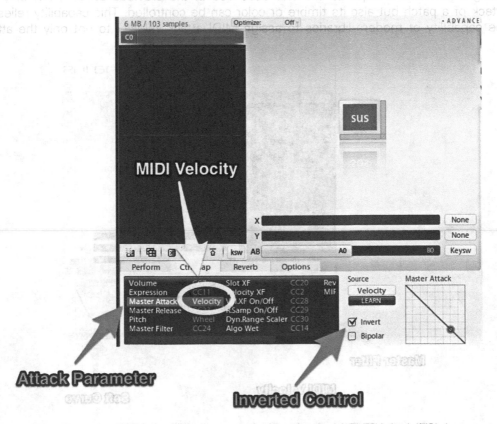

Figure 4.24 Example of the attack parameter controlled by the MIDI velocity in VSL Instrument

that is faster and fairly effective. This approach uses the MIDI velocity data to control the attack of each note. In most software samplers (or at least in all the most advanced ones), the filter sections of the synthesis engine are pretty sophisticated, allowing you to assign a series of MIDI data (CCs, velocity, aftertouch, etc.) to a series of modifiers (the amplifier envelope, the filter envelope, etc.). What this means in practical terms is that you can have the MIDI velocity of a part assigned to control the attack stage of the patch: a higher velocity will trigger a faster attack, while a lower velocity will translate into a slower attack (Figure 4.24). Notice how in VSL I inverted the control so that higher velocity (more intensity) will translate in shorter attacks. Some of the patches bundled with orchestral libraries might already have this assignment in place, but not all. Take advantage of this technique to virtually re-create a realistic phrasing.

Listen to Examples 4.3 and 4.4 to compare the same woodwind ensemble part rendered without and with the velocity-controlled attack parameter, respectively (Score 4.5).

4.10.2　Filter and Timbre Control

Using a similar technique to the one described in the previous section, not only the attack of a patch but also its timbre or color can be controlled. This capability relies on the capability of modern libraries to assign MIDI attack velocity to not only the attack

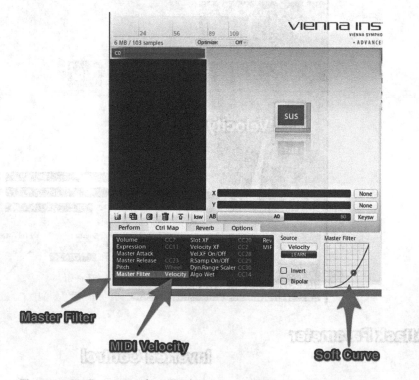

Figure 4.25 Example of a velocity-controlled filter in VSL Instrument

section of the amplifier but also to the filter. We learned in Chapter 2, when discussing the most popular synthesis techniques, that a synthesizer has a modifier section that is basically a filter capable of changing the waveform generated by the oscillators. In a sampler, the waveform is a complex wave that was recorded (sampled) from a real instrument. By carefully setting the filter to control the high frequencies and by assigning it to be controlled by the MIDI On velocity of the notes played, we can virtually re-create the response of an acoustic instrument to different dynamics. In most cases when playing a woodwind instrument at softer dynamics, you will produce a slightly duller and more muffled sound, whereas when playing it at louder dynamics you will obtain a brighter and sharper tone. While some of this variation can be reproduced fairly effectively through multisampling techniques, the addition of a velocity-controlled filter can improve the realistic effect of a virtual woodwind ensemble and/or solo instrument. As in the case of velocity-controlled attack, most advanced libraries implement this control as a standard feature. In order to achieve the best results for your productions you need first to set the filter of the patch to a low pass with a medium slope. This will have the effect of darkening the sound (cutting high frequencies), depending on the cutoff frequency of the filter. Next, set the cutoff point depending on your needs (a lower cutoff point will result in a darker sound). Finally, set the velocity to control the cutoff frequency of the filter. Lower velocity will result in a lower cutoff frequency that translates into a darker tone, while higher velocities will open the filter by having higher cutoff frequencies, which will result in a brighter sound (Figure 4.25). Notice how in VSL I have used a softer curvature in order to have a smoother application of the filter.

4.11 Performance Controllers

To get the most out of the sequenced woodwind parts, there is a series of MIDI controllers that, if used correctly, can help you achieve more realistic results. Some of these controllers will sound familiar since they were used when sequencing the string orchestra and some of the instruments in the rhythm section. Since CC#2 (BC) has been discussed above this section I will concentrate on other controllers, such as portamento (CC#5, 65, and 84) and soft pedal (CC#67).

The ability of wind instruments to slide into a note or to connect two subsequent notes with a sweep represents an important sonic characteristic of this family of musical instruments. Being able to reproduce such effects in a virtual environment is important in order to have a complete tonal palette. The portamento parameter of a synthesizer allows us to reproduce this technique accurately. This parameter (sometimes called "glide") controls the speed and the rate of the sweep between two subsequent notes. The first step toward using portamento in a patch is to check whether your synthesizer or sampler is equipped with this feature (most of them are). Next, set the portamento to a value that feels natural and not artificial. Playing subsequent notes at various intervals in different ranges of the patch will give a good idea of how the portamento will sound. Use this technique to get an overall value of the effect; you will be able to change it in real time later during the performance.

Once you have set the overall level, start playing the patch and part you need to sequence and fine-tune the overall portamento setting. Since the amount and settings of portamento need to vary depending on a particular passage or phrase you will play, you will need to have full control over its parameters during and after sequencing of the part. This task can be effectively achieved using three MIDI CCs: 5, 65, and 84. If possible, use a MIDI controller with assignable knobs in order to program these CCs and be able to change the portamento settings in real time. Through CC#65 you can turn portamento on or off all together: off (values between 0 and 63) and on (values between 64 and 127). By using CC#5, you can control the rate (speed) at which the portamento slides between the two subsequent notes. Lower values will give you a faster slide, while higher values will translate into a slower slide. Use this controller to tailor the glide quickly to a specific passage. The main problem with these two CCs is that they allow you to change how the slide is produced, but they don't give you control over the starting point (or note) of the slide. The portamento occurs always between note 1 and note 2. This is a problem if you need portamento between two notes that are separated by a big interval. The solution is CC#84. By setting a value for this controller, you specify the note from which the portamento will start. Note 2 will mark the end of the portamento slide. This can be extremely useful in constructing a precise slide when and where you need it. In practice it works like a grace note mark in a score, where you can specify from and to which

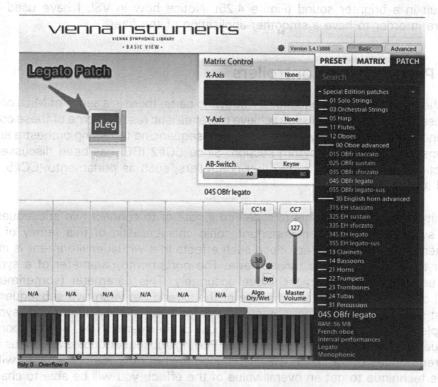

Figure 4.26 An oboe legato patch in VSL.

note the slide will occur. Like any other controller, CC#84 ranges between 0 (C2) and 127 (G8). A value of 60 returns C3 (middle C on the keyboard). It may take a little bit of practice to master CC#84, but the results can be very accurate. Keep in mind that, while almost all hardware synthesizers are programmed as default to respond to these controllers, software samplers are not and you may need to assign a controller manually to the portamento settings. Another factor to consider is that, no matter how good the samples of a library are, a synthesized portamento will always sound somehow artificial. Ideally, you want to use patches and samples that were recorded with natural portamento articulations. Most of the major orchestral libraries these days provide very realistic patches that feature natural legato glides (Figure 4.26).

There are two other MIDI messages that can be effectively used to add expressivity to woodwind parts. The first one is soft pedal (CC#67), which lowers the dynamic level of a patch while pressed (values between 64 and 127). It acts as a quick volume damper for soft passages. I recommend using it to improve the dynamic range of a part without using the usual volume and expression. Another MIDI message that can become handy in sequencing wind instruments in general is aftertouch. As this message can be virtually assigned to several parameters of a patch, it is very flexible. Generally, it is assigned as default to vibrato/modulation, but it can be equally effective to control dynamics if assigned to the amplifier section of a sound generator in a synthesizer or sampler. In either case, you can have aftertouch as an extra "weapon" at your service when trying to create a realistic rendition of a woodwind instrument. While it may be less useful when controlling vibrato, you can assign it to volume in order to accent single notes in a solo instrument part. In general, it is better not to use the generic modulation wheel to create vibrato effects for woodwind instruments. The result will be artificial and synthetic. Try instead to use patches that have vibrato key switches or alternate between two different patches, one with and one without vibrato. If the only option you have is the modulation wheel, then you could use very light vibrato settings that feature a slow period and a very low depth.

4.12 Performance Extended Controllers

In addition to the aforementioned controllers and MIDI messages, there are three extended controllers that can be particularly useful when sequencing for the woodwind section. These controllers are discussed in a different section because they are mostly implemented as default only on hardware synthesizers. They can, though, in some cases, be assigned to specific functions and parameters in a software synthesizer or sampler too. They are particularly targeted to give control over the shape (envelope) and color (timbre) of a synthesized patch. In particular, CC#72 and 73 give access to, respectively, the release and the attack stages of the amplifier envelope of the oscillators. In simple terms, they allow you to control the attack and release time of a patch. While other techniques can be used successfully to achieve similar results through the use of CC#11, the advantage of these two controllers is that they are specifically targeted to attack and release. In the case of a library, where these CCs are not assigned to any particular function, you need to assign them manually to the attack and release stages of a patch (Figure 4.27).

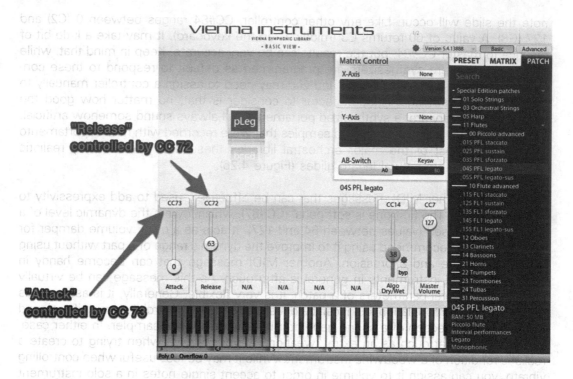

Figure 4.27 Example of CC#72 and CC#73, controlling, respectively, the release and attack of a flute patch in VSL

With this flexible approach, if you have a MIDI controller with assignable knobs/faders, you can use them to control in real time the attack and release of the part you are sequencing. This option can be particularly helpful when you need to quickly adjust the shape of the envelope of a patch. By storing these settings permanently on your MIDI controller, you can always have two assignable knobs/faders managing attack and release. In practical terms, you can either use CC#72 and 73 while you are recording your part (in real time) to smooth out a sustained note or to switch quickly between legato and staccato passages. You can also insert these controllers after sequencing the part by overdubbing a second pass with only the controllers, or by using the pencil tool and inserting them in the graphic editor of the sequencer. The latter method allows for more precision and efficiency. Make sure to use them smoothly and always remember that, as in most of the MIDI editing situations, you want the final effect to be felt but not heard.

Another useful controller is brightness (CC#74). It allows you to change the cutoff frequency of the filter section of the synthesizer playing the patch. By doing so, you can fine-tune the frequency response of the patch. Higher values assigned to the controller will bring the cutoff frequency higher, resulting in a brighter sound, while lower values translate into a darker and more muffled tone. As is the case for most of the situations analyzed so far, this controller is implemented by default in hardware synthesizers, but it needs to be manually assigned in most of the software synthesizers available on the market. CC#74 can be very useful and effective to create a wider contrast between sections characterized by soft dynamic passages (from pianissimo to mezzo piano) and sections that feature louder dynamics, such as forte and fortissimo. In order to use this controller successfully, make sure that the filter section of the synth engine is assigned to a low-pass filter. This will translate a higher cutoff frequency to a brighter sound and lower cutoff frequency to a darker tone. In a real acoustic wind instrument, lower dynamics usually generate a more muffled and intimate sonority that features a lower content of high frequencies. A similar effect can be re-created by lowering the cutoff frequency of the low-pass filter through CC#74. In an acoustic situation, louder dynamics express a sharper and brighter sound, an effect that can be re-created by opening (raising) the cutoff frequency of the filter by programming higher values of CC#74. You can overdub such controllers after having sequenced the part and programmed other controllers such as volume and expression. Be fairly conservative when applying CC#74 changes; you want some very subtle alterations in terms of tone, nothing too drastic. Listen to Examples 4.5 and 4.6 on the website to compare a flute part sequenced without and with CC#74, respectively.

All the CCs mentioned in the last paragraphs can be sent and programmed from any MIDI keyboard controller that provides assignable knobs and/or faders. While this option represents definitely a valuable choice, a BC is highly recommended. You will be able to achieve a much more cohesive and musical performance for virtual wind instruments by controlling dynamics, brightness, and attack/release through such a device. As mentioned earlier in this chapter, you can also gain control over all these aspects of a MIDI performance through more-sophisticated (and complicated to use) MIDI controllers such as the Synthophone, the AKAI EWI series, or the Yamaha WX series. These MIDI controllers not only allow you to assign breath intensity to any CC, but also feature several other interface elements (such as wheels) that can be programmed depending on the situation.

4.13 Special Sequencing Techniques for Woodwind: the Saxophone Section

Saxophone parts are among the most difficult to reproduce with a virtual ensemble. With the availability of more-sophisticated multisample libraries and more-advanced morphing technologies, the gap between a real saxophone ensemble and a virtual one is becoming smaller. Nevertheless, there are some difficulties that are not easy to overcome when rendering saxophone parts in a sequencer. This is particularly true for jazz, pop, and avant-garde scores. Most of the challenges come from the inability to re-create fully the expression and tonal nuances of these acoustic instruments. The soft vibrato, the delicate warm notes at low dynamics, and the intense slides into a note are just some of the effects that make the saxophone such a special addition to any arranger's palette. How can we effectively re-create such complex tones and colors in a virtual environment? While in the past synthesis techniques such as physical modeling have provided some promising tools, they have never completely fulfilled the need for a more realistic way of sequencing these instruments. Recently, though, a combination of PM and sample-based synthesis has been rising to the top as the best options for sequencing saxophones and wind instruments in general. In particular, a company called Sample Modeling has developed a new way of combing the realism of sample-based sound reproduction with the adaptability of PM. While PM by itself cannot reproduce the entire spectrum of articulations and tones of a saxophone, it can very effectively re-create specific inflections and sonorities such as tonguing, vibrato, and breaths. The sample-based component provides the main backbone of the sound with its realistic tone, while the PM layer adds the necessary nuances and articulation that would be too complicated to reproduce through samples only. Listen to Examples 4.7 and 4.8 to compare a saxophone section part sequenced respectively with a generic sample-based library and with the Sample Modeling library (Score 4.2).

A few other sample-only saxophone libraries (such as VSL) that can be used to render saxophone parts fairly realistically are available on the market.

If you think that neither sample-based nor PM synthesis can address your sequencing needs when it comes to saxophones, there is another option: phrase-based sampling technology. This alternative approach is less flexible than sample-based and PM synthesis, but it can generate some pretty astonishing results for certain styles. This technique is based on the principle that phrases and licks, instead of single note samples, can be used to create longer passages by stitching them together. While this can work very effectively for specific situations, this approach may be too limited and restrictive for the modern composer. Nevertheless, having this technique available in your bag of tricks can sometimes save the day. Among the phrase-based plug-ins for saxophones, the excellent Liquid Sax 2 by German software company Uberschall deserves a mention. Listen to Audio Example 4.9 to hear how a phrase-based saxophone library can be used effectively for your productions.

4.13.1 Blending Acoustic and MIDI

No matter which library or synthesis technique you use to sequence saxophone parts, it is highly recommended that you add one or two real instruments to replace some of the lines in a saxophone section. In a typical saxophone section of two altos, two tenors, and one baritone, if possible, replace the first alto with a real instrument. This will add some sharpness and clarity to the virtual section. If budget and time allow, also replace either the baritone or the first tenor to add realistic phrasing in the lower or the mid-range of the section, respectively. The main idea here is to create a "sandwich" of acoustic and virtual instruments, where one supports the other. Take a look at Table 4.3 to compare the different options for layering acoustic and MIDI saxophones.

This will create a blend of sonorities where acoustic and MIDI are impossible to tell apart. For the woodwind in general, and the saxophones in particular, you have to pay particular attention to blending the acoustic and the MIDI tracks together seamlessly.

Table 4.3 Different options for layering acoustic and MIDI saxophones

Saxophone	Option1	Option 2	Option 3
Alto 1	Real	Real	Real
Alto 2	MIDI	MIDI	MIDI
Tenor 1	Real	Real	MIDI
Tenor 2	MIDI	MIDI	MIDI
Baritone	MIDI	MIDI	Real

A saxophone section works very much like a cohesive and intertwined unit in which the voices support each other. Reaching the perfect blend when using acoustic and MIDI sources can be challenging. There are two main aspects to consider in order to reach the best combination possible: tone color and acoustic environment. The first can be matched through the use of light equalization. Make sure that neither the acoustic nor the MIDI tracks come across too bright and too dark in comparison with the other ones. If necessary, use a shelving equalizer to correct any obvious frequency imbalance between the voices of the saxophone section. To match the acoustic response of the environment of the different saxophone of the section, use a convolution reverb. If you use mainly wet samples then record the acoustic instrument as dry as possible (close miking), listen carefully to the ambience of the samples, and try to match it, through the convolution reverb, on the acoustic instruments tracks. If you use dry samples then you can afford to record your acoustic tracks with a bit more ambience and try to match it to the MIDI tracks. When adding one or more real saxophones to a virtual section, make sure to record the real instrument (or instruments) first if possible. Then sequence the MIDI tracks, trying to match the original acoustic phrasing (dynamics, accents, articulations, etc.) as accurately as possible. It will make a big difference if you follow this order, since it would be a mistake to sequence the MIDI parts first and have the acoustic instrument "boxed" inside a static and stiff phrasing.

4.14 Final Touches

Before discussing the mixing techniques for the woodwind section, your attention should be drawn to a few sequencing techniques targeted specifically at improving the final results for your productions. A very important aspect of creating realistic woodwind parts is the ability to generate enough variability and "human" feel so that the listener will have a hard time distinguishing between real acoustic instruments and virtual ones. With good samples and patches, an adequate controller, and advanced use of CCs can definitely produce good results, a few other tricks can bring your productions even further in terms of quality and realism. Let's take a look at them.

A neat trick is to use light detuning to re-create the sensation of a human performance. Samples and synthesized patches tend to be too perfect in terms of tuning. This can create the impression of a static and sterile ensemble, taking out the life of a performance. We can artificially bring back some of the natural excitement of a live performance through the use of controlled detuning. The detuning can be implemented horizontally or vertically. In the first case, you will apply a little bit of detuning on a track-by-track basis. Choose a slightly different value for each track of the section (not more than ± 5 percent), trying to keep the overall detuning balance close to 0. This option will create a smoother and more cohesive blend between the instruments. For an even more realistic effect, use slightly different variations over the span of the entire piece. Program the changes with increasing values toward the end to re-create the effect of players becoming more fatigued and therefore slightly losing control over their instruments' intonation. Be very careful not to overdo it, otherwise you will make your virtual ensemble sound like an amateur rehearsal group! You can apply the detuning directly from the sampler/synthesizer by changing the "fine-tune" control. This option will create a static detuning that can sometimes defeat the purpose of creating more variation in the MIDI parts. A better solution consists in assigning a MIDI CC to the fine-tuning parameter of the sampler and programming subtle changes along the track. This option will give you full control over the amount and the position of the detuning. Experiment with different settings before committing to one, and remember that you can always go back and edit the controller values to retune specific areas. Listen to Examples 4.10 and 4.11 to compare two woodwind ensembles sequenced, respectively, without and with detuning applied. A similar effect can be achieved without the use of MIDI CCs, but instead through the use of the velocity-to-pitch parameter of the sampler or synthesizer. This option works very much like the one presented when discussing how to control the attack stage of the amplifier envelope of a patch via MIDI velocity. This time, instead of linking the velocity to attack we use it to control the tuning variation. Depending on the synthesis engine used, you should be able to assign velocity to control the detuning of a sound. In Figure 4.28, you can see how I have assigned the "pitch" parameter to velocity. Notice how I scaled down the pitch range to "15" in order to make the pitch variations as subtle as possible.

I recommend programming the parameter so that higher MIDI velocity will translate into slightly sharper intonations, and lower velocity values will trigger slightly flatter pitches. This combination is particularly effective since it reflects what happens in an acoustic wind instrument where usually a stronger stream of air will bend the pitch slightly sharp and a softer air stream will bend the pitch flat. As explained in the previous section, make sure that the pitch variations are minimal and they are felt more than actually heard by the listener.

Another technique that I recommend implementing, in order to create more variation in terms of tuning, involves the use of a round-robin system. This option is only available for advanced libraries, such as VSL. The idea behind this approach is a multistep variation based on predetermined tuning curves that are applied cyclically to each note played. This function is sometimes called "humanize." VSL Instrument Pro comes with several presets, which range from "finding tune" to "out of tune" (Figure 4.29). Each micro curve is applied based in its position in the round-robin slot and note on a specific note/

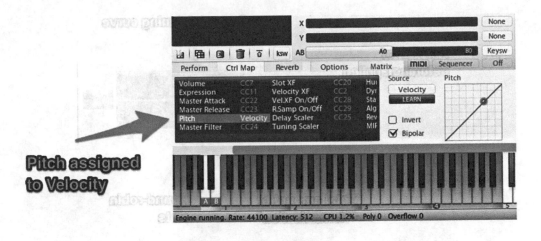

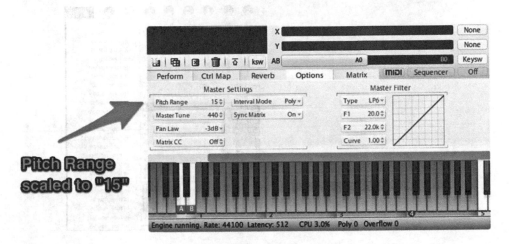

Figure 4.28 The "pitch" parameter assigned to velocity in VSL

pitch. This creates excellent variations that help reproducing how a real WW player would perform.

4.14.1 Performance Noises

Before discussing specific mixing techniques for the woodwind section, one more aspect related to sequencing for these instruments should be mentioned, which once again will increase the realism of your productions within a virtual ensemble. As for the string orchestra, the guitar, and the bass, the addition of background and performance noise represents a quick way to humanize a MIDI solo part or section. The same principle holds true for the woodwind instruments, where the addition of performance noises

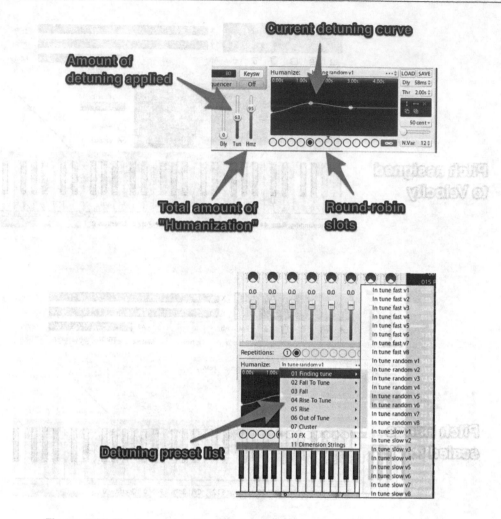

Figure 4.29 The humanize function and preset curves in VSL Instrument Pro

such as breathing, page turning, and keys can bring new life to a production. Table 4.4 describes the most common performance noises you can use when sequencing for woodwind instruments.

Some orchestral libraries come with a bundled selection of wind instrument performance noises. While those selections provide a decent starting point, you should record your own and create an original collection that you can use in your productions. When adding and mixing the noises, make sure to keep them very low in the mix and to use them sparingly. You don't want to turn your ensemble into a rustling, noisy bunch of undisciplined, unprofessional musicians! Listen to Examples 4.12 and 4.13 to compare a woodwind section sequenced first without performance noise and then with the addition of performance noises (Score 4.6).

Table 4.4 Summary of woodwind instruments performance-related noises that can be used in a MIDI rendition to infuse life into virtual performances

Type of noise	Instrument	Comments
Breathing	Any woodwind instrument	The addition of breathing noise, if carefully placed, will make your virtual parts much more real. Place a different breathing noise every time a phrase breaks right before a new phrase begins. Make sure to have enough noises to create variations so they don't become repetitive and artificial
Key noise	Any woodwind instrument	This effect can improve dramatically the realism of your parts. Make sure to have the right key noise for the right instrument. Usually saxophones and flutes tend to be more metallic while oboe and clarinet are more "woody"
Mouthpiece	Saxophones	If used accurately, these noises can create a nice, warm, and intimate sound. Use them at low dynamic and especially on the lower range of the instruments
Page turning	Any woodwind instrument	Use them with balance and mix them extremely low. Particularly useful if used to re-create the impression of a live recording in a classical context

If you have followed and experimented with the techniques and sequencing options explained in the previous sections, you should by now have a well-sequenced and solid production ready to be sent to the final step: the mix stage. This stage is crucial to put the final touches to your virtual ensemble. So, let's move to the mixing room.

4.15 Mixing the Woodwind Section

The mix stage for the woodwind section can be a challenging one for several reasons. The acoustic range and the sonic differences of the instruments belonging to this set of instruments are fairly wide and therefore the equalization stage can play an important role in terms of cohesion and blend. For the same reason, the panning of the single instruments can widely vary depending on the size and on the composition of the ensemble. Let's analyze each stage of the mixing process separately in order to get a better idea about settings and options.

4.15.1 Panning the Woodwind Instruments

The panning settings for woodwind can vary depending on the size and composition of the section. For a classical ensemble you can use the diagram in Figure 4.30 as a starting point. Here, the clarinets and flutes occupy the left side of the stereo image while the oboes and bassoons take the right side. How hard you pan left and right depends mainly on the context. For a smaller woodwind ensemble, which is featured in the piece, a wider panning will achieve more separation (Figure 4.30), while for a section inside a bigger ensemble such as a symphony orchestra keep the panning much more conservative to leave space for the other instruments of the orchestra (Figure 4.31).

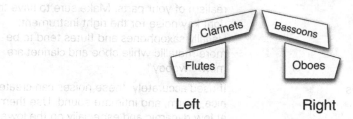

Figure 4.30 Panning settings for a woodwind ensemble. Notice how the ensemble takes the entire stereo image since it is featured in the piece.

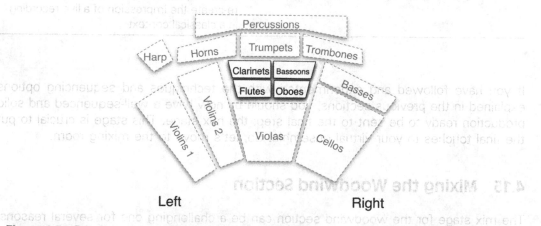

Figure 4.31 Panning settings for woodwind ensemble in an orchestral setting. Notice how in this case the stereo image is more conservative since the ensemble is part of the full orchestra.

If the woodwind section includes saxophones (the standard five-piece saxophones section) and it is featured as the main sonority, then you could take advantage of the full stereo image and spread the instruments to gain more separation and clarity. Keeping the basic panning position used for the simple woodwind section, "squeeze in" the saxophones as shown in Figure 4.32.

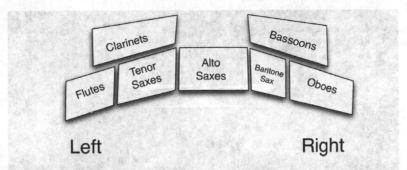

Figure 4.32 Panning settings for woodwind ensemble with the addition of the five saxophones: two alto, two tenor, and one baritone. This panning setting gives a cohesive and solid ensemble, yet retains clarity and separation

The pan settings for a solo saxophone section can vary depending on the composition of the section. Since one of the most common saxophone sections features two alto, two tenor, and one baritone, this particular formation will be discussed. The most traditional and widely accepted disposition over the stereo image has the two altos in the middle (with the first alto usually positioned slightly to the right), the first tenor to the left of the second alto, the second tenor to the right of the first alto, and the baritone to the extreme right (Figure 4.33).

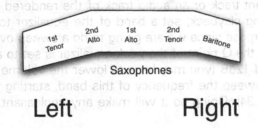

Figure 4.33 Panning settings for the saxophone section

4.15.2 *Equalization for the Woodwind Instruments*

The equalization stage for the woodwind section represents an important moment for the clarity and separation of the mix. Instruments such as clarinet, oboe, bass clarinet, and bassoon can sometimes easily "step on each other" and create confusion and muddiness. Equalization can also be used to smooth out some of the "reedy" edginess of instruments such as oboe, English horn, and clarinet. The first step toward successfully equalizing woodwind instruments is to check for any unwanted resonances that may jump out. These resonances are typical of reed instruments and even in sample-based libraries they can spoil a single sonority or, sometimes, an entire mix. These unpleasant resonances are often caused by less-than-perfect microphone placements or bad room acoustics. To identify these unwanted frequencies, you can use either your ears (usually

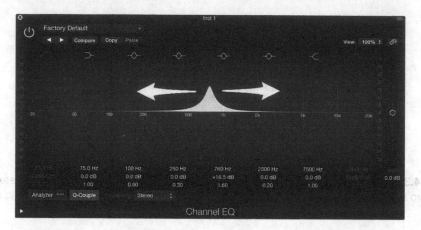

Figure 4.34 Example of an equalization sweep in Logic Pro for an English horn part using the Channel Equalizer in Logic Pro (Courtesy of Apple Inc)

the better option) or a spectrum analyzer. If possible, use both in order to verify your information with more accuracy. It is preferable to equalize not just a generic sound but the actual part that will be performed, since equalization settings can widely vary depending on the register and articulation used. To do so, after sequencing the part, insert a multiband equalizer on the channel where the woodwind part was recorded (you can use either the software instrument track or an audio track of the rendered version). Loop the part and play it back. During playback, set a band of the equalizer to "peak." Any available band will serve the purpose since we are going to do a sweep over the entire frequency range. Make sure that the Q point of the peak equalizer is set to a fairly high value. Now boost between 10 and 12dB (you may have to lower the volume to avoid damaging the speakers) and gently sweep the frequency of this band, starting from 20Hz and moving up to 20kHz (Figure 4.34). Doing so it will make any unpleasant frequency (resonance) stand out.

When you find any unwanted resonance, stop and set the gain to a negative value (usually around -3 or -4dB will be enough). Be very conservative with this technique, otherwise you will end up taking out too much of the character of the instrument. As with any equalization process, always compare the signal with and without equalizer in order to check whether you are going in the right direction. Repeat these steps for any other resonances that you feel are overpowering the natural signal of the instrument. You can also double-check the empirical results by using a spectrum analyzer to create a graph of the frequency range used by the instrument (or instruments) you are working on. To do so, insert a spectrum analyzer plug-in on the track playing the part and start looking at the frequency response of the part you sequenced. It is important to analyze the entire length of the part in order not to be tricked by a specific frequency or a series of frequencies related to a specific range. If your sequencer or plug-in allows, switch the freeze function on so that you can look at the frequency response over a longer period, as shown in Figure 4.35.

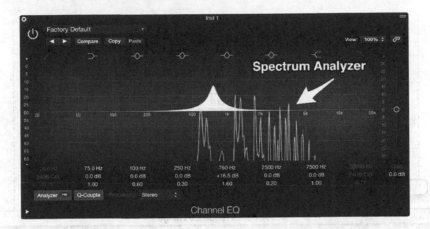

Figure 4.35 Example of a spectrum analysis of an oboe part in Logic Pro using the Channel Equalizer plug-in. Notice how the frequencies between 1 and 2kHz are more accentuated. This could be the cause of some unwanted resonances (Courtesy of Apple Inc)

By combining the results of the equalization sweep and the spectrum analysis, you will be able to pinpoint accurately and correct annoying resonances for your woodwind instruments. Keep in mind that this technique can be applied to any equalization situation and for any other instruments.

Once you have taken care of the unwanted resonances for each single instrument, it is time to use equalization to balance the frequencies of the woodwind section as a whole. The principle here is similar to the one presented in Chapters 2 and 3. Each instrument of the woodwind section features a certain frequency range that is typical of that particular instrument. Through equalization we need to try to prioritize it by gently rolling off frequencies that do not belong to that range and that may overlap with other ranges that are featured by other instruments. Figure 4.36 shows the main ranges of the fundamental frequencies featured by the most popular woodwind instruments. Each instrument covers a fairly specific range. The only exceptions are the bassoon and bass clarinet, which have fundamental tones spanning similar ranges. In a full woodwind arrangement, you should use equalization to cut frequencies that can be featured by other woodwind instruments and emphasize those frequencies that are specifically highlighted by the instrument you are equalizing.

The same can be said for the saxophone section, as shown in Figure 4.37.

Do not go overboard when using equalization for balancing the frequency range of the woodwind section. In general, it is better to cut than to boost, as this will create a more natural frequency correction. Start fairly conservatively with every filter you use and then push the gain slowly up to the desired point. Over-equalizing may lead to a fake and artificial sound.

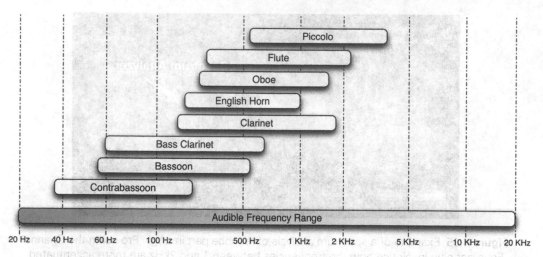

Figure 4.36 Fundamental frequency ranges of the most popular woodwind instruments

4.15.3 Reverberation

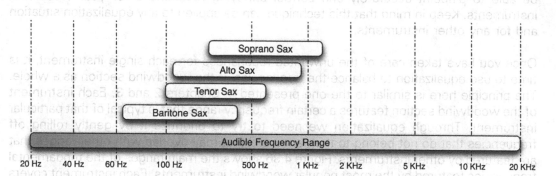

Figure 4.37 Fundamental frequency ranges of the saxophones

The final mix stage for the woodwind involves the addition of natural ambience through reverberation. As for most virtual acoustic instruments, a touch of reverb will add the final shining touch and bring alive the sequence as a whole. Woodwind instruments in general are fairly easy to place in space through reverberation. The entire family, including the low-end bassoon and bass clarinet, sit well in the mix with a reverberation time between 1.7 and 3 seconds. For entirely acoustic and classical productions, you can use a slightly longer reverberation (between 2.5 and 3 seconds), while for pop and studio orchestra productions shorter settings may be used (between 1.7 and 2.2 seconds). Since the woodwind section is contained in a fairly small space on the virtual stage, you can, for the most part, use only one reverb type and setting. If the section is featured and you use a more spread pan setting, as discussed earlier, two slightly different reverberation settings can be used for the left and right parts of the stage. To differentiate between the two sides, you can have small variations in terms of reverberation time (± 0.2 seconds), diffusion, predelay (± 10ms), and early reflections level. If possible, always use convolution reverbs instead of synthesized reverberation. As mentioned in the previous chapters, these reverbs provide the most natural and realistic reverberation for the virtual orchestra.

4.16 Summary

Continue to listen to woodwind instruments in all contexts: traditional orchestra, studio orchestra, marching band, theater productions, chamber ensembles, and in a solo capacity. With the knowledge you now have from reading this book, you will understand better how to listen. Transcribe simple solo passages you hear on a recording and study orchestral scores from Bach to Bartok. Practice locating the ranges and registers of the instruments on the keyboard. To write for woodwind, it is most helpful to think of the keyboard merely as a representation of pitches, rather than a piano. When playing at the piano, after the initial performance, strive to place the sound in your head and substitute the piano timbre with the woodwind of your choice. Independence is important when writing for woodwind. Voicings are not always playable and homophonic texture in general is most challenging. Once mastered, the woodwind will serve you well with regard to orchestral color and musical expression.

Sequencing for woodwind instruments involves several techniques and tricks to achieve the most realistic and accurate results. The first step is to choose the best possible sounds and patches available. Sample-based libraries provide some of the best options. The saxophones are among the hardest sonorities to re-create successfully in a virtual environment. For them, sample-based, PM synthesis, and, more recently, sample modeling will give the best results. Make sure that you can count on more than just a basic sustained articulation for each instrument of the woodwind section. Take advantage of techniques such as multisampling and KS to get the most out of the virtual woodwind ensemble. As in the case of the string orchestra, the addition of one or two real instruments to the sampled ones can vastly improve the quality of your orchestral renditions. This is particularly true for saxophones. While a keyboard controller can give a decent result, it should be combined with a BC to program volume, expression, and vibrato changes accurately. Make sure to have convincing phrasing and pauses for the "virtual musician" to breathe. You can also use wind controllers such as the Yamaha WX series, the AKAI EWI series, or Synthophone to get even closer to a real wind instrument player.

Additional use of CCs and editing techniques can further improve your woodwind arrangements. The goal here is to create enough variation in the woodwind parts to overcome the relative stillness of repetitive samples. Start by using CC#7 in conjunction with CC#11 to reshape the dynamic contour of a part and to avoid monotonic passages that stay at the same volume for several bars. You can either sequence the part, and in real time use CC#11 to shape its dynamic, or record the part first and then, in a second pass, overdub the CC#11 automation. If time allows, use CC#11 (by splitting a part in "up" and "down" notes on two different MIDI channels) to shape the attack and release of each note of a part. You can reach similar results by taking advantage of the velocity-to-attack parameter available in most software samplers. Through this option, you can have the MIDI attack velocity of a part assigned to control the attack stage of the patch: higher velocity will trigger a faster attack, while lower velocity will translate into a slower attack. Using a similar technique identified as velocity-to-filter, you can control not only the attack of a patch, but also its timbre or color. With a low-pass filter, lower velocity will result in a lower cutoff frequency that translates into a darker tone. With higher velocities, you will open the filter by having a higher cutoff frequency that will result in a brighter sound.

The performance controllers that you should use include portamento (CC#5, 65 and 84) and soft pedal (CC#67). Through CC#65, you can turn portamento on or off all together. By using CC#5, you can control the rate (speed) at which the portamento slides between the two subsequent notes. CC#84 allows you to specify the note (or notes) at which each time portamento will start the slide. Use CC#67 (soft pedal) to quickly control the dynamic of a part. Aftertouch can also be effective to control several parameters virtually, depending on the situation. Among the performance extended controllers, you should be familiar with CC#72 and 73, which give access, respectively, to the release and the attack stages of the amplifier envelope of the oscillators. CC#74 (brightness) can also be effective in changing the tonal shape of an instrument, avoiding unrealistic repetition and stillness. It allows you to change the cutoff frequency of the filter section of the synthesizer playing the patch. By doing so, you can fine-tune the frequency response of the patch.

The saxophone section is particularly challenging in a sequencing situation. A combination of PM and sample-based synthesis can help you construct more realistic parts. Phrase-based sampling technology has become more popular in recent years. This technique is based on the principle that phrases and licks, instead of single note samples, can be used to create longer passages by stitching them together. The addition of one or two real saxophones will be a big help to the virtual section. In a typical saxophone section of two altos, two tenors, and one baritone, replace the first alto with a real instrument and, if possible, either the first tenor or the baritone.

To further improve the realism of a virtual woodwind section, a series of final touches is recommended. A light detuning, for example, can bring life back to a solo part or a section. The light and subtle use of performance noises such as breathing, page turning, and keys can bring new life to a production.

The mix stage for the woodwind mainly involves the use of panning, equalization, and reverberation. Pan settings can vary depending on the size and composition of the woodwind ensemble. For a classical ensemble, the clarinets and flutes usually occupy the left side of the stereo image while the oboes and bassoons take the right side. For a smaller woodwind ensemble featured in the piece, use a wider panning, while for a section inside a bigger ensemble, such as a symphony orchestra, keep the panning much more conservative. The same principles can be applied to the saxophone section. The equalization stage is particularly important for woodwind instruments, since sometimes their frequency ranges can easily overlap and create muddiness. First, check for any unwanted resonances that may jump out for each individual instrument, using either a frequency sweep or a spectrum analyzer. As a second step, use equalization to balance the frequencies of the woodwind section as a whole. When it comes to reverberation, the woodwinds are very flexible and easy to place in the virtual stage. The entire family, including the low-end bassoon and bass clarinet, sit well in the mix, with a reverberation time between 1.7 (pop and studio orchestra) and 3 seconds (classical ensembles). To create the illusion of a more sophisticated space, use two slightly different reverbs for the left and right side of the stereo image. If possible, always use a good convolution reverb, as it will definitely make a big difference at the final mix stage.

4.17 Exercises

Exercise 4.1

(a) Take a popular but playful melody that you know (in your head) and write it down, using slurs and articulation symbols that illustrate which notes are long, short, or accented. Add a grace note where desired for a more playful expression.
(b) Sequence this melody using a clarinet sound using portamento, aftertouch, attack/release, and automation to render each articulation and grace note.

Exercise 4.2

(a) Assign the previous melody to each of the following woodwind instruments: flute, oboe, and bassoon, placing the melody in the most appropriate register for each instrument so the character and mood are preserved. Feel free to change the key if necessary.
(b) Sequence each instrument on separate tracks using the automation to control the overall dynamic and attack/release of each note.

Exercise 4.3

(a) Write a melody at the piano that can be performed on oboe or English horn. Transpose the written part for the English horn.
(b) Sequence this new melody for the English horn using CC#73 and 72 to control the attack and release of each note. If your library doesn't respond to these two controllers, use the velocity-to-attack technique of the sampler to control the attack of each note. Use also CC#11 to control the overall dynamic and the phrasing.

Exercise 4.4

(a) Write a musical phrase for the bassoon that stays below middle C on the piano. Transpose the part for the bass clarinet so the two instruments will sound in unison.
(b) Sequence this melody for both bassoon and bass clarinet, including dynamics, breath pauses, and control over individual attack and release of each sustained note (using

any technique learned in this chapter). Add performance noises such as breath and page turning noises.

Exercise 4.5

Sequence Score 4.3 (found in Appendix A) regarding each of the three orchestrations. Pay particular attention to panning, equalization, and reverberation.

Exercise 4.6

Write out a series of diatonic seventh chords in root position (i.e. Cma7, Dmi7, etc.) and enhance this "block voicing" by adding a fifth note an octave below the top note of the chord. Next, take the second note from the top of each chord and drop it down an octave to create a series of drop 2 voicings.

Exercise 4.7

(a) Using the set of voicings from Exercise 4.6, assign the notes to the standard instruments found within the typical saxophone section. If a top note from any of the chords exceeds the range of the alto saxophone, you will need to use the soprano saxophone.
(b) Sequence the progression for saxophones using attack/release control, performance noises, and the right reverberation, equalization, and panning assignments.

Exercise 4.8

(a) Take a popular tune that evokes the quality of a lullaby or a reverent sentiment and, using natural breakpoints within the melody, assign the various phrases to solo woodwind instruments of your choice. The main reasons for your choice should be with regard for tone color, dynamics, and expression.
(b) Create a sequence of this arrangement, completely mixed and polished, making sure to use the following techniques:

  (i)   performance noises;
  (ii)   attack/release control;
  (iii) use one real acoustic instrument (make sure to create a nice blend with the MIDI tracks);
  (iv)   use reverb, equalization, and pan to mix the sequence.

Exercise 4.9

(a) Using mostly intervals in thirds or sixths that move in parallel form, write for woodwinds in pairs. Feature each color group by having them emerge in imitation (e.g. you could start with the flutes, then have the clarinets enter, then the oboes, and finally the bassoons).
(b) Sequence your piece using all the techniques learned in this chapter.

5 Writing and Sequencing for the Brass Section

5.1   General Characteristics

Brass instruments are best known for producing a powerful and regal tone quality. In comparison to the lighter string and woodwind sections, they are unmatched in this ability, but they are also capable of creating a warm and lush tone quality that can blend somewhat with lighter instruments. In general, the brass should be respected with regard to their endurance. The physical exertion is greatest with this group of instruments in contrast to the less taxing woodwind and strings. Consequently, it is wise for the music arranger to avoid overusing the brass. An abundance of whole notes or any extended legato passages is very difficult to sustain physically as they require large amounts of air support. This concern becomes even more critical when brass parts are orchestrated in a high register. A good rule to follow is, "save the brass until you need them and, when used, look for an opportunity to give them a rest so they may be used again." With regard to the timbre of the brass, there is some diversity but not as much as the woodwinds. Most notably, the application of mutes can provide the most vivid contrast from the stereotypical regal brass quality. When the instruments are "open," their natural sound is essentially categorized within two broad timbral colors. The most commonly used brass instruments are outlined below:

Bright timbre	Dark timbre
Trumpet	French horn, flugelhorn
Trombone	Euphonium, tuba

Every brass instrument has within it a darker and brighter tone, but the horns and tubas certainly offer a more robust and darker tone in general. Conversely, these particular instruments have a less piercing attack than the brighter instruments.

5.2 The Brass Section in the Orchestra

The brass section within the standard orchestra is set up (in score order) as follows:

- four French horns
- three trumpets
- two tenor trombones
- one bass trombone
- one tuba.

As with the woodwinds, the first chair in each color group is the principal player, who is usually assigned any solos. The other chairs provide harmony, unison and octaves, and general support. The orchestral brass sound is different from the more commercial jazz and popular styles of music. It is a bit lighter and more refined as it fits with the context of the lighter woodwind and strings. This difference in sound and style is perhaps most noticeable in the trumpet section. Trumpet playing in jazz is much more muscular (usually requiring an extended upper range) and at times raucous as the style of music demands.

5.3 The Brass Section in the Jazz Band

In the jazz band, the brass section is smaller overall, omitting the French horns (replaced with saxes) and the tuba. There are basically two color groups: trumpets and trombones. Each section has a minimum of three players and, more usually, four. Within the trumpet section, the first trumpet chair is referred to as the lead player. This musician is responsible for playing the highest, most challenging and physically enduring notes of a musical composition. It is therefore recommended that the music arranger be respectful of this by not overwriting and overtaxing the player. In most professional situations, there is usually one other player within the section who can also play lead, offering a respite for the primary lead player. The remaining members of the section function as ensemble players. Usually there is one specific chair that is designated for improvised solos (the second or fourth chair). The trumpet section can also "double" on flugelhorn, which offers a mellow or lush sound. At times, there may be only one player to use a flugelhorn for a solo passage or to join with the darker trombones or saxophones.

The trombones represent the lower brass section in the jazz band. As with the trumpet section in jazz music, there are usually four trombones. Whether there are three or four players, the last chair is usually reserved for the bass trombone, providing some access to the notes that the absent tuba can so easily play.

Sometimes a French horn (or two) is added to a jazz band as an extra color, especially for more orchestral-style jazz. In lieu of this addition, a flugelhorn can provide to some degree the flavor of the horn section, especially when combined with a soft saxophone or muted trombone.

5.4   Creating Sound

As with the woodwinds, brass instruments require wind or breath from the performer. All of the instruments are made of brass piping and each has a metal mouthpiece, which is essentially a cup with a rim at the top for the player's mouth to blow into. Comprising the basic sound of each brass instrument, the overtone series plays an important role regarding available pitches. A bugle, which is essentially a trumpet without valves, most readily represents the principle of sound regarding the overtone series. The player blows air into the bugle, establishing a certain pressure with his or her embouchure (the formation of the mouth and its facial muscles) in conjunction with the stomach muscles to help guide and push the airstream. As the velocity of air is increased, the resistance against the pipe is overcome and the pitch begins to rise via the notes of the overtone series. Thus, pitch accuracy is determined solely through a process of hearing the pitch internally and taking aim physically to capture the desired sound. It is the wise arranger who respects the precarious nature of producing specific pitches on brass instruments in contrast to the other instrument groups. The written parts should be relatively easy to hear. Wide leaps, especially in erratic or quick motion, should not be a normal procedure. Also realize that it is more difficult to ascend melodically, particularly in a high register, as there is greater physical resistance within the horn.

5.4.1   Tonguing

Articulations via the use of the tongue are crisp and responsive on brass mouthpieces. Brass players use mostly single-tonguing, but can also use double-tonguing and even triple-tonguing (Figure 5.1). These techniques are used primarily for fast, repeated notes, but may also be used wherever the player feels it is appropriate.

5.4.2   Slurs

The slur is also used extensively in brass performance (for review, refer to this topic in Chapter 4). The instrumental technique is similar to the woodwinds (using valves instead of keys to change pitch), with the exception of the trombone. However, the brass player also can slur notes within a specific overtone series, though this requires more skill since pitches are derived solely with the airstream and embouchure.

Figure 5.1 Tongue technique for brass

5.4.3 The Glissando

A glissando, or gliss, is essentially a sliding (or smearing) up or down from one note to another. On trombone, this happens automatically because the trombone uses a slide instead of valves. When the slide moves in a downward direction with notes moving in the same direction, the sound created by the player will automatically result in a gliss.

The same effect occurs when the slide moves up, along with notes that are ascending. (If you've ever played a slide whistle, you will understand this principle.) Trombones are often called upon to provide a loud, raucous gliss that can certainly add comic spice or a weird Doppler effect. However, most music does not require this, so the trombonist must implement a technique called legato tonguing to avoid playing a gliss. The general rule is: whenever the slide and the notes move in the same direction, a legato tonguing must occur. This light tonguing procedure provides a defined attack and all professional trombone players are adept at it. But the arranger should respect the physical task by avoiding excessive speed. Although there are trombonists who have amazing dexterity, it is still wiser to write excessively fast-moving passages for instruments that are better equipped for that task, most notably the woodwind and then the strings, followed by the valved brass.

5.4.4 Articulation Markings and Effects

The brass follow the same markings found in the woodwinds chapter regarding staccato and legato markings with and without accents. There are additional markings indicated here in Figure 5.2 that are more specifically related to brass (and sometimes saxes) within the style of jazz and commercial music. The fall-offs are obvious, but the "doit" is a smear that ascends into infinity after the initial pitch is heard. The term, when spoken in an animated way, represents the sound that is emulated. The "rip" is essentially a smear into the attack of a note and is somewhat like the doit but reversed. It begins with an inaudible note, smears upward and culminates at the written note. The "bend" is merely a dipping of the pitch immediately after the pitch has been established and rising back to the original pitch. The "shake" is similar to a trill but is done solely with the embouchure, creating a deliberately more ragged sound.

Figure 5.2 Common articulations and expressions for brass in commercial music

5.5 The Specific Instruments

The following brass instruments (presented in score order) will be discussed here with regard to tone, technique, range (including registers and transposition), and musical function within an ensemble.

5.5.1 The French Horn

Tone: The French horn (or horn) is a particularly beautiful brass instrument. Depending on the register and volume, the quality can be noble and soaring, dark and threatening, or warm and mellow. The shape of the horn's tubing is round (looking like a coiled snake), eventually expanding into a flared bell that faces away from the audience. This accounts in part for its somewhat muffled tone quality. At the other end of the tubing is the mouthpiece. Near the mouthpiece, at the top of the tubing, are three valves. Since the tubing is longer and the bell of the horn is larger than the trumpet, it provides a lower, warmer, and more robust tone quality. Conventional orchestration books state that it takes two horns to equal the strength of one trumpet or trombone. (In live performance this rule may apply since the bell of the horn faces away from the audience, but in recording, or with any microphone usage, this rule need not apply.) Consequently, the horn section is set up in two pairs, with each pair reading one written part. The first horn and the third horn play the higher notes and the second and fourth horns play the lower notes. This may seem confusing at first, but consider that most conventional horn writing features a two-note harmony, with each note being played by two horns. Essentially, the pair of notes is assigned to horns 1 and 2 (high and low) and then duplicated for strength in horns 3 and 4 (see Score 5.1 in Appendix A, Example 5.20 on the website). In bar 5, the horn section divides into four-part harmony and the high and low horns continue in the same fashion. Since two parts are placed on one staff (divisi), it is easier to read when the notes are assigned this way. Today there are many different configurations of brass sections so, depending on budgetary and artistic needs, horn writing may be more varied. For a "section" sound, a minimum of two horns is required.

The horn is also capable of producing a loud, raspy effect known as bouché (stopped). Its sound stereotypically suggests a sinister mood. To achieve this, the player must insert his or her hand deeply inside the bell to choke the sound and create somewhat of a metallic rattle. This technique alters the pitch upward by a half-tone so the player must transpose the written part down a half-tone to produce the desired written pitch. To request this effect, the arranger can simply write the + symbol above the pitch and the player will do the rest. Since the player is transposing, the written part should not be too intricate (this sound is best used on a single sustained tone for effect or a rather slow-moving line). To cancel the effect, simply write the o symbol over a pitch, or write "open" or "normal."

Technique: Although the horn has a larger bore than the trumpet, the mouthpiece is smaller. As a result, the instrument is less accurate so demands on the player are greater. Although there are some unusual musical examples where the horn is required to perform an intricate passage (e.g. "Jupiter" from Gustav Holst's suite *The Planets*), it is more natural for the instrument to speak at a slower speed. Short notes on the horn are more blunt than on the trumpet, but long notes are excellent. In contrast to the other brass instruments, vibrato is generally not used.

Range: The range of the horn is extensive, depending on the skill of the player, but the extreme low and high ends of the range are not commonly used in most commercial music situations. The best range is the middle register, spanning two octaves from its written G below the staff to the G directly above the staff (Figure 5.3). This practical range offers the most accuracy, the most beauty, and the widest range of expression and dynamics. The horn reads treble clef and is pitched in F, meaning that its written C sounds a fifth lower on the note F. Consequently, the music arranger must write the part a perfect fifth higher.

The characteristics of its registers are as follows:

- Extreme low register: dark, cold, gruff, hard to control, slow response;
- Middle (practical) register: best register, offering the most flexibility and control; warm and robust, but becomes louder and more intense as it ascends;
- Extreme high register: strained, hard to control.

Musical function: Its location sonically within the brass family is similar to the violas in the string section, essentially bridging the gap between the higher trumpets and the lower trombones. However, the horns are placed on the conductor's score above the trumpets and below the bassoons because they also are used quite frequently in conjunction with the woodwinds. (The standard instrumentation of a woodwind quintet is flute, oboe, clarinet, bassoon, and horn.) The horn section often provides harmonic accompaniment, including "pads" (sustained notes) or rhythmic figures. However, arrangers love to assign a melody to a section of horns as well. The horn is also excellent as a solo instrument, but the background instruments should not be too heavy or loud. Sometimes, a series of solo phrases may be passed to other members of the horn

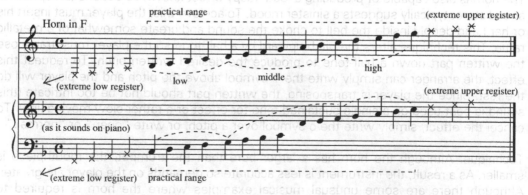

Figure 5.3 Range, registers, and transposition of the (French) horn in F

section in accordance with range; a higher phrase would be played by the first horn and a lower phrase would be played by the second horn.

5.5.2 The Trumpets

Tone: The trumpet looks and sounds very much like a bugle and is the highest instrument group within the brass family. There are several types of trumpets, but most sound similar within an ensemble, with the exception of the piccolo trumpet. The sound is relatively even throughout the registers, but a loud dynamic in the low register is limited and a soft dynamic in the high register will have significant intensity. Notes in the uppermost range will need more volume.

Technique: The trumpet is more agile than the lower brass because its registers are higher, and because it has three valves that are used in various combinations to transpose the "bugle" (or overtone series) into different keys.

Range: The most common trumpet (especially in jazz and pop styles) is the one in the key of B♭, requiring a pitch transposition. Since this instrument sounds a major second (or whole tone) lower, the arranger must write the part a whole step higher. (The treble clef is always used.) Its range extends from written low F♯ upwards to high C (Figure 5.4).

The notes beyond that point are usually considered the extended range, reserved for lead players who specialize in the performance of these very difficult notes. A specific top note cannot be readily determined as this depends on the individual, but a written G is normally the highest note to be used. Most arrangers avoid using these notes in any great frequency owing to the physical demands of the player and the sonic barrage that may wear on the listener. The best part of the trumpet range is from low E to high C (slightly over one and a half octaves). These notes sound most pleasing as they avoid the extreme registers and are easiest to play technically because the valve combinations (fingerings) fall most comfortably within the hand.

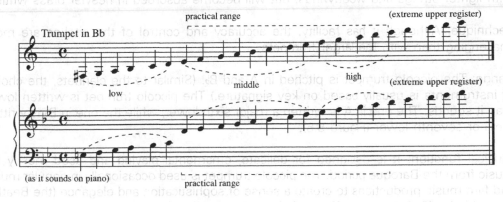

Figure 5.4 Range, registers, and transposition of the trumpet in B♭

Figure 5.5 Comparison of the transpositions for the trumpets in B♭ and E♭.

Orchestral trumpet players use other trumpets pitched in different keys. The C, D, and E♭ trumpets offer the player another option, similar to the clarinetist who might use a clarinet in A instead of the one pitched in B♭. For the brass player, a key signature is not the only consideration. The trumpet in E♭ can aid the player significantly where high notes are concerned (Figure 5.5). This example shows a musical passage written in the extreme high register for a trumpet in B♭. Notice how the same notes, when assigned to the E♭ trumpet (written a minor third lower from the sounding note on piano), are easier to perform since they fall within a slightly lower range.

5.5.3   The Piccolo Trumpet

Tone: The piccolo trumpet offers a unique timbre in contrast to the mainstream trumpet, almost sounding like a toy. Its volume increases as it climbs into its high register, but it is still rather pinched in contrast to the larger trumpets. It sounds fuller when placed with lighter strings and woodwinds, but will become absorbed in heavier brass writing.

Technique: Although it has facility, the accuracy and control of the pitches are more challenging than with the larger trumpets.

Range: The piccolo trumpet is pitched in A and B♭. (Similar to the clarinets, the choice of instruments is usually based on key signature.) The piccolo trumpet is written lower than it sounds. The one in A is written a major sixth lower, while the one in B♭ is written a minor seventh lower (Figure 5.6).

Musical function: It is designed for delicate, ornamental playing and used mostly for music from the Baroque period. The piccolo trumpet is used occasionally in popular music and film music productions to create a sense of sophistication and elegance (the Beatles used this instrument quite effectively in their recording *Penny Lane*).

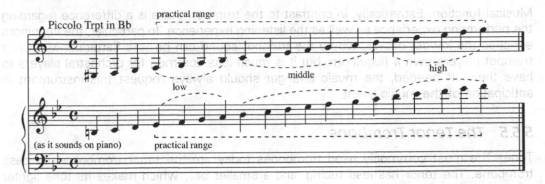

Figure 5.6 Range, registers, and transposition of the piccolo trumpet in B♭

5.5.4 The Flugelhorn

Tone: In contrast to the trumpet, the tone is warmer and a bit fuzzier, so the instrument works better in music that is more mellow. Its tone is formed with a bell that is larger than the trumpet but smaller than the French horn, and the slightly longer tubing is more square in appearance in comparison to the more streamlined trumpet. Although it is more lush, the instrument is a bit more difficult to control regarding intonation.

Technique: This instrument is played exactly as the trumpet (same fingerings but usually with a slightly different mouthpiece) and is in the same key as the B♭ trumpet, making the two instruments interchangeable at a moment's notice.

Range: The range of the instrument theoretically is the same as the trumpet, but ideally should be limited to a top written note G above the staff. Beyond that, the flugelhorn loses tone quality and sounds less comfortable. It is best used for playing in the lower and middle registers (Figure 5.7).

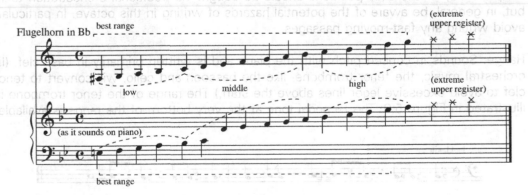

Figure 5.7 Range, registers, and transposition of the flugelhorn in B♭

Musical function: Esthetically, in contrast to the trumpet, there is a difference regarding the performing experience as well as the listening experience. In particular, the flugelhorn enriches the lowest octave, where the trumpet sound can be less flattering. Most jazz trumpet players own a flugelhorn, but it is much less common for orchestral players to have them. If needed, the music arranger should always request this instrument in anticipation of the music event.

5.5.5 The Tenor Trombone

Tone: The most commonly used trombones today are the tenor trombone and bass trombone. The tenor has less tubing, and a smaller bell, which makes its tone lighter and brighter. Its volume and tone control are quite broad; it can be warm and mellow or provide a loud, biting quality.

Technique: The trombone's mechanics are the most primitive of the brass family. A long metal slide (tube) is attached to the main tubing. The slide has seven positions that essentially elongate the tubing, in effect transposing the overtone series into different keys. The positions, however, are not marked or felt as would be the frets on a guitar neck. They are more like the finger positions on a fretless string instrument, requiring the player to find the note using aural skills in conjunction with the physical ones. The slide certainly requires more physical motion in comparison to valves, potentially slowing the response time. The other factor that complicates the matter is the need for a legato tonguing technique (this can be reviewed in Section 5.4.3, pertaining to the glissando). Yet another concern pertains to the lower register. To produce some notes here, the slide must be extended to the fifth, sixth, and seventh positions, which become physically more difficult to accomplish. A classic example of working in the lower register is illustrated below (Figure 5.8). What looks like an easy musical idea to perform on tenor trombone is, in actuality, very cumbersome. The Bb is produced via the slide's first or closed position, and the B natural, only a half-step higher, is produced via the slide's seventh position. To do this, the player must first have his hand (on the slide) close to his mouth (in first position) and then extend his arm completely (to reach the seventh position). A deeper study of the slide positions should be sought via traditional orchestration texts but, in general, be aware of the potential hazards of writing in this octave. In particular, avoid writing any fast-moving passages.

Range: Sounds in concert pitch with the piano and is written primarily in bass clef. (In orchestral music, the tenor trombone, like the bassoon and cello, will convert to tenor clef to avoid excessive leger lines above the staff.) The range of the tenor trombone is illustrated in Figure 5.9. The Bb pedal tone at the very bottom of the range is available,

Figure 5.8 Problematic performance on the tenor trombone

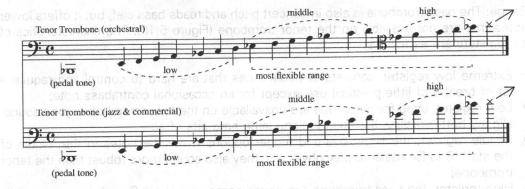

Figure 5.9 Clef protocol for the tenor trombone

but not very practical as it is difficult to control regarding expression. There is a gap from this note to the low E that begins the low register of consistently available notes for performance. The middle and high registers offer the most technical flexibility and are best for melodies. Many jazz trombonists have extended the upper range beyond its practical top note of high B♭. This register is most appropriate for an accomplished soloist, so arrangers should be aware of the capabilities of the individual performer.

Musical function: The tenor trombone usually plays the upper trombone parts within the trombone section. As part of the section, the trombones are great for harmonic padding and for rhythmic harmonic support, and are capable of a wide dynamic range. The trombone is used extensively in the jazz band in section work and for solos. Jazz players make frequent use of the upper range of the instrument, but use only the bass clef. Similar to flute players, they are quite used to reading four or five leger lines. Perhaps the most famous trombonist of the big band era who featured himself playing solo melodies in this register is the great Tommy Dorsey.

5.5.6 *The Bass Trombone*

Tone: The bass trombone's tubing is longer and the bell of the instrument is larger in circumference, making the overall sound a bit darker and fatter. In its upper register, its darker sound suggests a closer relationship to the French horn. In the middle and high registers, the volume increases and is generally more robust than the tenor trombone.

Technique: More air is required and the response time is generally slower than the tenor trombone, especially in the lower register. The slide features the same seven positions as the tenor trombone, but there are also two triggers (executed with the thumb of the left hand) that channel the air through longer tubing and effectively transpose the instrument into F and E, enabling the player to reach the lower notes without extending the slide as far. (The music writer does not need to be concerned with transposition; the part is written and always sounds in the concert key.)

Range: The bass trombone is also in concert pitch and reads bass clef, but it offers lower notes that are not available on the tenor trombone (Figure 5.10). The characteristics of its registers are as follows:

- Extreme low register: consists of pedal tones that are hard to control and require a lot of breath; of little practical use except for an occasional contrabass note;
- Low register: used for notes that are unavailable on the tenor trombone; good sound, slightly better facility, but still requiring large amounts of air;
- Middle register: the larger bell and trigger options make the notes in the middle of the staff (B to Eb) easier to play physically; they also sound more robust than the tenor trombone;
- High register: the bass trombone can sound somewhat like a French horn, especially when the bell is covered with a beret (or some similar cloth); excellent for melodies;
- Extreme high register: possible with some players for virtuosic solos but rarely used for standard ensemble purposes.

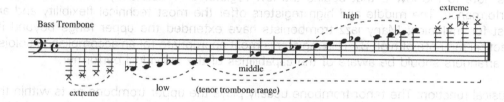

Figure 5.10 Range and registers of the bass trombone

Musical function: The instrument offers its strongest contribution in the bass register, but is also quite effective in the middle and high registers, where it supports the upper trombones in harmony and can also contribute melodically. Its versatility makes it a very handy instrument to use when there are only a few musicians in the brass section.

5.5.7 The Tuba

Tone: Tubas come in many different sizes and are more versatile than one might imagine. In general, the tuba's timbre can be more readily grouped with the French horn's tone character: warm and lush. Because they are more mellow than the trombones, tubas can play at a softer volume in the low register. As it ascends into the middle and upper registers, the tuba's dynamic contour actually thickens and its tone can protrude through the ensemble. This aspect is very good for significant melodic lines, but be careful when scoring the instrument for accompaniment.

Technique: Similar to the trumpet, the tuba uses valves to transpose the overtone series, making it more agile than its lower brass counterpart, the trombone. As with the bass trombone, the tuba requires large amounts of air, particularly in the lowest register. Tubas are pitched in various keys, but the writer need not be concerned with transposition (all tubas read and sound in concert pitch). The size of the tuba does matter and usually the player will choose the appropriate instrument for the music. Consequently, it is a good idea for the music arranger to notify the player regarding the characteristics of the music so he or she can plan accordingly.

Range: Figure 5.11 shows the range of the tuba. Notice the use of leger lines in the bottom octave. Unlike the double bass, there is no register transposition to avoid the use of leger lines. Convention stipulates that leger lines should always be used as opposed to writing 8va bassa. The bottom end of the range can vary depending on the type of tuba. Some tubas have a fourth valve that enables the player to access the lowest notes. In general, the practical range is from low G (the lowest one on the piano) to the B♭ on top of the bass clef staff.

The characteristics of its registers are as follows:

- Extreme low register: these notes are not always available and, although easier to perform than on the bass trombone, still require a lot of breath; of little practical use except for an occasional contrabass note;
- Low register: excellent for bass notes; warm and supple in tone with more agility than the bass trombone;
- Middle register: good for bass notes, better for melodies; the tone thickens but is most agile here and offers the most control with volume and expression;
- High register: used mostly for melodies; can sound strained; a smaller tuba should be used if this register is featured extensively;
- Extreme high register: possible with some players for virtuosic solos but rarely used for standard ensemble purposes.

Musical function: There is usually only one tuba in a brass section (except in the case of a marching band) since its presence is clearly established in relation to the other instruments. Most music is not heavy enough to support more than one player. The most typical parts are in the low register and the primary function is to play a bass line (as in a Sousa march), but sometimes the higher register may be featured. In lighter circumstances, a common double is for the player to use a euphonium.

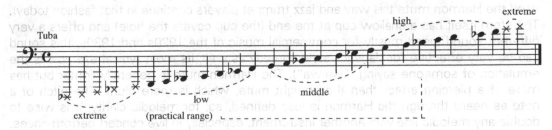

Figure 5.11 Range and registers of the tuba

5.5.8 The Euphonium

Tone: The euphonium sounds much like a hybrid of the larger tuba and the smaller horn. It offers a lighter tone than the tuba, but is still mellow and robust. The instrument fits more comfortably in the player's lap as it is closer in size to the French horn, although the shape of its tubing is more rectangular. As with the tuba, its bell points upward.

Technique: this instrument's higher register placement, along with its three valves, allows it to perform with the dexterity of a trumpet.

Range: Most similar to the range of the tenor trombone, it is primarily written in bass clef and sounds in concert pitch. However, unlike the trombone, the volume of sound thickens as it ascends into the upper register.

Musical function: The euphonium is great for melodic passages or any high-register tuba parts. It is used mostly in concert bands, but do not discount its value elsewhere as it can be quite effective in small brass sections when the need for doubling is required. Most professional tuba players own a euphonium (or tenor tuba), but the arranger should make an effort to notify the player regarding its use.

5.6 Mutes

The use of mutes with brass instruments is quite valuable. Mutes can drastically change the sound of the brass, making them softer to create different moods and enabling them to interact and blend more readily with the lighter woodwinds. Mutes also offer a wider spectrum of tone color. They are used mostly with trumpets and trombones. Below is a list of the most commonly used mutes, along with their characteristics.

Straight mute: A metal mute that is cone-shaped, it preserves the original open sound of the brass somewhat with the tone becoming more nasal. When played softly, this mute evokes the feeling of distance. When played loudly, this mute can offer a crisp, raspy tone.

Harmon mute: A metal mute that is shaped more like a donut. The hole in the middle appears when the stem of the mute is removed (the great jazz trumpeter Miles Davis used the Harmon mute this way and jazz trumpet players continue in that fashion today). The stem itself has a shallow cup at the end (the cup covers the hole) and offers a very different sound (used mostly for commercial music of the 1920s and 1930s, this sound can be very effective for slapstick-style comedy as well; its most typical example is the emulation of someone saying, "wa-wa"). The Harmon mute is generally lighter but has more of a piercing attack than the straight mute, which is more blunt. The pitch of a note as heard through the Harmon is less defined, so, for melodic clarity, it is wise to double any melodic line with another instrument, especially in live concert performances. In general, its characteristic tone offers a glitzy buzz that works well as a highlighter.

Cup mute: This fiberglass mute offers the warmest tone quality. It has a cone shape like the straight mute, but has a larger chamber (the part of the mute that sits in front of the bell), which makes the tone less nasal and consequently more resonant. This mute offers distinct clarity regarding individual pitches and only begins to "buzz" at the forte level when in the high register. It is excellent for blending with flutes and clarinets.

Bucket mute: This mute is a round cylinder about 5–6 inches long and about 4–5 inches wide. Inside the cylinder is a soft cloth material that also dampens the sound. Attached to the cylinder are three metal arms that clamp onto the bell of the instrument. This mute deadens the normal brass tone, making it darker and warmer but also somewhat veiled. Trumpet players today usually use flugelhorns for a darker tone, but a bucket mute is the closest substitute if a flugelhorn is not available. Conversely, if the bucket mute is desired, the flugelhorn can substitute. Another handy trick to create this effect (especially if there is no time to change instruments or attach the mute) is to have the trumpet or trombone player play into the music stand (if there is one). The bucket mute on trombone provides the same effect, but the trombone will sound more like a French horn. This mute is rather cumbersome to attach and detach, so extra time should be provided to accommodate this.

Plunger mute: This mute is essentially the same type of plunger that is used to unclog a toilet. The stick is removed and the player holds the rubber plunger with the left hand. The player's wrist anchors the bottom of the mute to the bell of the horn and, as the hand pivots, the mute moves to and from the bell. The opening and closing effect creates a "wa" sound, similar to the Harmon mute with its stem. This effect provides a sense of talking. This technique was used extensively by Duke Ellington in his jazz band. To indicate the closed position, use the symbol +. For the open position, use the symbol o.

The symbols should be placed directly above the notes. When both symbols are placed next to each other over the same note, the "wa" syllable is heard.

5.6.1 *Mute Preparation*

When mutes are desired, it is important to know that the insertion of the mute takes time to accomplish. In addition, the intonation of the instrument changes slightly (the professional player is accustomed to this, but ample time will allow for any adjustments to the instrument as needed). Trumpet players will have a slight advantage as they can perform using only one hand, but trombone players need both hands so they must have rests in order to grab and insert a mute (a minimum of five seconds for trumpet players and 10 seconds for trombonists is advisable). Of course, time is also needed when the mute needs to be removed, but this is usually an easier task. To indicate this, simply write "no mute" or "open."

5.7 **Orchestrating the Brass**

Voicing for brass is less problematic than with woodwinds since the instruments are of more similar timbre. However, there is still a concern as the blend is not as homogeneous as with the strings. The spread of the individual voices in a chord structure can be much closer than the woodwind and, in fact, usually works better. The instrumentation and the register will ultimately determine how a voicing needs to be constructed. Figure 5.12 shows a big band brass section (the kind you would hear behind the great twentieth-century singer Frank Sinatra) with four trumpets and four trombones. The trumpets are voiced in a block format and the trombones play the same voicing an octave lower (the stem direction will help distinguish both sections of instruments). This compact voicing provides a tight, bright, and punchy ensemble sound that flows nicely within the tempo and rhythmic feel. The melody, played by the lead trumpet, is placed strategically in a register that is high enough to sound exciting but not too high to avoid strain. The same melody (but an octave lower) also works well for the lead trombone as it places the instrument in the highest part of its practical range. When the lead trumpet ascends into its highest octave (in the last bar), the trombone voicing, if doubled exactly an octave lower, becomes less practical as a result of the extremely high range. The chord voicing used in the trumpets should then be rearranged for the trombones to avoid this problem. The results of this redistribution place the notes from the third and fourth trumpet parts (an octave lower) into trombone parts 1 and 2. The top two voices of the chord (essentially found in trumpets 1 and 2) are redistributed for the third and fourth trombones. There are many other ways to voice the brass in this style of music, but a lack of space will not allow any further study here, so an arranging book that focuses on this style should be consulted.

Score 5.2 in Appendix A (Example 5.18) shows the brass within the context of a fanfare style. The writing texture here is more polyphonic, beginning with one line and expanding ultimately into a full harmony. The piece is scored for a brass quintet and the distribution from the piano sketch to the individual instruments is rather obvious.

Score 5.3 shows how the same piece could be scored for a full orchestral brass section. The extra instruments afford the opportunity to dramatize the sound considerably.

Figure 5.12 A Sinatra-style voicing for a big band brass section

No new notes are added, but each line is doubled at the unison, with the exception of the bass line that the tuba plays at the octave with the bass trombone. More careful observation will show that the horn 1 part is doubled with the third trumpet. This not only thickens the horn line but, more importantly, adds punch from the brighter brass instrument. The same effect is established with the doubling of the horn 2 part by the first trombone. One final observation concerns the first and second trumpets: through most of the piece they are in unison, but in the last six beats split into harmony. (This occurs primarily to accommodate the expanding harmony, but the avoidance of unison in the highest register is sometimes good as it can prevent potential intonation problems and strident tone.) This loss of power in the melodic voice is reinforced an octave lower by the horn section.

With a full complement of instruments, there are more possible approaches to orchestration. The most important things to consider are: (1) register placement (each instrument is capable of performing well technically and sounding good regarding tone production and dynamics); (2) voice-leading (all parts are melodic, with any wide leaps being resolved comfortably); and (3) good balance within each instrument group (each section, when playing in counterpoint or harmony, has a homogeneous sound).

The next example features the brass in a solemn style of a traditional hymn. Here the brass is warm and lush. The sound features primarily the lower brass, although the trumpet still has the melody in its lower and middle registers (see Score 5.4, Example 5.12).

The piano part, as one would find in a hymnal, is written to be played at the keyboard. A direct transcription for brass may not necessarily be best since some of the musical lines have been compromised in order to conform to the physical capabilities of the pianist. Although the blend in the brass may forgive these compromises, it is better to score the brass in a way that offers each player a strong and, in this disciplined musical style, properly resolved melodic progression.

The trumpet part carries the melody and this may be done by a soloist or a section playing softly in unison. The horn 1 part also has the melody in unison with the trumpet, which provides darker hues and warmth. Cue notes have been written for the bass trombone as it may be too heavy for both the bass trombone and the tuba to play the bass line in this lighter context. Since the tuba is warmer, the bass trombone is asked to rest but, with the cue notes given, it may play the tuba part if needed.

5.7.1 Orchestrating Brass and Woodwinds

Since brass instruments are heavier than woodwinds, the brass must either curtail power and volume to match the weaker woodwinds or the woodwinds must be scored for maximum strength to compete successfully with loud brass. Sometimes the woodwind instruments can be heard, even in quieter registers, when the heavier brass instruments are in mutes. In Chapter 4, one example demonstrated how the woodwinds and brass might work together in an antiphonal way. The following example shows how the brass can accompany the woodwinds in similar fashion (see Score 5.5 in Appendix A, Example 5.6 on the website). The trumpets and trombones are muted while the horns provide a counterline to the woodwinds theme. The horns are strongest in this particular ensemble, but will not overpower the winds since they are not in a high register. They will also not interfere as their melodic line is gentle and simple.

5.8 Sequencing for the Brass Section: an Introduction

Compared to the woodwind instruments, brass instruments offer a wider dynamic range. This makes them slightly harder to render in a MIDI environment. Brass can drastically change color, function, and timbre depending on their range and dynamic. The use of mutes increases their multifaceted nature even further. Their chameleonic nature makes them hard to sequence consistently and effectively over their entire range. Brass sections can be used in a variety of styles, ranging from classical to pop, from jazz to R&B. For this particular reason, while some libraries and patches can easily work well in a classical setting, they may not work at all in a more contemporary style, and vice versa. With this in mind, let's start learning more about sequencing for the brass instruments.

5.8.1 Synthesis Techniques and Library Options

Because of the wide dynamic and color range of the brass family, a series of synthesis techniques can be used successfully for the virtual ensemble. The grade of effectiveness can vary depending on the instrument and on the musical style of the parts being sequenced. The main synthesis techniques that we learned so far are in general the most popular: lookup-table synthesis (a variation of wavetable synthesis) sample-based, physical modeling (PM), and hybrid (sample modeling). Table 5.1 lists the main applications, styles, and sonority that can be rendered most efficiently by each technique.

Lookup-table provides a useful source for contemporary brass sections that can be used effectively for styles such as pop, R&B, and funk. It has the advantage of offering a wide selection of patches some with "falls" and other basic articulations. The solo selection is limited, however, and it rarely offers enough realism to be used in a high-level production. Sample-based synthesis is the most versatile option. Multisample-based libraries can bring you as close as possible to real brass instruments and all their dynamic nuances and colors. Different articulations and timbres are usually available: sforzando, crescendo, and mutes can be rendered well through a professional sample-based library. If there is one area in which PM can really shine it is that of brass solo instruments. PM and sample modeling, by re-creating the different components of a brass instrument (e.g. mouthpiece, bell, pipes, valves, pistons) through the use of mathematical algorithms, can reproduce most of the different inflections and articulation of an acoustic instrument. PM patches can be even more realistic if layered with sample-based instruments. By using sample modeling libraries, you can take full advantage of all the standard MIDI CCs and use the advanced performance CCs (such as CC#2 breath controller) to control realistic vibrato,

Table 5.1 A detailed look at the use of some of the most popular synthesis techniques and their applications for the brass instruments

Synthesis technique	Suggested instruments	Styles	Comments
Lookup-table	Trumpet section French horn (solo and section) Tuba	Pop R&B Funk	Wavetable is well suited for modern sonorities and parts where an overall brighter and lighter timbre is required. In most cases it is not recommended for classical sequences where the brass instruments are particularly exposed
Sample-based	Trumpet (solo and section) Solo flugelhorn Trombones (solo and section) French horn (solo and section) Tuba	Any	As we learned earlier, sample-based is the most flexible and consistent technique to reproduce a variety of sonorities. Its versatility, combined with a multisample patch, can deliver extremely accurate renditions
Physical modeling	Solo trumpet Solo flugelhorn Solo trombone	Pop R&B Funk	PM is particularly recommended to re-create solo passages where the use of specific articulation is crucial
Sample modeling	Solo trumpet Solo flugelhorn Solo trombone	Any	This technique is specifically indicated for productions where the realism of the patch needs to meet highly expressive passages and articulations

pressure, pitch, and noise effects. Listen to Examples 5.1 and 5.2 to compare, respectively, a two-trumpet part rendered with a sample-based patch and the same part rendered with the Wallander Brass library that uses a combination of additive synthesis/behavioral modeling technology.

When it comes to choosing a library of sounds that can give good results in terms of sonorities, realism, and flexibility, there are several options available. Since brass instruments can be used in a variety of styles and configurations, you have to have clear in your mind exactly what you are looking for. In terms of style, there are three main categories: classical, pop (including rock, R&B, and funk), and jazz. Each category calls for slightly different tones, colors, and articulations. In addition, you have to consider whether you need solo instruments or sections. Table 5.2 summarizes the most common libraries and software synthesizers available on the market, along with their stylistic features and ensemble characteristics.

When choosing a library, keep in mind also the type of sequencing that you will be doing in terms of ensembles and sections. Some libraries offer great brass section patches, but they lack the versatility and complexity that is required when rendering solo parts. You will need at least two libraries, one used for solo passages and the other for ensembles. For simple and powerful unison section phrases, use a solid section patch that has the advantage of providing, without any particular programming effort, the richness of a real ensemble. For more complex section passages where divisi occur or where the lines move independently inside the section, use individual solo instruments to render each line. Make sure not to use the same patch for the different lines. This would create a flat and monotonous sonority that would sound artificial. Use instead a different solo patch for each line (even better, use different solo instruments from different libraries). This will re-create a more natural-sounding section. The final effect will be the same one created by different players with different instruments. If you have only one library with solo instruments try to use all the different variations that it provides for that particular instrument. If your library doesn't provide such an option, later in this chapter you will learn how to differentiate a single patch in order to use it in a multiline section.

5.8.2 *What to Look For*

A good and flexible brass library has to meet some basic requirements in order to represent a versatile source of sounds for the virtual orchestra. As was the case for strings and woodwind, repetition and lack of variation can make sequences sound artificial and lifeless. There are four main areas in which a brass library needs to feature a good amount of options and variation: dynamics, articulations, colors, and mutes. Having a series of choices for each of these categories will allow you to create convincing and realistic renditions of orchestral passages.

Since the dynamic range of brass instruments is fairly wide, your library should include at least four different variations, such as *p*, *mp*, *mf*, and *f*. Depending on your writing style, you might want to add also a *ff* dynamic, especially for instruments such as trombones and trumpets. Multiple dynamics can usually be implemented using either key-switching (KS) or multilayer techniques. In some libraries, individual patches for each dynamic level are also provided. In general, try to use libraries that provide all three techniques mentioned above. This will allow you to choose the best option depending on the passage you need to sequence. Multisample-based dynamic patches are usually indicated for a quicker and more seamless integration of different dynamics. In this case, the MIDI velocity (or a dedicated MIDI CC) allows you to vary constantly the overall timbre of a part. KS has the advantage of quickly differentiating between passages that are supposed to be played at different dynamics, giving you also the possibility of overdubbing the KS at a later time.

Table 5.2 A selection of the best brass libraries available on the market

Library	Platform	Sonic environment	Style	Solo/ensemble	Comments
Wallander Brass	VST, AU, RTAS	Dry	Mainly classical and jazz, but it could be used in some mellow pop arrangements	Both	Excellent library at an affordable price
Sample Modeling	VST, AU, AAX	Dry	Jazz and pop	Solo instruments but with a variety of timbres	Fantastic realistic library, very expressive and versatile
EastWest Hollywood Brass	VST, AU, AAX	Wet	Classical, studio orchestra	Solo and sections	A complete brass selection with extremely playable and smooth patches with a comprehensive choice of articulations
CineBrass (by CineSamples)	Kontakt Player plug-in (VST, RTAS, AU, AAX)	Dry/Wet	Classical and film	Solo and sections	Great library for big impact brass. Excellent for orchestral film scores
Vienna Special Edition	VST, AU, RTAS, AAX	Dry	Classical	Solo and sections	Very versatile and comprehensive library with a variety of articulations and sonorities. Perfect for orchestral renditions
Chris Hein Horns Pro	Kontakt Player plug-in (VST, RTAS, AU, AAX)	Dry	Jazz and pop	Solo and sections	This is a very flexible and great sounding library. Perfect for pop and light jazz

continued . . .

Table 5.2 Continued

Library	Platform	Sonic environment	Style	Solo/ensemble	Comments
Vienna Super Package	VST, AU, RTAS, AAX	Dry	Orchestral	Solo and sections	This is the ultimate orchestral brass library with a very comprehensive set of sounds and articulations
Broadway Big Band	Kontakt Player plug-in (VST, RTAS, AU, AAX)	Medium-dry	Jazz	Solo and sections	This is a fantastic library for any professional jazz and Broadway shows big band productions
Vir2 Mojo	Kontakt Player plug-in (VST, RTAS, AU, AAX)	Dry	Pop and jazz	Solo and sections	Excellent versatile library, particularly indicated for pop and Latin productions. It features a complete set of articulations
First Call Horns	Kontakt Player plug-in (VST, RTAS, AU, AAX)	Wet	Pop, funk, R&B	Solo and sections	This is a must-have collection for pop and contemporary productions. It features a variety of articulations and sonorities

Having a good number of articulations to choose from is also very important for a successful brass rendition. Brass instruments are capable of a fairly wide range of articulations and having them available in your library can really make a difference. Table 5.3 lists the most important articulations and their dynamic variations to which you should have access in your library.

A good brass library palette also needs to offer a variety of colors and tonal shades to represent the complexity of a full section effectively. For this reason, always try to have the most variegate palette possible. Use at least two different libraries to render the different instruments inside a brass section. A typical application would be the assignment of each part/instrument to a different library. This can be done randomly, in a cyclic way (by assigning every next instrument to a new library), or by choosing the right sonority from each library according to each part (e.g. the brightest patches would be assigned to the lead parts, and the darker and mellower ones to the fourth or fifth parts). Try to create as much variety as possible.

Strictly related to these topics is the issue of mutes. For brass instruments, mutes play an important role in adding variety and novelty to the sound palette. As you learned in the first half of this chapter, the addition of a mute to a brass instrument can add a whole new set of sonorities and acoustic effects. In the same way, when using a virtual orchestra, a completely new set of samples is required to represent such sonorities realistically. With the exception of some PM synthesizers, the rendition of a muted brass instrument (no matter which types of mute you use) cannot be realistically rendered without using a series of dedicated samples. Therefore, when shopping for a brass library

Table 5.3 A selection of the most common articulations and advanced dynamics options that should be available in your brass library

Articulation	Comments
Staccato	This articulation should feature not only a short staccato but also slightly longer variations such as 0.3- and 0.5-second staccatos
Sustained	For a more flexible sound you should have sustained patches with and without vibrato
Fortepiano	This is particularly effective in dramatic passages where the realism of the brass needs to drive the sequence
Sforzando	This articulation is essential to achieve variations and realistic brass passages
Crescendo-diminuendo	While this sonority can be somehow effectively reproduced through the use of MIDI CCs #7 and 11 starting from a basic sustained patch, the use of a real sample with this dynamics will definitely add realism to your parts
Flutter tonguing	While this technique is most likely used on a flute, it can be sometimes applied to the trumpet too
Falls and scoops	For any jazz or pop/funk production, you need to make sure that you have access to falls and scoops at different speeds

make sure that it includes an adequate number of patches with mutes for each section of the brass family. Among the mutes that are more often found in good sample libraries are the Harmon, the cup, and the straight mute. Having at least one or two mute patches for each section, especially for the trumpet and the trombone sections, will increase the realistic rendering of the brass ensemble and add a series of colors to your arrangements. Usually, classical libraries come with brass instruments sampled with straight mutes, while jazz libraries feature the plunger mute.

5.9 MIDI Controllers for Sequencing the Brass Section

To input brass parts in a sequencer, several MIDI controllers cover a fairly wide spectrum of techniques and technologies. As in the woodwind section, brass instruments require a controller with a good amount of flexibility and expressivity. The most common one is the MIDI keyboard controller. It can be effective mainly for long, sustained passages, especially for the lower range of the brass family such as tubas, trombones, and French horns. It is useful for sequencing pad-like parts where the dynamic range is not particularly wide. Because of the fairly limited action of the keyboard, it is hard to render accurately parts that run over a wide dynamic range. A keyboard controller is also limited in terms of expressivity. Even with the addition of dedicated faders and knobs, it is difficult to send MIDI CCs to sequence cohesive MIDI parts featuring slides, falls, and slurs. When using a regular keyboard controller, it is easy to forget about the breathing pattern of the virtual brass players. This can lead to sequence parts that tend to run for a long time without breathing pauses, making the part sound unnatural and artificial. As with the woodwind, a quick fix to this potential problem is to take a deep breath at the beginning of the phrase being sequenced and blow as though playing a wind instrument. When you run out of breath it is time to insert a very short pause in the phrase. A much better option is represented by the addition of a MIDI breath controller (BC), used in conjunction with a MIDI keyboard controller. The BC gives you a much higher degree of control over the parts you sequence. Because of their fairly wide dynamic ranges, brass instruments need to be controlled from a MIDI device that can easily render all the different degrees of dynamics, from *pp* up to *fff*. The short action of a keyboard controller has insufficient run to allow you to interact accurately and adequately with such dynamic range. A BC allows you to interact precisely with a brass patch by controlling the dynamic through MIDI CC#11. The normal range (coarse) of this controller (ranging from 0 to 127) can be managed very precisely, with airflows ranging from extremely light to vigorous. The advantage of the combination of keyboard and BC is also demonstrated by the fact that you will be able to sequence parts that are realistic in length, since the patch will sound only when air is blown into the BC; when you stop blowing, a CC with value 0 is sent and therefore virtually stops the sound. There are only a few options when it comes to breath controllers. Yamaha used to manufacture the legendary BC3A (now unfortunately discontinued), TEControl makes the USB MIDI Breath Controller, which gives you the same functionality of the Yamaha model with the advantage of having built-in USB connectivity (Figure 5.13).

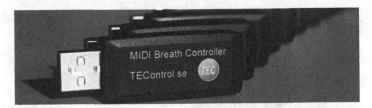

Figure 5.13 MIDI Breath Controller by TEControl (Courtesy of TEControl)

You can always use other controllers, such as the wind controllers mentioned in Chapter 4. They have the advantage of giving full control over notes, volume, bend, and vibrato of a patch. All the techniques demonstrated when discussing the woodwind instruments also apply to the brass section when using a wind controller such as the Synthophone, the AKAI EWI (Figure 5.14), or the Yamaha WX. These instruments are particularly effective when sequencing parts that require a high level of expressivity, such as solo or improvisational parts or ensemble parts featuring bending, sliding, and different types of vibrato. In the same category as these controllers is the excellent Morrison Digital Trumpet (MDT), which is unique in the class of MIDI controllers (Figure 5.14). The MDT uses the same fingering and techniques of a normal B♭ trumpet. It features three thumb controllers (operated with the right hand) and two switches that can, in total, send 10 MIDI CCs. A similar controller, called the MIDI EVI, is available from MIDI control specialist Nyle Steiner. This stylish and futuristic controller is based on a self-contained trumpet-like instrument that can also be retrofitted with a wireless MIDI transmitter. It can transmit CC#7, 11, 2, and 5, and aftertouch.

Figure 5.14 The Morrison Digital Trumpet (top) and the AKAI EWI 4000 (bottom), played by wind MIDI controller specialist Matt Traum (Courtesy of Matt Traum and AKAI)

Figure 5.14 Continued

As mentioned in the previous chapters, one of the drawbacks of these wind controllers is that they require you to be proficient with the instrument to start with. In the case of the MDT, you need to be fairly skilled in playing the trumpet to take full advantage of its potential. Nevertheless, it is an incredible tool that can be used to sequence not only brass parts but also strings, woodwind, and synth parts.

5.10 Sequencing Techniques for the Brass Instruments

Most of the sequencing techniques that you learned in the woodwind chapter also apply to the brass instruments. Therefore, the focus here will be on particular differences and advanced techniques that apply specifically to this family of instruments.

The first step, after having chosen the most appropriate library and patches, is to sequence the actual parts. No matter which MIDI controller you choose, it is important that you sequence each part on individual MIDI tracks. You basically have to treat your virtual score in the sequencer as the real score on paper. Playing one line at the time allows you to concentrate on expression, phrasing, and dynamics, exactly as a real player would do on a single part. Having individual MIDI tracks for each part allows you also to program automation with a higher degree of accuracy.

Having inputted the actual parts in your sequencer, it is time to start using some of the skills and techniques learned so far and apply them to the brass sections. With any type of wind instrument, the main secret of a successful MIDI rendition is the correct reproduction of its expressivity and variation in dynamic. While woodwind instruments use dynamics mainly to control expressivity at a higher level, meaning mainly at the phrase level, brass instruments in general tend to focus their expressivity to control the dynamic of single notes at a lower level. This approach makes it extremely important to use, in the virtual brass ensemble, automation in volume primarily through MIDI CC#11. With the accurate use of these controllers, the performance of a brass MIDI part can be effectively shaped and brought to life. Remember that MIDI parts and sample-based patches tend to be static and rigid. Volume and expression changes can add that missing spark. If you use only a keyboard controller to sequence the parts then you will have to do a little bit more work. Using a BC or a wind controller will make your life easier and lead to better results in general. Let's learn how to bring your brass parts to life!

Sequencing the brass parts from a MIDI keyboard only has the disadvantage of requiring more work at the editing level. The first step you have to take is to find the overall volume level at which you want a specific part to be. Then use CC#7 to "print" that level on each track of the brass instruments. The process is similar to the one learned for the woodwind. This CC#7 level will work as a master volume for each individual track. Now sequence the part or parts, remembering to have each part on a separate track and MIDI channel. Since we are assuming in this example that you don't have a BC, CC#11 will have to be inserted through a MIDI slider or knob. The keyboard controller should have at least one of these that can be assigned to output any type of CC MIDI message. CC#11 can be inserted either while performing or after the performance using the overdub mode of your sequencer. Another option is to take advantage of the pencil tool of the sequencer. Use the slider assigned to CC#11 to shape the dynamics of each part according to the score you are sequencing. Try to give a natural breathing flow to the part.

The main goal, when using automation, is to create two different types of dynamics. One is a macro-dynamic level that controls the overall crescendo and decrescendo of the part (this is a more traditional automation), where the score dictates the changes. The other one, which may be called micro-dynamic automation, targets single notes and their attack, sustain, decay, and release (more on this in the next section). The macro-dynamics can be inserted either by using the pencil tool of your sequencer of choice or by recording them (or overdubbing them) in real time using a fader or knob that can send MIDI data from the MIDI keyboard controller. When inserting CC#11 graphically with the pencil tool, use an exponential curve (Figure 5.15), as this shape is the most accurate in re-creating the dynamic crescendo of brass instruments, where the first part of the blowing (the weaker) creates a slightly less powerful sonority than the second one (the stronger).

For other instruments, such as woodwind and non-wind instruments, the regular linear curve can be used when inserting CC#11 automation data graphically. If you prefer to insert the automation data in real time, you could use a BC (more information on this in a second) or a MIDI fader/knob first, then use the graphic editor to smooth out and clean

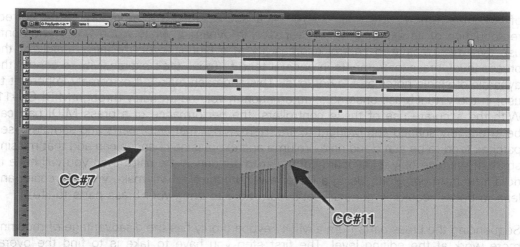

Figure 5.15 Example of a French horn part with dynamic automation inserted graphically (CC#11) using an exponential curve. Notice also the CC#7 inserted at the beginning of the part functioning as main volume (Courtesy of MOTU)

the recorded curves. A BC may be seen as essential when sequencing wind instruments in general and brass instruments in particular. From a MIDI keyboard it is very hard to get the feel of how hard it is to blow into a piece of metal for a sustained period and come out with nice and smooth melodies or with edgy and powerful tones. By simply pressing on the keys you won't be able to render a brass ensemble realistically. By assigning CC#11 data to a BC you can experience the same effort that it takes to generate sounds from a brass instrument (or at least close to it). A BC allows you to select not only the type of data you can send (in this case, CC#11 expression), but also the shape of the curve used to translate the blow into data. Program the BC to respond using an exponential curve for brass instruments and a linear one for other lighter wind instruments (such as flute, oboe, and clarinet).

As you have probably realized by reading the previous chapters, there are really two main parameters that we can control when it comes to render realistic dynamics for acoustic instruments. There is loudness (we can generally call it "volume"), and there is the sample switching/cross-fade that it is associated with the sonic characteristics of the instrument at different dynamics. As a default, many libraries have the former assigned to MIDI CC#11 and the latter controlled by MIDI velocity. These settings work well when we sequence passages that features very rhythmically active phrases. Since velocity controls the sample switching, we can effectively render the dynamic shape of a phrase by playing each note (let's say sixteenth notes) with the necessary velocity. This approach doesn't work, though, for long sustained notes. Imagine a French horn part where a long sustained note needs to be performed with dynamics ranging from *pp* to *ff*. Since we have only one MIDI velocity (at the attack of the note), we can access the sample switching feature using velocity. While we can use MIDI CC#11 to change the loudness of the note, this will not have any influence on its sonic characteristic. This can be particularly problematic for brass parts. In this case I recommend using a different MIDI CC (usually CC#1 is the most common) to control the sample switching. By doing so, you will control the loudness via CC#11 and the sample switching via CC#1. While it is possible to use the same CC for both, I prefer to use two different CCs in order to have a finer control over the two parameters. Some libraries (such as VSL Instrument) allow you to turn on or off the sample switching triggered by velocity so that you can have it on for fast rhythmic passages and off for long, sustained notes.

Listen to Examples 5.3 and 5.4 to compare, respectively, a French horn part sequenced from a keyboard controller (without any automation) and a keyboard controller in conjunction with a BC.

5.10.1 *Micro-level Automation*

As discussed for the woodwind (Chapter 4), micro-automation at the note level can be used to bring the realism of virtual parts to a higher level. This will add a greater level of variation since each note will have a slightly different envelope. Whereas for the woodwind the main focus was on the attack and release of a note, for the brass we should concentrate primarily on sustain and release. In the case of brass instruments, this technique should be reserved for use on long, sustained notes; leave short and medium-length notes without micro-automation.

The technique is similar to the one discussed in Chapter 4 for the woodwind. After sequencing the part, move every other note onto a new MIDI track and assign it to a different MIDI channel/device. To create more variation, select a slightly different sonority for the second track (make sure to select a patch that is very similar to the other one). At this point you will have the original part split between the two MIDI tracks. This step is necessary in order to avoid MIDI automation data conflict between subsequent notes. Now, using the pencil tool, insert slight changes in volume using CC#11. Be very gentle and do not overdo it. This technique is only indicated to add a bit more life to virtual brass parts and not to create very noticeable changes. The main areas of a sustained note that you should work on are the sustain and the release sections of sustained notes as shown in Figure 5.16.

The advantages of splitting a part over two or more MIDI channels are many. You can create more variation by assigning two slightly different patches of the same instrument to the two MIDI channels. This will create a much more natural feel and, since the notes that will be assigned to each patch will vary depending on their order in the phrase and not on the note itself, the part will sound much more realistic. This technique can be improved if you spread a single phrase or part over more than two MIDI channels, each with a different patch assigned. The splitting technique can be improved further by changing slightly some of the parameters of each patch such as tuning, filters, or attack. Some specific detuning techniques will be discussed in more detail later in this chapter.

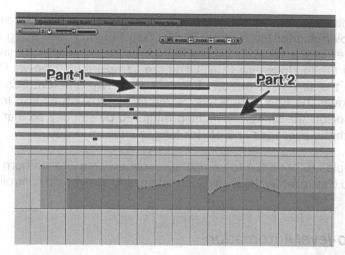

Figure 5.16 Example of micro-automation applied to a French horn part where the sustain and the release parts of the two sustained notes have been altered to add a more realistic effect (Courtesy of MOTU)

As mentioned in Chapter 4, when discussing woodwind, a valuable technique to create more variation in wind instruments is the use of the velocity-to-attack function of the sampler or synthesizer. While, in most cases, the attack parameter doesn't have a huge impact on adding realism to brass instruments, it is nevertheless a valuable technique that can improve sequenced virtual parts, especially solo passages. The principle is the

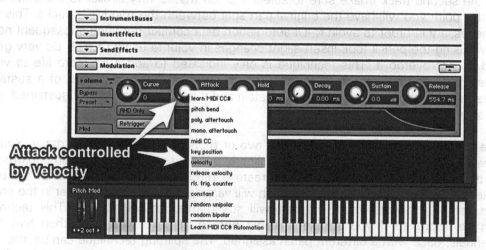

Figure 5.17 In Kontakt any modifier (CCs, MIDI Data, etc.) can be assigned to virtually any parameter of a patch. In this case, MIDI velocity was assigned to the attack section of the amplifier envelope (Courtesy of Native Instruments)

same as for the woodwind. Advanced software samplers and synthesizers allow you to assign the MIDI velocity data of each Note On message to control not only the level of the note (loudness), but also several other parameters of a patch. In this case the aim is to control the attack time of the amplifier envelope to add variation to the way in which the waveform of the patch is played back. Lower velocities will provide a softer attack, while higher-velocity values will result in sharper attacks. This is particularly useful in re-creating how a real wind instrument reacts to a softer or a harder blow. Figure 5.17 shows the velocity-to-attack parameter in Kontakt by Native Instruments.

Depending on the sophistication level of the sampler or synthesizer, this technique can go a step further by using MIDI velocity to control not only attack but also other stages on the amplifier envelope such as sustain or release.

5.10.2 Quantization

Quantizing orchestral parts is always a challenging task. How to quantize them is one of the most asked questions at presentations and demonstrations I give. It can be tricky to strike the right balance between accuracy and natural flow of the parts. Use a quantization setting that is too tight and the virtual orchestral will lose the natural groove performance that is typical of a real acoustic ensemble, while an ensemble that is quantized too little (or not quantized at all) can sound unprofessional and sloppy. This problem is particularly evident when sequencing for brass instruments, since they usually play parts that can range from pad-like and fairly open to rhythmically challenging and busy. To strike the right balance, follow some basic rules that, in general, can also be applied to the other sections of the orchestra.

Try to play the parts as accurately as possible. Nothing can replace a good, solid performance. If you don't feel comfortable in sequencing fast passages, you can slow down the tempo of the sequence while recording a part and then speed it up later. Try to be particularly careful and precise when sequencing slow passages, since they usually require a more melodic interpretation that cannot be re-created through quantization. Leave slow, melodic phrases or pad-like passages unquantized to preserve the natural performance. For more complicated and rhythmically challenging parts, never use 100 percent quantization strength. If you do so, the parts will sound mechanical and fake no matter how good the samples are. Music is a combination of sound, rhythm, and human groove, and all three fundamental elements have to be accurately translated to the virtual ensemble to achieve the best results possible. By starting with 50 percent strength and -60 percent sensitivity, you will get a light quantization that will simply tighten up the part slightly. The results of these settings vary greatly depending on the accuracy used when the parts were recorded. The sloppier the original performance, the higher the quantization strength and sensitivity you will have to use. After the first quantization round, if you feel that your parts are still a bit rhythmically off use a much lower strength setting (between 10 percent and 20 percent) to move each MIDI event gently toward the quantization grid selected. This last step can be repeated until the part sounds the way you want it. This incremental quantization technique is very useful since it allows you to improve the rhythmic positioning of the notes slowly, without drastically changing it in a single shot. Table 5.4 lists the main quantization scenarios for brass parts.

If you decide to use some quantization for your brass parts, do not use exactly the same quantization values and settings for every track. Try to use slightly different variations of

Table 5.4 Most common quantization settings for the brass section. These principles can be used also for strings and woodwind sections

Part type	Quantization settings	Comments
Long, sustained notes, pad-like (i.e. French horns, trombones, tuba)	Do not quantize unless absolutely necessary	If the part doesn't sound right rhythmically, try to play it again instead of quantize it
Medium rhythmic complexity, counterpoint-like part (i.e. trumpet, trombones)	Use incremental quantization starting with mild settings and use second and third quantization rounds to improve the quantization if necessary	Start with conservative quantization settings such as 50 percent–60 percent strength and –60 percent sensitivity. For the following quantization rounds, use between 10 percent and 20 percent strength with 100 percent sensitivity
Busy rhythmic ensemble parts	Start with a more decisive quantization and use additional round with milder settings to improve the quantization even further	The first quantization should be set to 70 percent–80 percent strength and 80 percent sensitivity. For the following quantization rounds use between 10 percent and 20 percent strength
Solo parts with simple and consistent rhythmic activity	Try to preserve as much as possible the original performance in order to keep the natural phrasing and flow. If necessary quantize the part directly with small incremental rounds with strength between 10 percent and 20 percent	
Complex solo parts with a variety of rhythmic subdivisions	Try to preserve as much as possible of the original performance; if necessary, try to re-record the hardest passages until you get it right. Use quantization only if absolutely necessary	These are the hardest parts to quantize. If you have to use quantization, start with mild settings and use second and third quantization rounds to improve the quantization if necessary

strength and sensitivity to create more variety. Do not use very different settings, but instead add or subtract between 1 percent and 5 percent for both strength and sensitivity. This is another reason why it is important to sequence each part of a section on different MIDI tracks/channels, since it will give you much greater control over the quantization of each individual part. Listen to Examples 5.5 and 5.6 to compare, respectively, a brass and woodwind section quantized with the same settings for all the parts (100 percent strength and 100 percent sensitivity) and the same section quantized with different settings and variations for each part (Score 5.5).

5.11 Performance Controllers for the Brass

Performance controllers allow you to expand the flexibility of a sampled or synthesized patch to gain a higher control over their performance in a virtual environment. As with all the wind instruments, brass can take advantage of a series of these controllers. Among them, CC#2 (breath controller), portamento, CC#67 (soft pedal), and aftertouch are particularly recommended. Let's take a look at some of their main applications when sequencing for brass.

5.11.1 Breath Controller

The importance of using a BC device to sequence brass parts cannot be stressed enough. Some of the most basic applications of such controllers were discussed earlier in this chapter, and some of the more advanced options and advantages of using a BC device in conjunction with CC#2 will be demonstrated. We already learned how to use the BC to send volume control changes such as CC#7 and expression CC#11. The natural application of a BC device is to send CC#2 messages. This controller is usually not assigned to a specific parameter but instead, in more modern software and hardware

Figure 5.18 The new Breath MIDI controller by TEControl with the addition of the bite, nod, and tilt controls

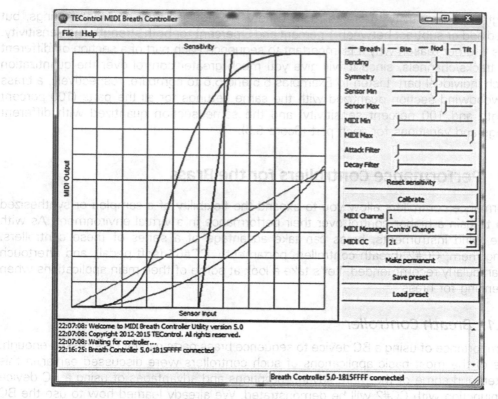

Figure 5.19 The software editor for the TEControl MIDI Breath Controller

synthesizers, it can control a variety of parameters. When used in combination with brass patches, you can assign it to parameters such as vibrato and pressure. Its flexibility depends on how sophisticated the synthesis engine that receives that CC is. Physical modeling is among the types of synthesis that allow greater flexibility in terms of live sound manipulation and variations. Listen to Examples 5.7 and 5.8 to experience how CC#2, controlled through a BC, can be used to change the vibrato and pressure of a brass patch in Wallander Brass. Another advantage of using a BC is the flexibility that the controller itself provides. The TEControl USB MIDI Breath Controller can be easily programmed through its own bundled software to send aftertouch, any CCs, pitch bend, and even SysEx. This is particularly useful since often you may want to send CC#11 to control expression and then use it to overdub CC#2 or any other controller. In addition, you can control the shape (linear, exponential, or logarithmic) of the curve used to translate your breath into the MIDI messages.

5.11.2 Portamento

Portamento is another performance controller that can be useful when sequencing brass parts. The ability to slide into a note is a typical sonority of the brass instruments. Although it is better to use real samples to re-create an accurate rendition of smears and scoops for piston instruments such as the trumpet, for slide instruments such as the trombone you can use portamento effectively to reproduce a realistic effect. To take full advantage of the portamento feature, you need to use all three portamento controllers: CC#5, 65, and 84. CC#65 allows you to turn on (values between 64 and 127) or off (0–63) the overall effect. CC#5 controls the speed, or rate, of the portamento effect: lower values will give a faster slide, while higher values translate into a slower slide. With CC#84, you set the note from which the slide of the portamento will start: it ranges between 0 (C2) and 127 (G8). A value of 60 returns C3, meaning middle C on the controller. Since it can be challenging to change the value of CC#84 in real time, this technique can be done off-line by penciling in the correct value for each smear, slide, or scoop. First, insert where you want the portamento to be on and off with CC#64 (Figure 5.20), then insert the rate data CC#5, and finally insert CC#84 to set, for each slide, the starting note. Keep in mind that CC#84 gets reset after each new Note On message; therefore, even if you want to set the next slide to the same note as the previous one, a new CC#84 must be inserted.

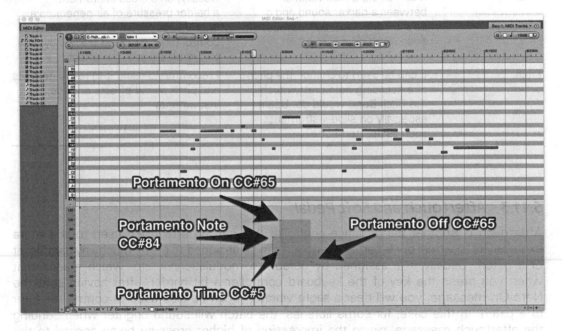

Figure 5.20 Use of Portamento parameters

It is important to underline that CC#5 and 84 are normally part of the standard set of controllers for hardware synthesizers. For software samplers and synthesizers there is usually no standard assignment and therefore they may not work as expected. Listen to Examples 5.9 and 5.10 on the website to compare a trombone part sequenced without and with portamento CCs, respectively.

Table 5.5 Useful applications the aftertouch performance controller

Parameter controlled	Application and instrument	Comments
Volume	Sudden dynamic changes for *pfp* or subito *f* passages. It is indicated for all the brass instruments	This is a technique to quickly create abrupt dynamic changes inside a part. It can be effective if used, especially on trumpet and trombone parts
Pitch modulation	Creation of vibrato effect. It is indicated especially for trumpet and trombones	If used on its own and assigned to the LFO's intensity it can create a decent vibrato (make sure to set the LFO's rate fairly slow). You can assign it to control the LFO speed and therefore, in conjunction with CC#1 (which would control the LFO's intensity), vary the speed of the vibrato
Filter	It can create a nice variation between a darker sound and a brighter one. Use it on trumpet and French horn	Usually, on a brass instrument, a harder pressure of air generates a brighter sound. By using aftertouch to control the filter cutoff frequency, you can re-create a similar effect
Pitch	It can be used to change the pitch of a sounding note pretty much like Pitch Bend would do. Use it especially on slide instruments such as the trombone	I recommend using it to re-create doits and pitch sound effects on slide instruments

5.11.3 Aftertouch and Soft Pedal

Aftertouch is another performance controller that can be effectively used to add more expressivity to virtual brass parts. Aftertouch can be assigned to pretty much any parameter of a patch. Usually it is assigned by default to modulation/vibrato, which means that when you press the key of the keyboard controller a bit harder, after having sent the Note On message, you will hear a slight vibrato. It can be also used to control the filter of a patch. In this case, for some libraries, the patch will sound brighter when sending the aftertouch message, giving the impression of higher pressure being applied to the virtual mouthpiece of the instrument. You can also use it to control the overall volume of a patch, creating the effect of a *pfp* dynamic that would be hard to create otherwise. You should experiment with aftertouch and use it to control the parameters that best suit your parts on a case-by-case basis. Table 5.5 lists the most common applications of the aftertouch message for the brass.

Figure 5.21 shows an example of aftertouch assigned to the filter cutoff parameter in Logic EXS-24.

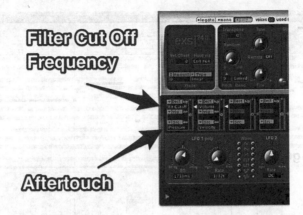

Figure 5.21 Aftertouch assigned to control the filter cutoff of a trombones patch in Logic EXS-24

Another performance controller worth mentioning in this section is soft pedal (CC#67). This controller can be inserted off-line with the pencil tool in the sequencer. If used correctly, it can create some quick change in dynamics often used in *subito piano* passages. While this controller may not be at the top of the list, it can be a useful addition to the arsenal of sequencing weapons.

5.12 Extended Performance Controllers: Attack and Brightness

To add even more realism to virtual brass parts, two other CCs can be particularly helpful in shaping up the color and expression of brass patches: CC#73 (attack) and CC#74 (brightness). Once again, these controllers are set as a default for most hardware synthesizers, but they need to be manually assigned for the majority of the software samplers and synths.

5.12.1 Attack Control

The control over the attack time can be useful when used to differentiate soft passages from loud ones. As for wind instruments, a softer dynamic involves a rounder and less edgy sonority. You can use CC#73 to gently smooth out the attack section of a waveform during **pp** and **p** passages. Use a shorter attack for louder sections (lower controller values) and use a slightly longer attack for a more intimate sound (higher controller values). If you use extremely high values (100 or higher), you can get a good crescendo effect without having to program volume changes. CC#73 can be used as a quick tool to give new life to a patch without having to reprogram it completely. You can use it, for example, to change slightly the attack of every other note in a random way in order to create more variety to a specific part. Use the random shape of the sequencer pencil tool to insert subtle changes in the attack of a patch and it will sound much more real

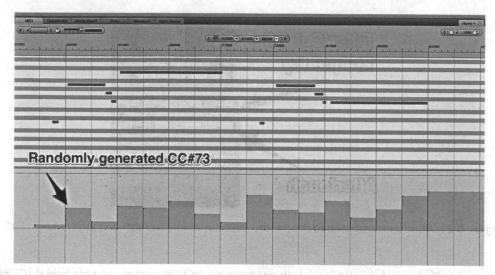

Figure 5.22 Example of CC#73 (attack) inserted with the pencil tool with a random pattern. Each note now features a slightly different attack, giving the part a more realistic effect (Courtesy of MOTU)

and credible than a similar patch that has all the notes with exactly the same attack time (Figure 5.22).

5.12.2 Brightness Control

A similar technique can be used with another important extended performance controller: CC#74 (brightness). As with the woodwind, this CC allows you to control the filter section of a synthesizer, specifically by changing the cutoff frequency of the voltage-controlled filter (VCF). Once again, for wind instruments in general but for brass in particular, being able to change the color of a patch is fundamental to create a realistic rendition. In acoustic brass instruments, softer dynamics generate mellower and darker sonorities. The same effect can be re-created through CC#74. Lower values will translate into a lower cutoff frequency and therefore a darker and more muffled sound, while higher values will generate brighter timbres. You can insert this controller either in real time or off-line after sequencing the part. If you choose the first option, if possible, use a BC to send it, so that a stronger airflow will create a brighter sound and a softer blow will create a darker sonority. Another option is to overdub the data during a second pass, which will usually translate into more relaxed and focused results. If you don't have a BC, the old-fashioned pencil tool is always available, but keep in mind that inserting the data graphically usually takes longer and gives less musical results. The attack and the brightness controllers can be combined for the ultimate virtual brass experience. Listen to Examples 5.11 and 5.12 on the website to compare first a brass part sequenced without the use of these controllers, and then the same part sequenced with the addition of CC#73 and 74 (Score 5.4).

5.13 Advanced Sequencing Techniques for the Brass: Detuning

The ability to re-create, in a virtual environment, the natural sonority of which acoustic brass instruments are capable relies not only on good samples, attack, or brightness, but also on an ability to create enough variation in terms of detuning. Some of the acoustic brass instruments, such as trumpet, French horn, and trombone, can be particularly hard to play in tune for a long period or in extreme registers (mostly the high end). A sample can only give a certain degree of realism, but for better MIDI renditions you will need to go the extra mile and try to reproduce artificially some of the natural detuning nuances. This can be done at two different levels: solo and section. No matter which technique or situation you are going to use for artificial detuning, always remember to apply it musically without overdoing it. Too much detuning can be disastrous for your productions.

5.13.1 Solo Instrument Detuning

For solo parts the technique is fairly straightforward. The goal is to create variation by having some of the notes slightly detuned from the rest. In order to be able to change the detuning over time, assign an open controller (refer to Table 1.2 for a complete list of available controllers) to the fine-tuning parameter of the software sampler or synthesizer. Next, sequence the part as normal. Now use the pencil tool to insert subtle changes in tuning. Use settings that do not go over ± 5 cents in order to keep the parts sounding solid and consistent with the other instruments of the arrangement. In addition, start with very mild changes and increase them further in the piece (usually a more tired performer, especially brass players, will start slightly losing control over the instrument). In addition, generally in brass instruments, very high notes tend to be harder to control and therefore they may sound more out of tune. For this reason, use slightly higher values for notes in the higher register. Listen to Examples 5.13 and 5.14 to compare, respectively, a trumpet solo part sequenced without and with small tuning alterations.

Randomized Pitch parameter controlled by CC#23

Figure 5.23 Example of random detuning controlled by MIDI CC#23 in Kontakt

me libraries provide a detuning option (random or round-robin-based) that can be extremely effective. In this case, you can assign to a MIDI CC the parameter that controls the amount of random detuning applied. This will allow you to accurately control the amount of detuning at any given time.

Be careful when using these automatic pitch variations, since they can quickly turn a professional production into an out-of-tune mess. Values between 4 percent and 8 percent should be used; anything higher than this can result in too much detuning.

5.13.2 Section Detuning

A similar technique can be applied to a full brass section. This will create a much more realistic blend between the single instruments forming the section. This technique is recommended if you plan to use all patches from the same library or the same patch for all the parts of the section. The principle is simple: slightly detune each part of the section by a few cents. The secret is to use this technique so that the overall tuning is centered on a perfectly in-tune part. For example, on a five-trumpet section leave trumpet 1 in tune, and detune trumpet 2 by 16 cents, trumpet 3 by 26 cents, trumpet 4 by 13 cents, and trumpet 5 by 23 cents. This setting will avoid having one part of the lines more out of tune than the other. Detuning settings can vary depending on the quality and accuracy with which the live samples were recorded. Start with more moderate settings and make sure not to overdo it in order to avoid unpleasant results. If you want to bring it a notch higher, try also varying the detuning of each part over time. A ± 1 cent variation in relation to the original starting detuning of each part will be enough to make each part more realistic. To create this variation over time, use the same technique described for the solo parts but, this time, apply it to each part in the section. Try always to create small variations over time and try to increase slightly the amount of detuning with the passing of performance time, as would happen in an acoustic situation.

5.13.3 Velocity-to-Pitch

Another valuable technique for implementing detuning in brass parts is the velocity-to-pitch function. Similar to velocity-to-attack, this method involves the control of the fine-tuning parameter of your library of choice through the MIDI velocity data. Most advanced libraries allow you to assign MIDI velocity to tuning. The main idea here is that the tuning of an instrument can vary depending on the velocity-on value sent from the MIDI controller. As a default, higher values (which means louder notes) trigger a higher detuning, and lower values (softer notes) will feature less or no detuning at all. This technique works well for brass since, in general, higher notes are louder and therefore a little bit more out of tune than lower and softer notes.

This approach has several advantages. It is much faster to implement than using a dedicated CC since the programming time is reduced to a minimum. In addition, it provides a good amount of natural variation since the amount of detuning is not related to a specific note. One problem with this approach is that, usually, louder notes will sound a bit more out of tune than softer notes and in the long run this could defeat the purpose of greater variation.

Listen to Examples 5.15 and 5.16 to compare a French horn part sequenced, respectively, without and with velocity-to-pitch variations. Pay particular attention to how the second example gives the impression of a much more natural and human-like performance.

5.14 Addition of Real Instruments

The addition of one or two real brass instruments blended with the virtual ones can have a big positive impact on a production, as was the case with the string orchestra. By simply substituting the top voice of each brass section of an ensemble with a real player you will achieve a more expressive and realistic effect. Let's take a look at different scenarios where this technique can be particularly useful.

Table 5.6 Different options for enhancing MIDI brass section with the addition of real instruments

Type of ensemble	Configuration	Comments
Small orchestral brass section	Trumpet—real French horn—MIDI Trombone—MIDI (or real) Tuba—MIDI	If possible, always replace the trumpet. If the budget allows, also replace the trombone
Large orchestral brass section	Trumpet 1—real Trumpets 2 and 3—MIDI French horns—MIDI Trombone 1—real Trombones 2 and 3—MIDI Tuba—MIDI	Substitute only the first chair of the trumpets and the trombones section. Alternatively, you can also replace the first French horn
Small pop and jazz ensemble option one	Trumpet—real Alto sax—MIDI Tenor sax—real Trombone—MIDI Baritone—MIDI	This option provides a fatter sound
Small pop and jazz ensemble option two	Trumpet—real Alto sax—real Tenor sax—MIDI Trombone—MIDI Baritone—MIDI	This option provides a brighter and top-heavy sonority
Latin ensemble	Trumpet—real Tenor sax—MIDI Trombone—MIDI (or real)	
Big band	Trumpet 1—real Trumpet 2 through 4—MIDI Alto sax 1—real Alto sax 2, Tenor sax 1 and 2, Baritone sax—MIDI Trombone 1—real Trombone 2 through 4—MIDI	

For orchestral brass sections where you have only one part/virtual player for each instrument (for example, one trumpet, one French Horn, one trombone, and one tuba), you should replace the top voice (trumpet) with a real player. It is recommended not to double the MIDI part since it could create some imbalance in the voicing. If you can afford two real players, use the "sandwich" technique I discussed in the WW chapter: replace the trumpet and the trombone with real players. For larger orchestral brass sections (two or three chairs for each instrument), you can add the real player to the section, but pay always attention to the overall balance of the voicing. The addition of one or two real instruments will add a shiny and edgy sonority to the MIDI parts. If possible (but this is not crucial), try to double also the first trombone, as this will add body and depth to the section. French horns in section do not usually need an acoustic doubling unless they are very exposed. For solo orchestral parts, try to use a real acoustic instrument if possible. It will help bring the entire production to a new level. For small pop and jazz ensembles (four or five horns), always replace the top line with a real instrument if possible. To achieve a more realistic effect and a slightly fatter sound, use the "sandwich" technique by substituting also the tenor saxophone. For a brighter sound, substitute the trumpet and the alto sax. If you are working with a Latin brass section, I recommend substituting the trumpet and the trombone with real instruments. For big band productions, where we are dealing with three larger sections (four trumpets, four trombones, and five saxophones), you should replace the lead instrument of each section (first trumpet, alto saxophone, and first trombone). Look at Table 5.6 for a summary of all the different possibilities for enhancing the brass section with the addition of real instruments.

5.15   The Final Touch: Performance Noises

As has been the case for every section and instrument discussed so far, before starting the mix process there are a few final touches that can be helpful in bringing a production to a higher level. First, double-check that all the most important sequencing aspects have been implemented. Focus especially on aspects related to the realism of the sequenced parts, such as correct range, breathing pauses, correct dynamics, attack and release, and automation. Play back and fine-tune each part, individually first and then as a whole section, making sure that your virtual brass section and soloists are as expressive and as real as possible.

Once everything has been checked, it is time to add the final touch: performance noises. These can be easily added at the end of the sequencing process. They can be either mapped to MIDI keys in a software sampler and inserted as MIDI notes or inserted as audio files on audio tracks. For brass instruments, performance noises are more effective for solo parts than for full sections. Keep the noises as low as possible; they need to be barely heard in a track. Give priority to breathing noises. They need to be strategically placed to fit the rhythm and the natural phrasing of the part. Placing them in the right spot is fairly easy. Remember where you left a little bit of space in between notes to create the natural flow and phrasing of the brass part you just sequenced? Well, that's exactly where you want to put your breathing noises. Other noises, such as mechanical

Table 5.7 A list of brass instruments performance-related noises you should consider to add more realism to your productions

Type of noise	Instrument	Comments
Breathing	Any brass instrument	Place these noises where you left space for breathing in your parts. Try to change noises here and there in order to avoid repetition. Use them especially for solo passages
Valve and slide noise (mechanical noises)	Any brass instrument	While brass are in general more quiet than woodwinds, I recommend using mechanical noises sporadically and in a very subtle way
Mouthpiece	Trumpet and trombones	These noises should also be used seldom. In conjunction with the breathing noises, these can add some extra realism to your brass parts
Page turning	Any brass instrument	Use them with balance and mix them extremely low. They can add a sense of live performance to your productions

noises (e.g. pistons and slides) or page turning, can be placed more sporadically where needed. For a full list of performance noise for brass instruments, refer to Table 5.7. Specific brass performance noises are hard to find in orchestral libraries, especially noises that are repetitive and that will give you enough variety for extended works. It is a good idea to create your own selection of brass performance noises, even if you are not a brass instrumentalist. Breathing noises are easy to create since you don't actually need a brass instrument to record them. The same can be said for page turning noises. For the mouthpiece, if you cannot get hold of a brass instrument such as a trumpet or trombone, you can re-create it by blowing into a bottle neck of an empty bottle and recording the noise of your lips vibrating against the glass. If you record these noises on your own, make sure to use a very sensitive and quiet dynamic microphone with a close positioning. Listen to Examples 5.17 and 5.18 to compare a brass section sequenced without and with performance noises, respectively (Score 5.2).

5.16 Mixing the Brass Section

Since the instruments of the brass family produce sonorities that can be fairly similar in timbre but different in range, the mixing stage for such instruments can present some difficulties. This is particularly true if you are mixing low-register brass such as trombones and tubas. For this reason, a balanced panning and a precise equalization can definitely improve a production. Let's take a look at how these steps can help your virtual brass sections.

5.16.1 Panning the Brass Instruments

Panning brass instruments on the stereo image with accuracy is an important step in the production process. For orchestral renditions, follow the traditional disposition of the classical symphony orchestra, with the French horns to the left (between 10 and 11 o'clock), the trumpets in the center, and the trombones to the right (between 1 and 2 o'clock). The tubas go to the right of the trombones and usually behind the string basses. Figure 5.24 gives a graphic representation of the panning of a brass ensemble inside the symphony orchestra.

If you are writing and sequencing for brass ensemble only, the overall pan settings can be opened up, giving the effect of a wider stereo image and of a closer positioning of the listener in relation to the ensemble. In this case, place the French horns to the left (between 9 and 10 o'clock), the trombones in the center, and the trumpets to the right (between 2 and 3 o'clock). The tuba can be placed in the center too, but you will have to program and apply a slightly different reverb to it in order to place the tuba virtually behind the trombones as shown in Figure 5.25.

The principle applied to the low-end brass instruments (trombones and tuba) is similar to the one used for the string basses when discussing panning for the string orchestra. When dealing with the brass as a separate ensemble, it is recommended that you place the low-end brass in the center of the stereo image in order to keep a more accurate balance.

Another type of brass ensemble that it is important to discuss in terms of panning is the jazz big band. Chapter 4 described how the saxophones are panned in such an ensemble. Now let's take a look at trombones and trumpets (Figure 5.26).

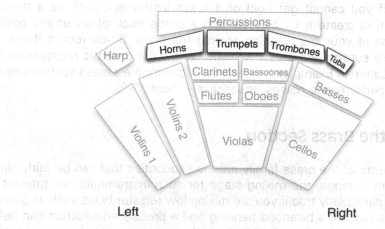

Figure 5.24 Traditional disposition of the brass section within the symphony orchestra

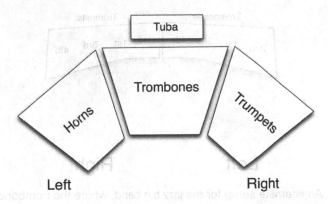

Figure 5.25 Pan settings for a medium-size brass ensemble

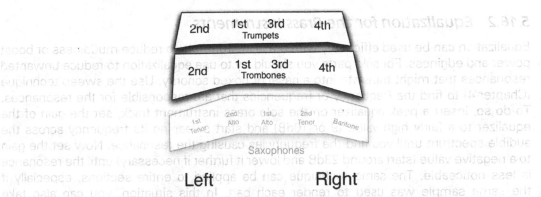

Figure 5.26 Standard panning for brass in the jazz big band layout. Notice how the first trombone and first trumpet are placed inside each section.

If you use these pan settings, make sure to take full advantage of the entire stereo image, spreading the sections completely. If this setup sounds too cluttered, you can use an alternate one that spreads the trombones and the trumpets left and right of the stereo image, respectively, as shown in Figure 5.27.

Notice how the alternate option balances the low frequency of the baritone sax (right) with the low end of the trombones (left) to achieve acoustic balance. Remember that these are starting points for conventional ensembles. Feel free to experiment with less conventional settings. The main thing to keep in mind is to strive for balance and realism in order to achieve the most realistic results.

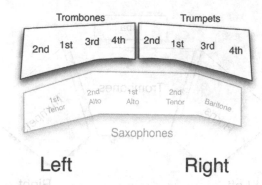

Left　　　　　　　　　Right

Figure 5.27 An alternate setup for the jazz big band, where the trombones have been moved to the left and the trumpets to the right

5.16.2　*Equalization for the Brass Instruments*

Equalization can be used efficiently with brass instruments to reduce muddiness or boost power and edginess. For solo parts you should try to use equalization to reduce unwanted resonances that might translate into a nasal or boxed sonority. Use the sweep technique (Chapter 4) to find the frequency or frequencies that are responsible for the resonances. To do so, insert a peak equalizer on the solo brass instrument track, set the gain of the equalizer to a fairly high value (9 or 10dB) and start sweeping its frequency across the audible spectrum until you find the frequencies causing the resonance. Now set the gain to a negative value (start around 23dB and lower it further if necessary) until the resonance is less noticeable. The same technique can be applied to entire sections, especially if the same sample was used to render each part. In this situation, you can also take advantage of equalization to diversify the sonority of each instrument slightly. Use one or two peak equalizers on each track of the section and boost or cut (by not more than 3dB) different frequencies for each part. You are basically trying to give to each player of the section a slightly different instrument. One could be slightly brighter (usually the first part of the section) or darker. Listen to Examples 5.19 and 5.20 on the website to compare a horn section sequenced, respectively, without and with the use of equalization to differentiate the instruments. Example 5.21 features a horn section sequenced with the use of equalization and two different libraries to separate the instruments (Score 5.1).

Equalization can also be used to make a section fit better inside a full orchestra mix. The principle is basically the same as for the other sections. Each instrument has a frequency range that is specific and characteristic. The goal is to enhance that range by lowering (if necessary) the frequencies that do not belong to it in order to eliminate, or at least reduce, possible conflicts with frequencies that are specific to other sections and instruments. Figure 5.28 highlights the ranges of the fundamental frequencies of the orchestral brass instruments.

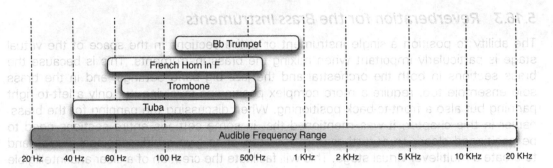

Figure 5.28 Fundamental frequency ranges of the orchestral brass instruments

Brass instruments and sections can be morphed in several different colors and attitudes, depending on the equalization applied. Table 5.8 lists some useful equalization settings for brass instruments.

Table 5.8 The principal frequencies and their applications for the brass instruments

Instrument	Frequencies	Applications
Trumpet	250–350Hz	It controls the bass register of the trumpet. Boost to fatten up the sound; cut to have a lighter sonority
	750–850Hz	It controls the nasal tone of the trumpet. Cut in order to smooth out the sound
	1200–1300Hz	Boost this range to have a more open trumpet sonority
	3500–4000Hz	It controls the transparency of the trumpet. Boost to have a lighter and more ethereal sonority
	7000–8000Hz	Boost to have an edgier and more breathy sound
Trombone and French horn	100–200Hz	It controls the bass register of trombone and French horn
	350–450Hz	Cut to avoid muddiness and the "boomy" effect
	600–700Hz	It controls the nasal range of these instruments. Cut to reduce unwanted nasal boxy resonances
	1300–1400Hz	Boost to get a more open and transparent sound
	2400–2600Hz	It controls the breathiness of the sound
Tuba	80–100Hz	Bass register. Boost to obtain a deeper sonority
	180–250Hz	Cut to reduce muddiness and "boomy" timbres
	300–400Hz	It controls the nasal range of these instruments. Cut to reduce unwanted nasal boxy resonances
	600–700Hz	Boost this range to have a more open sonority
	1100–1200Hz	Boost to have an edgier and more breathy sound

5.16.3   Reverberation for the Brass Instruments

The ability to position a single instrument or entire sections in the space of the virtual stage is particularly important when mixing the brass instruments. This is because the brass sections in both the orchestral and the jazz big band settings, and in the brass solo ensemble too, require a more complex positioning involving not only a left-to-right panning but also a front-to-back positioning. When discussing the panning for the brass, earlier in this chapter, it was mentioned that in some settings entire sections need to be positioned closer to or further from the listener to give a bidimensional illusion and to create a multilevel virtual stage. This will facilitate the creation of a clear and intelligible mix. The secret weapon used to achieve these goals is reverberation. The amount of reverberation applied to a certain instrument or section, in conjunction with volume, allows you to place them precisely in a bidimensional space. Before getting to specific settings for the three main ensembles described earlier in this chapter, let's take a look at the more generic reverb parameters for brass instruments. Depending on the style of music being sequenced, the reverb parameters can change considerably.

For brass solo instruments and sections alike, you should use a good convolution reverb, as was the case for the other orchestral sections. As a starting point, use a slightly brighter reverb for brass than the one used for woodwind. Brass instruments are usually darker and less piercing than woodwinds, and therefore a slightly brighter reverb can help bring out the edginess of the metal of these instruments. Use the equalization section of your reverb of choice to control the color of the reverb. Most reverberations are programmed to damp high frequencies faster, resulting in a darker (and more realistic) reverberation. For the brass, you can work with the equalization section to brighten up the color of the reverb tail just a little bit. The brass will come through the mix better and will add a nice edge to the arrangement. An even brighter reverberation can be used for pop brass sections and solo instruments, and a slightly darker and warmer reverberation for orchestral brass, especially for arrangements that feature low dynamics. In terms of reverberation time, you should be quite conservative for pop and funk sonorities in order to have a punchier sound. Usually times between 1.2 and 1.7 seconds are ideal for punchy and fast passages, especially for trumpet sections. For orchestral brass parts, you can use a longer reverberation time to get a more mellow and round sonority. In this case, you can start with settings between 1.8 and 2.1 seconds and increase them if necessary. A well-balanced reverb can also help you blend the brass sonorities together and achieve a much more cohesive sound.

As mentioned earlier, you can use reverberation to place accurately, on a virtual stage, the different brass sections of a symphony orchestra, a jazz big band, or a brass ensemble. The trick is to use slightly different reverberation times and volume settings to re-create how an instrument would sound if placed on different parts of a stage. Since this technique was described in Chapter 3, this section will focus on taking advantage of built-in reverberation engines that allow you to achieve the same result with much less effort. Some of the most advanced convolution reverbs available on the market have built-in features that allow you to position a sound source on the virtual stage by simply moving an icon (representing the sound source) on a tridimensional plan (the stage). In this category is the convolution reverb MIR-VSL. Its graphic interface is designed to allow you to place precisely in space the instrument or section on which you are working.

In the case of the brass instruments, this feature is particularly handy since their positioning inside the orchestra, the jazz big band, or the ensemble is crucial for a successful and realistic rendition. Figure 5.29 shows how this feature was used to place the brass section of an orchestra according to the traditional positioning of the symphony orchestra seating chart.

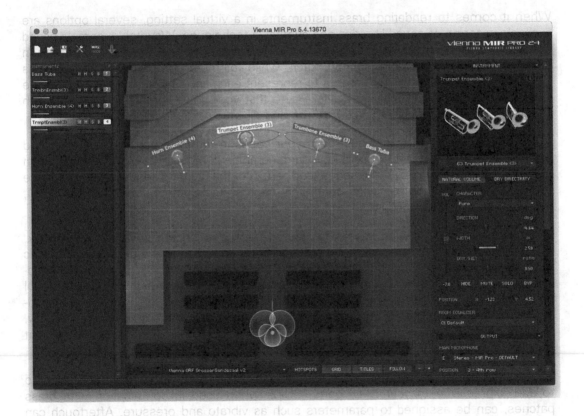

Figure 5.29 Positioning of the brass instruments for the symphony orchestra using MIR convolution reverb engine

These types of reverb can require a substantial amount of CPU power from your computer since they need to be inserted on each single channel of your virtual orchestra in order to take advantage of the individual positioning on the virtual stage.

5.17 Summary

In general, the listener's tolerance for the brass sound is less than for the other color groups. This sensibility should be respected not only for the audience but also for the performers. Brass players require the greatest amount of physical endurance, so it is best not to have them playing constantly. Remember that the brass can also be subtle or lush, and mutes can provide a completely different tone color. Various musical styles can have a significant effect on performance with regard to tone, articulation, and range considerations. Careful listening, transcribing, and score analysis will continue to develop this awareness.

When it comes to rendering brass instruments in a virtual setting, several options are available in terms of sound libraries and synthesis techniques. Wavetable, sample-based and PM are all viable options, depending on the style. For simple and powerful unison section phrases, use a solid section patch, while for more complex section passages (divisi) use individual solo instruments to represent each line. If possible, do not use the same patch for the different lines. Make sure that your brass library allows for enough variation in terms of dynamics (*p*, *mp*, *mf*, and *f*), articulations, colors, and mutes.

Choose a MIDI controller that best suits your needs in terms of performance skills and effectiveness. If you want to use a keyboard, combine it with a BC, which gives a much higher degree of control over the sequenced parts. For the ultimate sequencing experience of brass parts, you can use wind controllers such as the Synthophone, the AKAI EWI, the Yamaha WX, or the MDT.

When sequencing brass parts, always record each part on an individual MIDI track. Having sequenced the parts, take advantage of CC#7 and 11 to shape the overall dynamic of the phrase and the individual attack, sustain, and release of each individual note (micro-level automation). The quantization stage is crucial in achieving a high standard in MIDI rendition. First of all, try to play the parts as accurately as possible. Leave slow (pad-like) passages unquantized as much as possible to retain the natural performance flow. For faster and more rhythmic passages, start with 50 percent strength and –60 percent sensitivity and keep working with the strength if a tighter quantization is required.

Take advantage of performance controllers such as CC#2, aftertouch, and portamento to add expressivity to your MIDI parts. CC#2, when used in combination with brass patches, can be assigned to parameters such as vibrato and pressure. Aftertouch can be assigned to pretty much any parameter of a patch, although it is usually assigned by default to modulation/vibrato. Extended performance controllers such as attack (CC#73) and brightness (CC#74) can be used to manage the different sonorities generated by different dynamic levels. Use a shorter attack and brighter filter for louder sections, and use a slightly longer attack and darker filter for a more intimate sound at lower dynamic levels.

Detuning can be very effectively used to create natural micro-variations in pitch that occur quite often in brass instruments. Remember, though, that too much detuning can be disastrous for productions. For solo passages, use the pencil tool to insert an available open CC assigned to the fine-tune parameter of the synthesizer or sampler. Use settings that do not go over ± 5 cents in order to keep the parts sounding solid and consistent with the other instruments of the arrangement. For entire brass sections, detune each part of the section by a few cents, trying to compensate the 1 and 2 between the different parts so that the overall tuning is centered on a perfectly in-tune part. A similar effect can be achieved using the velocity-to-pitch parameter of the synthesizer or sampler.

The addition of one or two real acoustic instruments will greatly improve a rendition. Double the first part of a brass section to get a punchier and more lively sonority. For pop and jazz bands, where each section is formed by four or five instruments, you should double with real instruments the first and third parts of the section or, alternatively, the first and last (fourth or fifth) parts of the section. For a final touch, the addition of performance noises such as breathing, valve/slide, mouthpiece, and page turning will give a sense of live performance to your productions.

Panning, equalization, and reverberation will have a big impact on the final production. An accurate and balanced pan setting over the virtual stage will improve the mix in terms of intelligibility. Equalization can be used efficiently with brass instruments to reduce muddiness or boost power and edginess. Use equalization to reduce unwanted resonances that might translate into a nasal sonority. The amount of reverberation applied to a certain instrument or section, in conjunction with volume, allows you to place them precisely in a bidimensional space. A slightly brighter reverb should be used for brass than for woodwind. Brass instruments are usually darker and less edgy than woodwind instruments and therefore a slightly brighter reverb can help bring out the edginess of the metal of these instruments. In terms of reverberation time, lengths between 1.2 and 1.7 seconds are ideal for punchy and fast passages, especially for trumpet sections. For orchestral brass parts, you can use a longer reverberation time of between 1.8 and 2.1 seconds and increase it if necessary. If available, take advantage of convolution reverbs that allow you to position your instruments on a virtual stage, such as MIR by VSL or Altiverb by Audio Ease.

5.18 Exercises

Exercise 5.1

Write a slow-moving but bold melody for solo horn. Stay within a range that places the written horn part within the treble staff.

Sequence the above melody for French horn, programming a light horizontal detuning all the way through the piece.

Exercise 5.2

(a) Write a thoughtful and reverent melody for Bb trumpet. Stay within a range that places the written part within the treble staff.

(b) Sequence this melody for trumpet using automation (CC#7 and 11) to control the dynamics, and CC#73 and 74 to control attack and brightness.

Exercise 5.3

(a) Write a comic melodic statement for solo tuba using the register that offers the most flexibility and control.
(b) Sequence the part using a breath controller (if available) for the dynamics by sending control CC#11. Add a slight horizontal detuning and performance noises.

Exercise 5.4

(a) Write a two-voice chorale in homophonic texture for a pair of horns. Have the piece performed by four horns (two per voice) as well to appreciate the sonic difference.
(b) Sequence the piece for two horns with solo patches (on separate MIDI tracks) using the following techniques:

- vertical detuning;
- horizontal detuning;
- use one acoustic horn and blend it with the MIDI horns;
- using the up/down technique to control the attack and sustain of each part.

(c) Now repeat the previous steps but, this time, use French horn section patches and notice the differences between the two ensemble sounds.

Exercise 5.5

(a) Take a simple four-part chorale and transcribe it for four horns. (Try to use one whose bass line is not too low in range.) Remember, also, to have horns 1 and 3 play the notes in the chorale's treble staff and horns 2 and 4 play the notes within the bass staff. Then combine the notes for horns 1 and 2 on a single part and similarly for horns 3 and 4.
(b) Sequence the chorale using the following techniques:

- detuning (vertical or horizontal or a combination of both);
- using one acoustic horn and blend it with the MIDI horns;
- using the up/down technique to control the attack and sustain of each part;
- using reverb, equalization, and pan to mix the final sequence.

Exercise 5.6

(a) Using the music from Figure 5.12, create a sequence emulating a big band brass section. If possible, use a real instrument for the top trumpet line and for the top trombone line. Make sure to pay particular attention to blending the acoustic and the MIDI instrument smoothly together through equalization and reverb.

(b) Now compose and sequence a series of "block" voicings (close position) for big band brass (four trumpets and four trombones). Keep the lead trumpet part above the staff for a "shout" effect. If the first trombone part gets too high, restructure the trombone section voicing.

Exercise 5.7

(a) Write some three-voice harmonic structures in an R&B style for B♭ trumpet, sax (alto or tenor), and trombone.
(b) Sequence the progression using advanced quantization technique to get a nice and natural groove out of the horn section.

Exercise 5.8

(a) Write a melody in unison for two B♭ trumpets in mutes (one with a Harmon and the other with a cup), along with a flute and B♭ clarinet.
(b) Sequence the melody using all the techniques learned in Chapters 4 and 5. Use one acoustic instrument to smooth out the MIDI instruments. Pay particular attention to dynamics, attack, sustain, and release, phrasing, and mix.

Exercise 5.9

Orchestrate a major triad at a loud dynamic level for one horn, two B♭ trumpets, and two trombones. Also add the following woodwind instruments: flute, oboe, clarinet, and bassoon, to sound as balanced as possible.

6 Writing and Sequencing for Vocal Ensembles

6.1 Introduction: General Characteristics

Most commercial composers, arrangers, or orchestrators will, at some point in their career, have to write for singers. Unlike instruments, they are much harder to replace with digital samples. If necessary, the easiest context is when all that is needed are some ambient voices singing on syllables such as "ooh" or "ah." Today there are certainly many choir samples that can produce this effect. But, when lyrics are to be delivered in a convincing and expressive way, singers become essential.

When vocalists must be hired, there may be several contexts to consider: the soloist, the duo, the trio, the quartet, or the choir. Theatrical productions often include all of the aforementioned contexts. Sometimes there may be previously written vocal arrangements. In this case, the instrumental arranger can proceed in a more normal fashion. But when any vocal music (mixed with instrumental) must be written, it is paramount that the music writer address the vocalists first! The range of most instruments is clearly defined. Within the range, the quality and characteristics of individual players may require slight adjustments. But with vocalists this issue becomes murkier because the voice is a human instrument—not a mechanical one. Its range, tone, and technical ability are unique to each person.

Whenever there is a vocal soloist, the arranger must first determine the best key(s) and consider the singer's tone, power, and style. This becomes more complex when there are two soloists to accommodate. When there are three or four soloists, the ensemble of vocalists becomes a hybrid of solo, duo, trio, and quartet contexts. In this setting, it is best to avoid writing music that requires an extensive range. In general, the smaller the range of the melodic material the easier it is to accommodate several singers to sing the same musical role. This happens in theater productions where the music is written for the original actors/singers. After the initial run or tour, the producers and director will want to keep the show running into the indefinite future. There may even be simultaneous productions of the same show. So the music writer must assume that several different singers will eventually perform the same vocal part(s).

Another variable with singers is that their range changes over the course of their lifetime. Male singers who start their acting/singing career as young boys will, several years later, no longer be able to sing that particular part (or role). The onset of puberty creates a change in their register and ultimately their range. On the other end of the spectrum, women will also experience a drop in their range as they age. This can occur in men, too. It is common for a solo singer who has had arrangements created in mid-life (20s–30s–40s) to eventually, in their senior years, need an arranger to adapt the existing arrangements by lowering them to a more comfortable key. Because most singers experience range adjustments during the course of their lifetime, it is so important for the arranger to ask the vocalist about the key at that moment in their life.

Singers are not necessarily trained musicians. Although many vocalists are, the great majority—especially in the theater and entertainment worlds—are primarily actors who must learn to sing and dance. They may have significant notoriety and an incredible amount of natural ability but little or no formal musical training. Therefore, it is wise for the arranger to anticipate the latter scenario. They should assume that the singer does not read music or know the names of the piano keys and would certainly not have a clue about finding an appropriate key. The arranger must be prepared to help the singer find the right key by sitting at the piano and running through the piece. If the singer is asking for a composition, the writer might listen to some recordings of the singer to determine the individual's range and then begin to compose. But be careful when listening to recordings—particularly with singers who have had a 20+ year career. Today it is very easy to find YouTube clips of prominent artists, but remember that the singer's key at that time may not be ideal for the present time. Always confirm it with the singer.

Interpretation by the solo vocalist is another consideration. The lead sheet for a song is written deliberately in a generic rhythmic fashion. It is assumed that the singer will interpret the song in various ways—stylistically, tempo, groove, etc. Many times a singer will "lay back" or "anticipate," which basically means that the written music is manipulated in a more conversational fashion. This is important for the arranger to consider, as the "holes" between the phrases can change. It is customary for the arranger to write counterpoint or instrumental activity of any kind in between the vocalist's phrases. Unless the vocalist stays within the prescribed rhythms as stated on the printed music, the arranger's ideas may collide with the singer's manipulation of the phrasing. To anticipate the problem, it is a good idea for the arranger to get a recording of the singer interpreting the song and then work around that.

6.2 Vocal Timbre—Finding the Appropriate Singer(s)

Each vocalist has a unique "instrument" with regard to tone and ability. But there are certainly general distinctions that can be made.

Classically trained singers: Develop a tone that transcends their natural ability to produce sound. Their teachers develop the singer's natural anatomical instrument to expand the tone to a point where it can fill a theater without the use of microphones. This type of singing is required in all operas and most classical choral works. In most cases, the tone is exclusive to these styles and does not cross over into the popular styles of singing.

Popular singers: Have opposite tonal qualities from the classically trained singer. A significant minority have musical training but many or most are strictly natural talents. Their tone production is naturally produced and most, if not all, need the aid of a microphone to help boost their sound. In some cases, particularly in more intimate jazz and cabaret styles, the singer often sings with a whisper tone. This contrast is similar to actors working in television versus theater. A television camera (and close microphones) can foster a much more intimate setting for an actor.

Theater singers: Embody the widest spectrum of tone. Although the vocalists in all musicals today wear small contact microphones, the style of the genre still requires singers with "big" voices. Depending on the style of the production, classical singers may be required or popular singers may be preferred. In some cases, vocalists may need to have hybrid qualities. There could even be a mix of singers—some with music training, some without. Sometimes the role (a domineering father, an innocent female teenager) may suggest whether the singer should have a classical tone or a more natural tone.

Studio singers: Performers in this genre need to have strong musicianship skills. They are the ones hired on recording sessions for radio and TV commercials, for film and video soundtracks, and sometimes as background singers for theater productions. They are excellent sight-readers who are capable of adapting to the musical situation by manipulating their voice with regard to style (except for the classical/operatic tone) and can sing with no vibrato when requested. They are great at creating a homogeneous blend with other singers and they are experienced at doubling their voice (via overdubbing) in unison or adding harmony parts. In simple styles (pop, country, etc.,), many can actually create harmony parts and sing them in the moment.

Stylistic singers: Unlike malleable studio singers, most other singers have a specific style that provides a special character: the opera singer, the country singer, the heavy metal singer, the jazz singer, the cabaret singer. Racial background can also be a factor. The character of American Gospel music would require the sound of the African-American voice—not only in timbre but also in musical interpretation. But this distinction is minimized with operatic training. As an experiment, listen to two female opera singers (one black, one white). Compare the minimal contrast to two female popular singers of these two racial backgrounds. I think you'll notice that the contrast is more striking in the popular style. Ultimately, in this category, finding the appropriate singer for your project is as unique as the director who must cast a play or a movie with just the right characters.

6.2.1 Vocal Timbre—"the Break"

With regard to range, there is an area of the voice that is known as "the break." Similar to the throat tones on the clarinet, the notes in and around "the break" are usually weaker than the greater surrounding notes. For singers it can also be tiring, so it is wise to not linger there. This spot is most commonly found around Bb or B (middle of the treble staff) in female singers, but it is best for the arranger to interview solo singers whenever possible with regard to their individual characteristics and abilities. As with all wind instruments, any vocalist needs periods of rest. Sustaining notes (especially higher ones) can be very tiring.

6.2.2 Vocal Timbre—"Chest" versus "Head" Voice

Below "the break" the vocal tone is strong and clear. This is called the "chest voice." Above "the break" it becomes harder to remain in "chest voice," so the singer usually converts to "head voice"—also known as "falsetto" in the classical genre. The tone in this part of the range is noticeably weaker, so all singers strive to develop and strengthen this area to create a more even tone throughout their entire range. That said, "falsetto" is appropriate for certain musical effects and styles. In fact, it is quite possible to bring the "falsetto" tone into the lower range. The sound is wispy and delicate. In general, instead of being naturally limited to either "chest" or "head" tones, the best singers develop both timbres globally throughout their range and utilize the choice as an option for greater musical expression.

6.2.3 Vocal Timbre—Vibrato versus Non-vibrato

As previously mentioned, the studio singer is the one who most likely has the ability to control vibrato or omit it completely. This is a valuable asset as some styles of music require the omission of vibrato—also known as singing with a "straight tone." There are some natural singers who can sing with no vibrato (and sometimes have difficulty trying to muster a nice vibrato), but many trained singers have a built-in vibrato and sometimes do not have the ability to control or omit it without noticeable difficulty. The need for "straight tone" singing becomes apparent when multiple voices are in harmony—especially more complex harmony. The "wobble" of a wide vibrato is simply too thick to create a clear and precise sound of a specific chord tone. Most vocal jazz ensembles and "jingle" singers (ones who sing for radio commercials) avoid vibrato. The singing of music from the Renaissance period also requires a suspension of vibrato in favor of a leaner and more pristine tone; within the murky environment of a cathedral this sound is much clearer.

6.3 Reading Music—Capturing Sound

For any aspiring musician who wants to become a professional, reading music is required and eventually mastered. But there are varying difficulties in accomplishing this task.

Pianists do not have to hear the notes before playing them. As long as their fingers hit the proper key, the correct note is heard. Similar to pianists, woodwind players press certain keys to capture particular pitches but must also be aware of intonation and register. The brass player's job is a bit harder. The valve combination or slide position does not guarantee that the specific pitch will be heard. The player must first hear the note and adjust the embouchure in conjunction with the mechanics of the instrument to eventually broadcast the correct pitch. With string players, this task becomes even more precarious. Although there are "positions" on each string, they are not labeled or felt (as frets would be on a guitar). The player must hear the pitch and simultaneously approximate the location on the fingerboard to create the proper pitch with good intonation. The vocalist's job is even more difficult than the string player's. He or she must hear the pitch out of thin air and then sing it. This is difficult enough as a soloist. Imagine how difficult this becomes in the context of a full choir where there are several simultaneous parts! Singing "a cappella" (voices alone) is much harder to accomplish than singing with instrumental accompaniment. The singers must retain their individual pitches and those pitches must be in tune with the pitches emanating from the other singers. This is very difficult to do! When writing for voices you must have a keen awareness of this process and write the music in such a way that makes each part as "singable" as possible (more on this later in the chapter).

6.3.1 Preparing the Vocalist's Entrance

Unlike writing for instrumentalists, the vocalist usually needs a reference pitch from the instrumental ensemble. This makes introductions mandatory; they establish the tonal center (and key) and help the singer find the first note. There are other critical points in an arrangement as well: for any modulations and after any instrumental interludes.

6.3.2 Accompanying the Singer

After addressing the primary function of determining the key and broadcasting the vocalist's entrance note, the arranger can begin to think about the process of creating a suitable accompaniment. The first thing to realize is that the singer is delivering a message through the lyrics that most, if not all, in the audience will understand. There is a story line that must be respected. So the first job of the arranger is to endorse the meaning and mood of the lyrics. Next, the arranger must be aware of the specific aspects of the singer—the tone, strength, and style—and strive to match and support them. Some singers have large voices; other have small voices. The orchestration must not overpower the singer. Although there are microphones (in popular music contexts) that allow an arranger to write dramatic and large orchestrations, the mics will not help the audience hear the words if the orchestration is too active during the singer's phrases. It is important to preserve a "sonic window" for the singer and have the instrumentalists work around the singer's phrases. Doing this also creates less competition for the singer with regard to finding the right pitches.

6.4 Choral Writing

The standard choir consists of what is known as an SATB format. The letters represent the soprano, alto, tenor, and bass voices. The letter B also refers to the voice of the baritone, whose range falls between the basses and tenors. Particularly in public school choirs, the range of the B part is actually closer to a baritone than a bass. This basic four-part format is easily expandable (with full choirs) by incorporating divisi parts within each section. Sometimes a score is expanded to additional lines for easier reading. For example, SSATBB would represent first and second sopranos, altos, tenors, baritones, and basses.

There is no prescribed number of personnel in a choir, although most of them typically have anywhere from 20 to 60 or more. Sometimes it is desirable to have a large number for epic works and a much smaller number for more of a chamber sound. For studio recordings, a small group of singers (four to eight) may be hired and then overdubbed to create a larger vocal ensemble sound. But to create a 30-voice choir in the studio you will need at least 10—12 singers to create a homogenous sound. Otherwise, even with overdubs, the unique timbre of the individual singers will remain.

6.4 The Vocal Score

Unlike instrumentalists, singers in any vocal ensemble read a full vocal score. In other words, each vocal part (SATB) has its own stave and each stave is placed in score order (from soprano on the top line downward to the baritone or bass part). If there is a piano part, the grand staff for the piano is placed below the vocal staves. It is used either for accompaniment in performance, or for rehearsal only. Even if the choral work is for an a cappella choir, it is a good thing to create a piano reduction if possible. The reduction contains all of the vocal parts inside the grand staff for the pianist to play. Sometimes there are isolated places where the parts do not conform to the hands of the pianist but the notes nonetheless are there for the pianist to do what is possible to help the choir hear the notes. Many times, the piano part may have an indication that says: for rehearsal only. Then, in the concert, the choir director may use a pitch pipe or play the opening notes on the piano for the choir's starting pitches. After that, the singers must now capture their own part while merging with the other vocal parts—all without the aid of the piano. Score 6.1 is an example of a traditional piece for choir. The score contains all four SATB parts, along with the grand staff reduction for the pianist to play. The pianist and each singer will read this score.

As mentioned earlier, singers need some way to find their pitches. A full vocal score enables each singer to see all the vocal parts. Many times, a singer can find an important pitch located elsewhere in another vocal part. The awareness of this can alert a singer, who then has a better chance of hearing that note as well. The piano reduction is also helpful to see the relationship of the pitches. For example, an alto reading her line and a tenor reading his line may not notice that their parts will rub at some point to create a tension in the music. But viewing the parts in the context of piano music may clearly show the location of both parts in close proximity. This prepares the singers mentally to expect some dissonance (or at least tension) to occur. Examples of this can be found in Score 6.1. The first location is in bar 8 on the word "head," where the soprano and alto voices sing an F and an E♭ respectively. Another rub with the sopranos and altos occurs in bar 16 on the word "hard." Several more occur on page three of the score.

With regard to clefs, the female parts (soprano and alto) always read in treble clef. The tenors usually read in treble clef but there is a small number 8 placed beneath the clef. This indicates that there is a register transposition. In other words, the written notes are actually written an octave higher from where the voice sounds. The baritone and bass singers always read in bass clef. Some publishers (for choral arrangements in the pop and jazz styles) merge the sopranos and altos onto one stave, with the tenors and basses on another staff. This is less than ideal as it forces the tenors to read in bass clef, which may be less familiar to them. Some publishers use this format solely for economic purposes. This writer advises against that practice unless a publisher forces the issue. It is cumbersome with regard to lyric placement and it limits the writing of divisi parts and polyphonic textures between the vocal parts.

Lyrics are always placed below the written music. Many times, because the lyrics below may be too congestive, dynamics and other musical indications are placed above the staff. Polysyllabic words require a hyphen between each syllable (notice the first word, "La-dy," in Score 6.1). Usually there is one pitch for each syllable, but quite often there can be two or more notes sung on one syllable. Where this occurs, the arranger must write a phrase marking above those specific notes. Similar to string parts where the phrase marking indicates the notes for one bow stroke, this alerts the singer that the notes within the phrase marking (known as a "melisma") correspond to one syllable. This also keeps the typing of the lyric intact. It is easy to spot a novice vocal arranger who perhaps wants several notes on a monosyllable word (let's say "though") and makes it look like "th-o-u-gh" underneath the set of notes. When done properly, the word "though" appears intact and then a solid line to the right of the word extends underneath the subsequent pitches. Score 6.1 shows a melisma occurring in the alto part in bar 16 on the word "hard." In bar 18, all parts are melismatic except for the soprano voice. In bar 20, the inner voices sing two notes on the single syllable "wed." Most digital notation programs will automatically add this line as you type the lyrics underneath the music.

6.5 Assigning Pitches

Now that the basics have been discussed, it's time to address the compositional process of assigning pitch. There are several aspects that should be respected:

- Create the melodic material in a way that captures the mood of the text (lyrics).
- Stay within a comfortable range for each voice as much as possible.
- Choose harmony that supports the mood but make each part as singable as possible.

6.5.1 Setting Text with Pitch and Rhythm (Creating a Melody)

Remember that, with vocal music, the music itself should be designed to accompany the text (lyrics). This means that the rhythm should be written so that the lyrics flow as naturally as if they were spoken. There will be pauses in mid-sentence but those pauses should be placed so that the smaller phrases or ideas are not disrupted. The following phrase is presented with bold type to suggest the natural accents that occur within the English language: "**La**-dy **mine**, most **fair** thou **art** with youth's **gold** and **white** and **red**." Score 6.1 shows how this phrase is presented in a musical context. (Music will, to a certain extent, force a more rigid presentation of the language because of its stricter tempo, but the overall sense of the accents and general flow of the language should be preserved.) Notice that "Lady" in this sentence is an anacrusis (a pick-up to the stronger syllable "mine"). How different the rhythm (and pitches, possibly) might have been if the lyric were "My lady." Notice the comma after "Lady mine." Any reader of this poetry would pause and then proceed with "most fair thou art." To capture the natural flow of the poetry, the music lingers on a half note followed by an eighth rest for a quick breath. If you say the last part of this phrase, you will notice that the duration of "gold," "white," and "red" are more significant, while the word "and" is passed through swiftly. This natural rhythm is captured with the longer dotted quarters for the more important "colors," while the quicker eighth notes represent the conjunction "and." A reader of this poetry may link more closely the second and third sub-phrases; but, in a musical context, lingering on the word "art" provides a nicer shape for the melodic line with a more graceful climb to the octave by stopping midway on the fifth degree (B♭) of the scale.

6.5.2 Set the Melody in a Comfortable Range

As mentioned earlier, every singer is unique but there is a general range (particularly within the choir) that every composer can assume will work. The melody is always the prime element for consideration. The smaller the range of the melody, the easier the task of arranging becomes. The melody in Score 6.1 is found solely in the soprano voice. The range of the melody extends from the C below the treble staff up to the E♭ in the top space. For soprano voices in a traditional choral setting, this is a good practical range. Once the melody has been created and assigned, it is time to add the harmony parts. These parts will usually fall in their respective comfortable registers too, but when the soprano voice ascends significantly it is wise to keep the voices in a more spread context. Please refer to the piano reduction in Score 6.1. Notice in the beginning the close proximity of the female voices. In bar 3, as the soprano leaps up to the octave, notice how the interval increases between the female voices. In bar 13, the gap is noticeably larger. The sound is gratifying as the lower three voices provide a lush background for the soaring soprano voices. (The specific ranges of each of the SATB voices are discussed later in this chapter.)

6.5.3 Writing Interesting Music with "Singable" Harmony

This aspect is perhaps the hardest to do. The composer should first choose harmony that portrays the emotion in the lyric. Then it is best to create the voice parts not only to capture the sound of the harmony but to also make the parts as easy as possible. Remember that the harmony parts that you write will actually become individual "melodic lines" for each singer. A keen awareness of this (extracting each line and singing it) will provide the proper perspective. To garner the greatest empathy for the difficult task that singers have in capturing pitch, take Score 6.1 and sequence it with each voice on its own track. Males can then try muting the tenor and bass voices and possibly the alto voice (one at a time) and sing those respective parts in a cappella fashion with the synthesized voices. Females can follow a similar procedure with the soprano and alto voices and possibly the tenor voice. After singing each vocal part, proceed to the next set of examples in Figure 6.1. You should notice that the easiest of the four is what most closely represents the vocal parts from Score 6.1. But hopefully the overall effect of the music doesn't sound as simple as it is to sing!

As you can see and hear, repeated notes work very well in vocal music. They are the easiest to sing and, unlike instrumental music, the possible dullness of many repeated notes is balanced with the interest from the lyrics. When a line must begin to move it is easier for the singer when the notes are diatonic. This is actually more important than distance. Figure 6.1 illustrates this point with four levels of difficulty, from easiest to hardest.

Try singing these examples yourself but *don't* play the examples on the piano first. Play only the first pitch in each exercise for a starting reference point. Then hear the succeeding notes in your head before you sing them. See if you can sing the entire phrase and then play the entire phrase on the piano. If it matches what you sang, then you're a good sight-singer. If you were unsuccessful, you now have a new appreciation for how difficult the task is for the singer. If you had difficulty with the entire phrase, try one note at a time and then play the note on piano. Ultimately, I think you will experience that example 2 is not as difficult as example 3, even though there are wider leaps in the second example. This is because all of the notes are diatonic in the key of C major. The excessively chromatic third example (which is still in C major!) with its smaller leaps is definitely harder to hear. Of course, music must become chromatic at times. In those places, it is best to resolve the chromatic note stepwise to the nearest diatonic neighbor tone unless a series of chromatic tones outlines a chromatic harmony. This occurs with the first two notes—E♭ and A♭ outline an A♭ chord; G and F♯ could suggest either Dsus4 to D major or a G chord; the F and D♯ suggest a G7+5 with the E sounding as the third of the C major chord.

Figure 6.1 Sight-singing examples

Observe a similar process in Score 6.1. The first chromatic note appears in the alto voice in bar 9. The B natural (existing inside the G dominant chord) serves as the leading tone to the relative key of C minor. The B pulls upward to the C as it should. Later, in bar 21, the altos have a series of C♭s. Here the chromatic note is derived from the parallel key of E♭ minor. It is heard as the sixth degree of that scale and it resolves as it should—down to B♭. So, you see, spelling is important.

6.6 The General Ranges of the Human Voice

The ranges of the human voice can be approximated in the traditional choir. Ideally, it is best to stay within the practical range as much as possible. Ultimately, it would be wise to look at a variety of choral music to confirm the general ranges as suggested below in Figure 6.2.

Within each range, the terms "low," "ideal," and "high" suggest the characterizations of viable performance. The low notes will sound dark but confident. The high notes are also manageable but should be respected and not overused. The extreme notes are the ones to write very cautiously. They are possible but this depends on the specific abilities of the performers. Age is also a factor. A young school choir's alto section may not be able to reach those extremely low notes. Elderly sopranos in a church choir (who are usually non-professional volunteers) may not be able to sing the extreme high notes.

Much like the instruments of the string orchestra, you will notice a significant overlap between the sections. Although every professional singer usually has a two-octave range, the extreme ends of the range are less flexible or limited to certain musical contexts. Similar to instrumentalists, they are capable of broader expression as a soloist and prefer to have the parameters limited when in an ensemble context. Notice that the general

Figure 6.2 General ranges of the SATBB voices

ranges for each section as indicated in Figure 6.2 only span approximately one and a half octaves. Staying within these more limited parameters will guarantee viable performances by choral singers.

Score 6.2 is written for a 30-second "jingle." It is set up much more informally and is basically a lead sheet that is designed for flexibility (Audio Example 6.1). Studio singers are extremely versatile and many decisions (by the producer, writer, or even the singers) are often made in the moment. Notice that there is no specific designation for SATB. The indications are for "gals" and "guys." There is a main melody and one spot for background vocals. The melody is sung mostly without harmony except for the last few bars. The range of the melody is a major ninth. Its location (between C below the treble staff up to the D line) suggests that a soprano voice would be more appropriate. A mezzo soprano (halfway between soprano and alto) would also work quite well. Of course, the style of singing for the melody must be non-classical, in chest voice, and with a straight tone. The guys sing in their natural register down an octave. The range is suitable for tenor or baritone voice. As the piece enters the repeated section, there is a color change where only the guys sing. The sopranos answer with the words "sun dried" in harmony. Notice the indication for "head voice." This effect creates a light, angelic tone that suggests the bright rays of the sun. (Although this part could be performed together with the main vocal part, it would more likely be recorded as an overdub for ease of performance and better control of the mix.) Then all the singers sing the "hook." In the final section, the vocal ensemble splits momentarily into four-part harmony in bar 11. Here is where experimentation may begin. The parts could be split evenly, with the upper two notes being sung by women and the lower two notes sung by men. Or, for a lighter sound, the upper three notes may be sung by the women. But this latter option would not be ideal for the harmony in the last bar. Since the A♭ below the staff becomes rather dark for the female voice, it would not support the lively mood. The best option here would be an even split between the women and men. For this scenario, at least four singers should be hired (two women, two men) to get a "section" sound in each timbre. Overdubbing for a thicker sound would be optional.

6.7 Complex Texture

Many choral pieces have added dimension within the standard four-part writing. Compared to a string quartet or sax quartet, there is the additional aspect of the lyrics to be considered. Sustained parts must have a syllable of some kind for the vocalists to sing on. Furthermore, melodic fragments must use sentence fragments that make sense when isolated. Score 6.3 shows some ways to achieve this effect. During the first phrase, the choir begins in the standard homophonic format. The basses in their deep register add darkness (suggesting night); the choir moving as one suggests the aspect of being "bound." Notice the very important addition of the apostrophe at the end of this phrase; the marking encourages the choir to take a moment longer to catch a breath for the next phrase but it also allows the listener to process the words along with the music. Especially in an a cappella format, the tempo automatically becomes more malleable to allow moments for easier breathing and the greater expression of the lyric. During the second phrase the texture changes considerably. The lower three voices sing on the syllable "ooh," which creates a smooth and lush pad, while the sopranos sing the melody. The soprano timbre suggests a contrast of light. The pads are soothing and pleasing (supporting the word "fond"). Notice the register change in the basses during the first part of this phrase. Their upper register also provides a more floating quality. The altos and tenors echo the words "bring the light," which provides a sense of transport into the latter half of the phrase, where the ensemble broadens into a much wider homophonic texture (suggesting "the other days around me"). The sopranos continue with the next phrase, but notice the quarter rest in the other parts. The solitary soprano note is striking and the re-emergence of the pad texture immediately following is even more pleasing. Notice the register of the basses (baritones) as they ascend into a very high range. They will need to (and should) transition into a falsetto timbre during this phrase. Also notice how the homophonic texture gradually thins to only two notes (E♭ and G) on the word "years." The isolated major third and the wistful falsetto of the men's voices support the sense of childhood and innocence. The next phrase returns to homophonic texture. The final phrase begins with an inversion of the earlier texture: now the basses take the melody while the other voices sing pads. During the final two bars, the sopranos initially claim the melody with the altos in thirds (portraying "cheerful hearts") but the basses claim the melody's final notes with the tenors in harmony. Their dark timbre portrays the "broken heart."

Complex texture is always more difficult for the pianist to perform. In this example everything is reduced exactly, with the exception of bar 4. At that point, the first three notes of the phrase correspond exactly to the vocal parts but the next three notes change slightly. The notes in the tenor voice needed to be moved up an octave to accommodate the limited stretch of the pianist's hand.

6.8 Writing for Vocal Jazz Ensemble

The vocal jazz ensemble is somewhat more complex than writing for traditional choir because the aspects of traditional writing and jazz writing are combined. Traditionally, the SATB format is retained and in some cases expanded to more parts. The piano grand staff is present and sometimes a bass and drum part is added. The singers read the full score while the same rules apply with regard to setting text properly and comfortably. The main differences are with rhythm, harmony, and style. Because of limited space, the reader may want to review the brass chapter for a reminder about jazz articulation and stylistic markings. They are useful in vocal jazz writing.

The big challenge is to write singable parts when using complex harmony. Most vocal jazz writing typically has extended harmony and excessive chromaticism. Unless the singers have some experience with the jazz style, they may have considerable difficulty hearing the voicings—even when good voice-leading is present. The rhythmic concept can also be difficult. Figure 6.3 shows some typical jazz harmony in four parts that expand to five. The swing rhythms and their articulation are emulated with special "words," known in jazz

Figure 6.3 The vocal jazz ensemble

as "scat" syllables. These syllables are critical to the sound and feel of the music. The example is designed to represent the vocal group emulating the sound of a big band. The voicings are very similar to how they would be arranged for a jazz horn section (much more on this in the next chapter). This exercise would be good to sequence and then try muting the individual parts and singing them with the sequence. You may find once again that the parts, although more difficult than the previous traditional choral writing, are still rather simple within the more complex jazz harmony. One final comment about range: the soprano voices ascend to a fairly high range. In this context they are emulating a lead trumpet sound—they would most likely use a powerful falsetto on the top note (E) of the phrase. When they are singing a lyric, it is usually best to work in a slightly lower range so they can remain in the more appropriate timbre of the chest voice.

6.9   Syllables for Background Texture

As seen earlier with Score 6.3, complex texture is often employed for greater dimension by creating various backgrounds within the vocal ensemble. Figure 6.4 shows the most common syllables and their musical effect. The examples in the first three bars are primarily used as "pads" to create a smooth background for a theme. They are found in both traditional and jazz styles. The rhythms in the last bar emulate a drum pattern in a traditional military style. The correlation of this last example to the "scat" syllables used in the jazz style is worth noting. In both scenarios, fabricated "words" are designed to emulate the rhythm and shape of sound to create greater musical expression.

The syllables in the first three bars have been organized in a deliberate way to create the effect of a crescendo. If you say them aloud, you will notice that "mm" is most

mm   ooh   looh   ah   bah   dum   da dum dee dah

Figure 6.4 Syllables for background texture

quiet. This is analogous to placing a mute over the bell of a brass instrument. The vocal cavity is closed and the sound resonates completely inside. The syllable "ooh" forces the mouth to open narrowly while the sound remains in the back of the throat. The syllable "looh" forces the tongue to engage, which thrusts the sound forward to create a subtle increase in volume as well as a light and more defined attack. The syllable "ah" forces the mouth to open wider and thus the sound increases and projects further. The syllable "bah" provides greater force and articulation. Choose these syllables carefully to create the appropriate texture with regard for dynamics and articulation. You can remain on only one syllable or mix them as you wish to create varying textures. Ultimately, you can use your imagination to invent another syllable that may better represent the musical aspect you wish to portray.

6.10   Sequencing and Production Techniques for Vocals

As we learned so far, adding vocals to an arrangement is a powerful way to increase the color palette that is available to you, and, of course, it is a necessity in the case of a song with lyrics. When it comes to sequencing vocal parts, though, the tools that we have available are unfortunately more limited than for any other instrument that we have analyzed so far. While the artificial re-creation of the vocal sound has been studied, researched, and even brought to market for many years, the complexity and flexibility of the vocal cords make it particularly hard for a synthesizer or sampler to re-create the expressivity of the human voice. Nevertheless, there are several tools and techniques that we can use to effectively incorporate vocal parts in our production. Let's take a look at some of these exciting possibilities!

6.11   Vocal Sample Libraries

One of the simplest and most direct ways to integrate vocal tracks in our sequences and productions is to use prerecorded sample libraries. The advantage of this solution is that vocal sample libraries are quick to implement and easy to use. The drawbacks, on the other end, are several. These libraries tend to be much less flexible and adaptable; this is true particularly for parts that are in the foreground. If you are planning to rely on a sample library for a lead vocal track, it is important that you manage your expectations in terms of what you can achieve. Key, tempo, and available phrases pose a big limitation to what you can do with phrase-based vocal sample libraries. Usually, these libraries have samples of different phrases assigned to different key triggers that you can combine and re-order in any way you want (Figure 6.5).

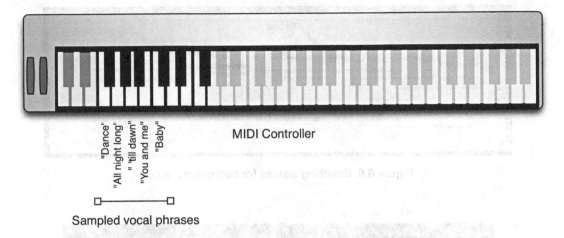

Figure 6.5 Example of a phrase-based vocal sample library

The techniques used to integrate these types of libraries in your production are very simple. Just sequence the MIDI key triggers on a MIDI track that outputs its data to the vocal sample library of choice. Usually, these types of libraries work well for beat-based productions such as electronic dance music (EDM) or ethnic parts. In Audio Example 6.2, you can listen to how a phrase-based vocal library can be effectively used to create a basic vocal lead track for an ethnic piece.

In this case, we need to pay particular attention to have the tempo and key of the song matching the key and the tempo of the samples. This can be tricky and can often limit what we can write. I recommend using this technique mainly for demos or for pieces that require limited use of vocal parts. For example, I like to use phrase-based sample libraries for TV commercials or short cues, where the turnarounds are often quick and the durations of the pieces are short.

6.11.1 Background Parts

For background parts where only ooh, ah, and mm are needed, sample libraries can produce fairly realistic results. The repetitiveness of these parts is ideal for static samples that are supposed to change very little over time. When sequencing these types of parts, there are two main things to keep in mind. First of all, you need to make sure to remember to implement breaks for breathing. As it was the case for woodwinds and brass, adding realistic breathing pauses is essential in order to re-create renditions of vocal background parts. While sequencing each part, try to sing it and time the breathing pauses in a musical way. Make sure to break the part/note when a short breathing pause needs to occur. To create the necessary breathing break, just leave anywhere from 200 to 400 ticks between two sustained MIDI notes (Figure 6.6).

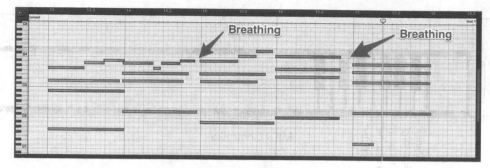

Figure 6.6 Breathing spaces for background vocals

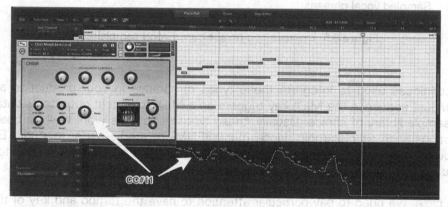

Figure 6.7 Controlling the vowel morphing parameter using MIDI CC#11 in Kontakt

To further increase the realism of background vocal parts, I recommend using a sampled patch that allows you to morph between vowels, therefore giving you additional control over the small variations occurring during change in dynamics and expressivity. In Figure 6.7, you can see how I have assigned MIDI CC#11 to control the "morph" parameter on a choir patch in NI Kontakt.

For best results, try to use mm or ooh settings for lower dynamics, and the open ah sound for louder passages. Listen to Audio Example 6.3 to hear how the changing of the vowel can affect a vocal background part.

To further enhance the realism of a background vocal group, we can use a similar technique learned when sequencing and producing for the string section. Layering a real vocalist on top of the MIDI parts can bring back to life a dull and artificial sampled (or synthesized) patch. In a typical SATB vocal ensemble (four vocalists in total), substitute the soprano MIDI part with a live singer. The additional harmonics and realism of the

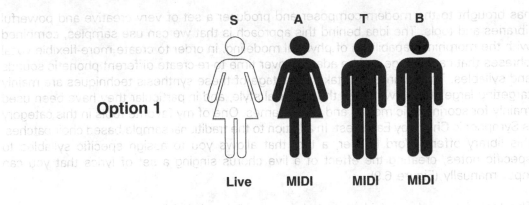

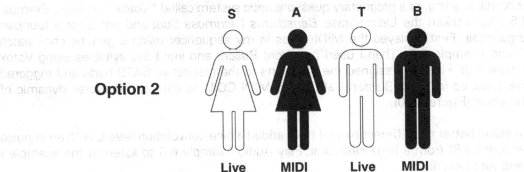

Figure 6.8 Available options for replacing vocal parts

live performance can increase the quality of the final result substantially. Since this is a small group (four singers), we are going to substitute the soprano MIDI part with the live singer (Audio Example 6.4).

If you have the possibility, I recommend replacing also the tenor part in order to create a "sandwich" of live and MIDI lines (Figure 6.8).

In this case, you will have the top female line and the top male line sang by live players and the other two parts played by MIDI.

6.12 Sample Modeling and the Evolution of Phrase-Based Sample Libraries

If you find that phrase-based libraries are too limiting, there is another option, which gives you more flexibility and can allow you to achieve impressive results for particular styles. The combination of sample-based and physical modeling synthesis techniques has brought to the modern composer and producer a set of very creative and powerful libraries and tools. The idea behind this approach is that we can use samples, combined with the morphing capabilities of physical modeling, in order to create more-flexible vocal phrases that can change and be adapted over time to re-create different phonetic sounds and syllables. The libraries that take advantage of these synthesis techniques are mainly targeting large chorus writing in the classical style, and in particular they have been used mainly for scoring epic movies and video games. One of my favorite tools in this category is Symphonic Choirs by EastWest. In addition to the traditional sample-based choir patches, this library offers Word Builder, a tool that allows you to assign specific syllables to specific notes, creating the effect of a live chorus singing a set of lyrics that you can input manually (Figure 6.9).

This technique makes it fairly easy to have your "virtual choir" sing convincing parts with any lyric that you need. The way you input the lyrics can follow traditional English spelling, phonetic spelling, or a proprietary quasi-phonetic system called "Votox." In Audio Example 6.6 I have taken the Latin phrase *Benedictus Lacrimosa Sunt* and set it for a four-part large choir. First, I played the MIDI notes in my sequencer using a generic choir patch (Audio Example 6.5). Then I used the Word Builder and input the syllables using Votox (Figure 6.9). Finally, I assigned the four parts to the respective SATB parts and triggered the syllables in Word Builder. I also used MIDI CC#1 to control the overall dynamic of the choir (Figure 6.10).

To blend better the different parts, I have added a nice convolution reverb with an impulse response (IR) from a large cathedral. Play Audio Example 6.6 to listen to the example I have just described.

This, and similar tools, are ideal for large choir productions where the budget doesn't allow for the hiring of a live large vocal ensemble. In order to improve the quality of these productions, we can use the layering technique we learned earlier for the four-part background ensemble. Layering a live soprano on top of the MIDI tracks can inject more life to the ensemble, while bringing the top line and the lyrics out a little bit more. Since we are dealing now with a large ensemble, we are not going to substitute the live singer with the MIDI, but instead we are going to add the live performance on top of the soprano MIDI track. Practically, we are slightly expanding the size of the soprano group. If you remember, this is the same technique that we used for the string section, where we triple-tracked a live violin and layered the takes on top of the MIDI violins. In fact, I recommend using the same technique, by triple-tracking a soprano singer in order to achieve a better blend between the MIDI and the live tracks as shown in Figure 6.11. If the budget allows for two live singers, you should record three takes each for the soprano and the tenor parts, therefore creating a nice stack of alternating textures (live, MIDI, live, MIDI), as shown in Figure 6.11.

In Audio Example 6.7 and 6.8, you can listen to the "Benedictus Lacrimosa Sunt" line with the addition of respectively a triple-tracked live soprano voice and a triple-tracked live tenor voice.

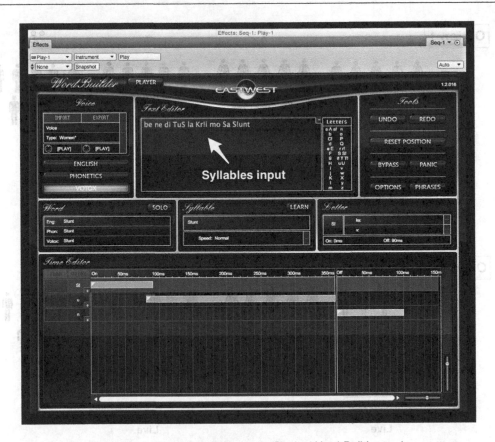

Figure 6.9 EastWest Symphonic Choirs Word Builder tool

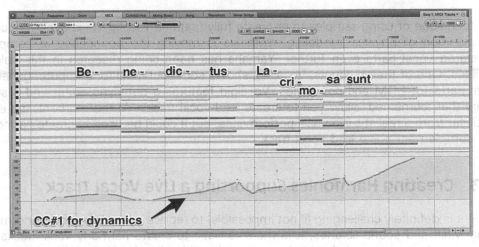

Figure 6.10 The four parts (SATB) assigned to EastWest choir and Word Builder, with MIDI CC#1 controlling the dynamics

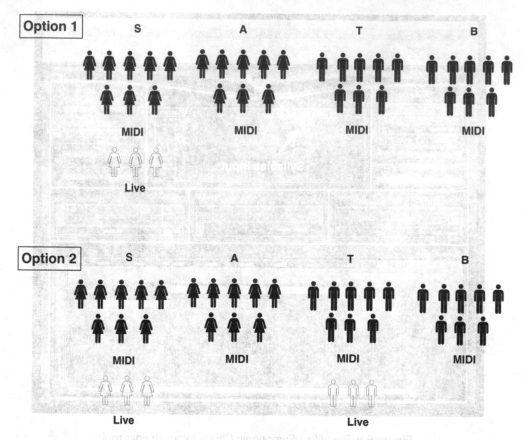

Figure 6.11 Layering techniques for large choirs

The technique of using samples to re-create fully featured vocal parts is evolving quickly. The addition of physical modeling synthesis is moving the boundaries of computer-generated vocal parts even further. Yamaha has been doing research in this field for decades. One of their products, Vocaloid, shows some potential for computer-generated vocal parts for EDM and pop styles. Vocaloid features a note/lyric editor and a sound library that, combined, allow you to create lead vocal tracks. At the moment, the effect is still far from a realistic live vocal performance, but it could be used to create harmonies that complement a live lead vocal track.

6.13   Creating Harmonies Supporting a Live Vocal Track

While it is definitely challenging (if not impossible) to replace a lead live vocal performance using MIDI and sound libraries, it is more realistic to create harmony vocal tracks that support and complement the lead one. Imagine this situation: it is midnight, you are done recording your ensemble, the lead part is done, and you are ready to put the final touches at the mix stage. Now the producer calls; he would like to add a couple of harmony parts that highlight the chorus sections of the song, and of course he wants the final production in the next two hours . . .

While the ideal solution would be to call the lead vocalist back in the studio, most likely you don't have the time. Trust me, this is a situation in which I have found myself several times! Fortunately, this is a situation that can be easily solved using some of the tools available in your DAW, or via third-party plug-ins. Let's learn how to create a two- or three-part harmony starting from a live vocal lead track!

6.13.1 DIY: Harmonizing a Live Vocal Track

To harmonize a prerecorded live vocal track, there are two main approaches: we can use a pitch shifter (or pitch tuner) and manually alter the pitches for the harmonized parts, or we can use a specialized plug-in that can automatically create two-, three-, or four-part harmonies, such as Antares's Harmony Engine EVO. While the simplest and fastest way to create harmonies is to use the latter, the former option can be cheaper; it also gives us a bit more control and flexibility when it comes to note choice and placement. The first step to create a vocal harmony is to have the live lead vocal on a single track. Make sure that this is the final take and that there are no tuning issues (I'll talk discuss how to correct pitch problems on a live vocal track later in this chapter). Then make a copy of the lead vocal track into a new audio track and name it "Harmony 1" (or something similar). The next step is to use your DAW built-in pitch analyzer (or a third-party one such as Melodyne by Celemony) and analyze Harmony 1's pitch content (Figure 6.12).

The result is a view that combines waveform and MIDI piano roll. Here you can easily see the pitch of each note in the Harmony 1 track (Figure 6.12). Now we need to decide where we would like to have the harmony part. Natural sections are the chorus of our song, or parts of a verse where there is a call-and-response structure, but feel free to

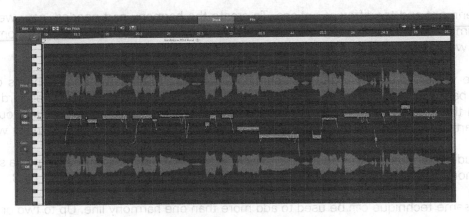

Figure 6.12 Example of live lead vocal pitch analysis in Logic Pro X

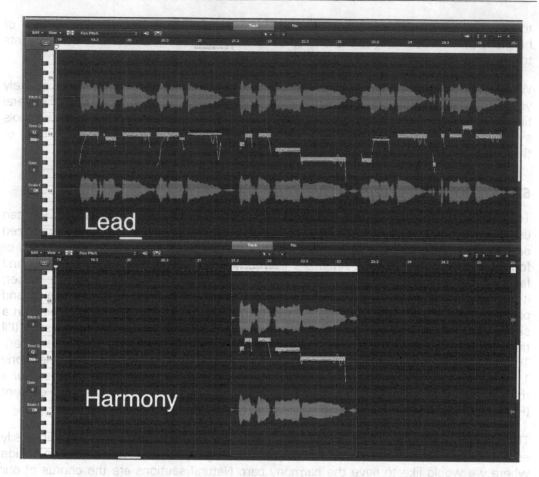

Figure 6.13 Original lead part altered in order to create the new harmony part

experiment and try different options. To create the harmony part needed, now we have to simply move the notes/pitches on the Harmony 1 track to follow the harmony part that we wrote (Figure 6.13).

Since we are basically using pitch-shifting algorithms to alter the original notes of the lead part, it is recommended to shift the harmony line not more than a major third away from the original pitch. Anything higher or lower than a major third could start sounding too artificial, though results depend on the type of audio signal we are working with.

In Audio Example 6.9 (courtesy of Ella Joy Meir, www.ellajoymeir.com) I created a simple harmony line starting from the original lead vocal track as shown in Figure 6.13.

The same technique can be used to add more than one harmony line. Up to two or three harmony lines can be easily added for a more complex texture.

6.13.2 Automatic Harmonizers

If you find that the previous technique is too time-consuming or too convoluted, fear not! There are several plug-ins available that allow you to automatically create harmony parts for you, starting from the lead vocal. An example of these plug-ins is Harmony Engine EVO by Antares (Figure 6.14).

The main idea behind an automatic harmony engine is that we can feed the live lead track and then the plug-in will create between one and four voices based on a series of parameters that we program. The simplest way to set the harmony engine is to set the harmonized voices to a fixed interval (for example, up a major third and down a minor sixth). Of course, if we have a changing harmony structure (different chords) behind the lead vocal, this system will most likely produce wrong harmony notes. For more-sophisticated harmonic progressions, we can program the harmonizer using "harmony

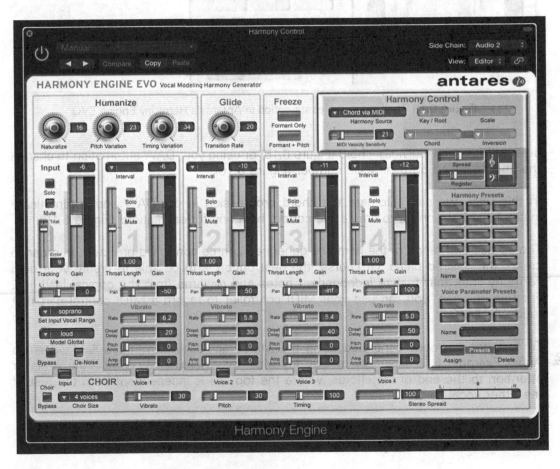

Figure 6.14 Antares's Harmony Engine EVO

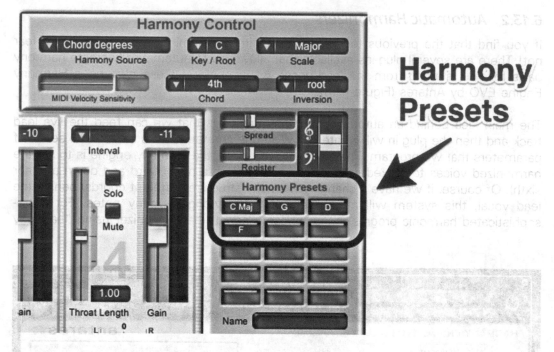

Figure 6.15 The harmony presets option in Antares's Harmony Engine EVO

presets" for each individual chord (Figure 6.15) or using your DAW automation to change the chords based on their scale number (Figure 6.16).

In this way, we can automate the chord progression in our DAW and automatically instruct the harmonizer to create the correct harmony for each chord (Figure 6.16).

The spread and range of the harmony generated can be controlled precisely, in order to create the desired blend. Of course, the best option is to be able to create your own harmony lines that feature independent notes and rhythms from the original lead line. For this purpose, you can use your MIDI controller to record a MIDI control part that feeds the correct pitches to the harmonizer (Figure 6.17).

To hear the harmony generated using the MIDI track shown in Figure 6.17, listen to Audio Example 6.10 (courtesy of Ella Joy Meir, www.ellajoymeir.com). Pay particular attention to the end of the chorus, where the top line is independent from the rest of the other vocal parts.

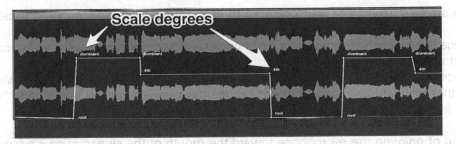

Figure 6.16 Automation lane in Logic Pro with the harmony preset switching sequence

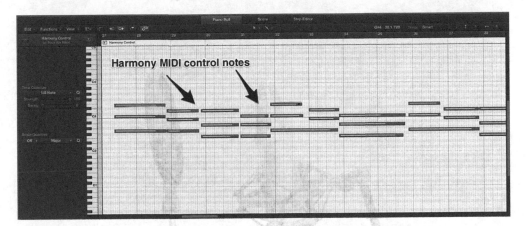

Figure 6.17 The MIDI control notes that are sent to the automatic harmonizer plug-in

6.14 Size Matters!

Often, you will find yourself in a situation where you need to make either the lead part or the background parts sound bigger and with more voices than you have available. Consider the situation where a producer asks you to expand the size of the background vocals, for example. Ideally, we would record a live choir, but budget and time restriction often do not allow for a large ensemble recording session. If this is the case, we can create the effect of a larger live choir by recording multiple takes of the same singer.

6.14.1 Recording Vocal Tracks

To record vocal tracks and achieve the best results, I recommend using some specific techniques aimed at minimizing unpleasant phasing effects, while at the same time creating the feeling of several different people singing together.

First of all, start with a solid lead vocal track that will serve as a guideline for all the other voices/parts. Here are some quick tips on how to effectively record the voice of a singer.

There are few "rules" or best practices to keep in mind when recording a vocal track:

- Place the microphone stand on a rug or carpet in order to minimize extra low frequencies and noises.
- Place the microphone about 12 inches from the singer as starting point. Moving it further will capture more room and reduce the plosives caused by "p" and "b."
- To further reduce plosives, you can use a pop filter (Figure 6.18).

Instead of pointing the microphone toward the mouth of the singer, raise it a few inches to be level with the singer's eyes, aimed at a slightly downward angle. This will further decrease the chances of annoying plosives occurring (Figure 6.19).

Figure 6.18 Condenser microphone with a pop filter for vocal recording

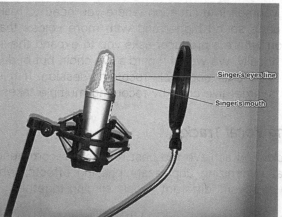

Figure 6.19 Ideal positioning of a microphone in relation to the singer

Figure 6.20 Primacoustic Voxguard (Courtesy of Primacoustic,
www.primacoustic.com/voxguard.htm)

If your tracking space is not ideally isolated and presents some problematic reflections, I highly recommend adding a portable mic isolator such as the Primacoustic Voxguard. Such a device prevents the microphone from capturing standing waves and unwanted reflections bouncing off the walls of the tracking space.

Place the singer about eight to 10 inches from the pop filter.

6.14.2 Multiplying a Vocal Track

Once you have a solid and well-recorded lead vocal track, it is time to expand the virtual ensemble. To do so, we are going to record multiple takes of the same part. Usually, recording between four and six takes allows us to give the impression of a mid-sized vocal ensemble. The main problem with recording the same voice over and over, though, is that we can easily run into phasing problems, defeating our goal of giving the impression of a larger ensemble. Phasing issues can easily add or subtract low frequencies, render the recording shallow and empty, and of course add unwanted chorus effects and artifacts. To avoid these issues, we need to follow some simple rules.

First of all, change the type of microphone between takes; this will help "characterizing" in different ways each performance. Switch between condenser, dynamic, and ribbon mics in order to create more variation. Try also to move the singer around the recording space; this will allow for the microphone to capture different reflections of the walls, injecting a series of differences in each take. In addition to the regular takes, record some tracks with the singer facing toward the wall with the microphone at his back. These are nice "filler" takes that help to increase the size and the realism of the vocal ensemble. Finally, have the singer perform with different attitudes for each take; some takes should be more aggressive, for example, while others more mellow and laid back.

Widening the stereo image of the virtual vocal ensembles is crucial in order to create the effect of a bigger group. Make sure to pan each take slightly off center from the previous one, so that we can achieve a wider spread over the left and right channels.

6.14.3 *Automatic Choir Multiplying*

Another way to enlarge the size of our vocal ensemble without actually recording a full choir is to use a "multiplier" plug-in. While this technique by itself is not ideal, it can nevertheless be a worthy alternative and an excellent addition to the previous multitake technique. There are several "doubler" plug-ins available that allow us to duplicate a vocal track in order to make it thicker and more present. These tools can be effective only when trying to re-create smaller ensembles. For larger ensembles, you need to use more-specific tools. One of my favorites for this type of task is again the Antares Harmony Engine. In addition to being able to create fairly sophisticated automatic harmonies (as we learned in the previous section), this plug-in is capable of multiplying a single vocal line to up to 16 voices (Figure 6.21).

With this tool, we can multiply the lead and the four harmony parts up to a choir of 16 elements. The effect is fairly realistic. Listen to Audio Examples 6.11 and 6.12 to compare,

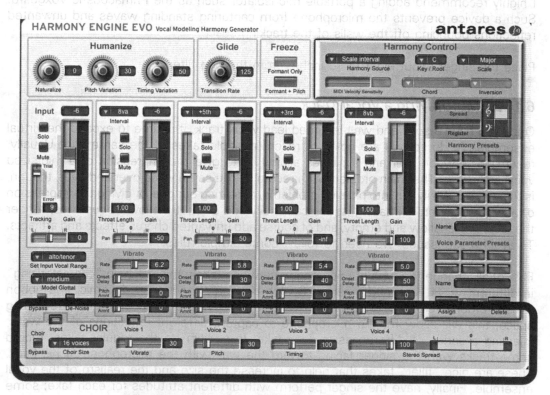

Figure 6.21 Antares Harmony Engine's "Choir" section

respectively, a solo lead vocal and the same lead vocal "multiplied" 16 times using Antares's Harmony Engine.

If you want to create the impression of an even larger ensemble, you can multiply also the harmony parts, as you can hear in Audio Example 6.13 (Courtesy of Stefano Marchese, www.stefanomarchese.org).

To create larger and even more realistic choirs, I recommend combining the two techniques described in the previous sections. Start creating a mid-sized to large ensemble utilizing a multiplier plug-in, then overdub three or four takes of a live vocalist using different microphones, positioning, and interpretation. The final result will be a very realistic large vocal ensemble recorded in your project studio!

6.15 Mixing the Vocal Ensemble

As we learned for the other instruments and sections, the mix stage can be crucial for a successful and realistic production. Vocal tracks and ensembles are no exception. With a well-calibrated equalization, compression, and a good reverb we can achieve great results!

6.15.1 Equalization

When equalizing vocal tracks, always aim for clarity, intelligibility, and separation between vocals and the supporting instruments. In general, try to use a nice multiband equalizer that is transparent and that provides excellent flexibility. You will find that, in most cases, you will use a peak equalizer in order to "surgically" work on specific frequency ranges. In Table 6.1 I have listed some of the most common types of equalizers.

Each type of equalizer has an easily recognizable symbol associated with it (Figure 6.22).

When equalizing vocal tracks, there are a few specific frequency ranges that can be effective to tweak in order to reach the desired results as shown in Table 6.2.

6.15.2 Compression

After applying the necessary equalization setting, I advise using a little bit of compression in order to achieve a more cohesive and solid vocal sound. Compressor is an effect that belongs to the dynamic effects family. The name of this family comes from the fact that the effects are designed to alter, in one way or another, the dynamic range of an audio signal. They are important because they allow you to control over time the ratios between high and low peaks in the dynamics of an audio track. A compressor allows us to reduce the overall dynamic range of a signal without reducing the overall output. In Table 6.3 I have listed the main parameters of a compressor.

Table 6.1 Characteristics and parameters of the most common types of equalizer

Type of equalizer	Description and parameters	Typical uses
Peak	It allows you to cut or boost frequencies around the center frequency. Center frequency: it determines the frequency to cut or boost Gain: positive gain boosts, negative gain cuts Q point: this determines the "shape of the bell" or how wide the area around the cutoff point is going to be: the lower the value the larger the bell and vice versa the higher the value the smaller the bell. The Q parameter can usually (but not always) vary from a value of 0.7 (equal to a two-octave frequency range) to a 2.8 (half-octave)	A Peak equalizer is extremely versatile. It can be used to pinpoint and cut/boost a very precise frequency or it can be used in a broader way to correct wider acoustic problems. It is usually utilized in the middle of the frequency range
Notch	It leaves all the frequencies unaltered except the one(s) specified by the center frequency that will be cut. Center frequency: it determines the frequency to cut Gain: positive gain boosts, negative gain cuts Q point: it determines the "shape of the bell" or how wide the area around the center frequency is going to be: the lower the value the larger the bell and vice versa the higher the value the smaller the bell. The Q parameter can usually (but not always) vary from a value of 0.7 (equal to a two-octave frequency range) to a 2.8 (half-octave)	Usually it is used in live concerts to reduce feedback since it allows you to choose a very narrow set of frequencies to cut drastically
High shelf	It cuts or boosts the frequency at the cutoff and all the frequencies higher than the set cutoff point. It has only two parameters: the cutoff frequency and the gain	It is usually used in the mid-high and high end of the spectrum. It can be effectively used to brighten up a track by using a positive gain of 3 or 4dB and a cutoff frequency of 10kHz and higher (be careful because this setting can increase the overall noisiness of the track). It can also be used to reduce the noise of a track by reducing by 3 or 4dB frequencies around 15kHz and higher

continued . . .

Table 6.1 Continued

Type of equalizer	Description and parameters	Typical uses
Low shelf	It cuts or boosts the frequency at the cutoff and all the frequencies lower than the set cutoff point. It has only two parameters: the cutoff frequency and the gain	It is usually used in the low–mid and low range of the audible spectrum to reduce some of the rumble noise caused by microphone stands and other low-end sources
High pass	It cuts all the frequencies below the cutoff point. It has only one parameter, which is the cutoff frequency	It is a very drastic filter. It is often used to cut very low rumble noises below 60Hz
Low pass	It cuts all the frequencies above the cutoff point. It has only one parameter, which is the cutoff frequency	It is a very drastic filter. It is often used to cut very high hiss noises above 18kHz. Use with caution in order to avoid cutting too much high end of the track

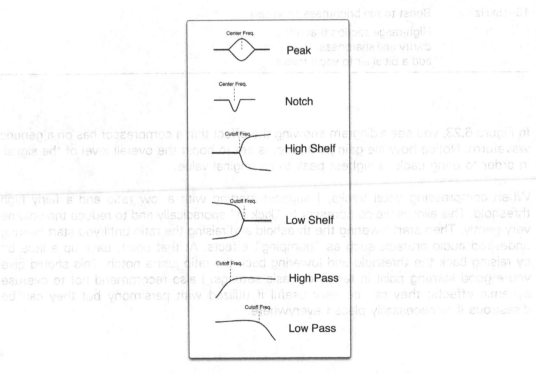

Figure 6.22 Graphic symbols for different types of equalizers

Table 6.2 Frequency ranges and their features used for equalizing vocal tracks

Frequencies	Application	Comments
20–120Hz	Cut to reduce rumble, noises, muddiness, and proximity effect	It is a good idea to always reduce this area by 4 to 6dB in order to lower the low frequencies noise
200–300Hz	Boost to add fullness to vocal tracks Cut to reduce "boomy" effect	Be careful not to boost too much of this frequency range in order to avoid adding muddiness to the mix
400–500Hz	Cut to reduce "boxiness"	If a tracks sounds a bit boxy, cutting 1 or 2dB in this frequency range can help
1–4kHz	Cut to reduce nasal effect and harshness	If a vocal track sounds too nasal, try to reduce this range a little bit
5–6kHz	Boost for vocal presence A general mid-range frequency area to add presence and attack	
7.5–9kHz	Cut to avoid sibilance on vocals A mid- to high-range area that controls the sibilance on vocal tracks	
10–15kHz	Boost to add brightness on vocals High-range section that affects clarity and sharpness. Useful to add a bit of air to vocal tracks	

In Figure 6.23, you see a diagram showing the effect that a compressor has on a generic waveform. Notice how the gain parameter is set to boost the overall level of the signal, in order to bring back its highest peak to its original value.

When compressing vocal tracks, I suggest starting with a low ratio and a fairly high threshold. This allows the compressor to "kick in" sporadically and to reduce the volume very gently. Then start lowering the threshold and raising the ratio until you start hearing undesired audio artifacts such as "pumping" effects. At that point, back up a little bit by raising back the threshold and lowering back the ratio just a notch. This should give you a good starting point in terms of basic settings. I also recommend not to overuse dynamic effects; they can be very useful if utilized with parsimony but they can be disastrous if unnecessarily placed everywhere.

Table 6.3 The parameters of a generic compressor

Parameter	Description	Comments
Threshold	It sets the level in decibels above which the effect starts reducing the gain of the signal	If the signal is below the threshold, the effect doesn't affect the signal. As soon as the level goes over the threshold the effect starts reducing the gain of the signal
Ratio	It sets how much the gain of the signal is reduced after the level goes over the threshold	It is usually set as a x:y value. With a setting of 1:1 the level is not altered; with a setting of 30:1 the level is highly compressed. With a setting of 100:1, the effect is considered a soft limiter. In certain limiters, the ratio parameter is omitted, implying a ratio set to infinite. In this case, the effect can be referred as a hard-limiter
Attack	It sets how quickly the effect reacts after the signal goes over the threshold	
Release	It sets how quickly the effect reacts after the signal returns below the threshold	
Knee	It controls the curvature during the transitional moment below and above the threshold point. A "soft knee" allows for a more gentle transition, while a "hard knee" generates a more drastic transition	This parameter is not found on every plug-in
Gain	It allows you to control the overall gain of the signal after compression/limiting	Since the clear effect that compression and limiting have on the signal is a reduction of amplitude, the gain parameter allows you to boost back the compressed signal by the amount specified. Some dynamic plug-ins featured an "auto gain" option, which automatically set the level of amplification so that the output, after compression, matched the input before compression
Side-chain input	The side-chain input allows you to trigger the dynamic effect of the track where it is inserted (for example, track A) through the signal coming from another track (for example, track B). To set it up, send part of the signal from track B to a bus (via an aux send), then set the input of the side-chain dynamic on track A to the same bus	Not all the dynamics plug-ins feature a side-chain input. Practical applications include, for example, compressing a bass track with the side-chain input set to the output of the bass drum, in order to add clarity and punch to the combination of the two signals

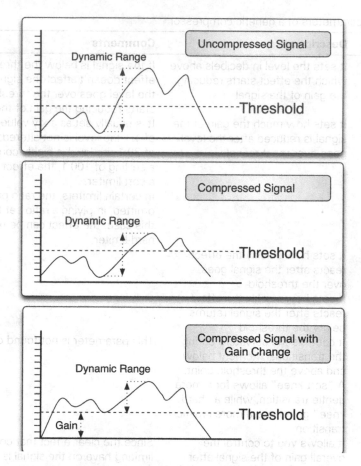

Figure 6.23 Example of a compressor

6.15.3 Reverberation

Another important aspect of mixing vocal tracks is the type and amount of reverb you apply in order to place them correctly in a natural ambience. Natural reverberation is produced by a build-up and complex blend of multiple reflections of the waveforms generated by a vibrating object (e.g. the string of a violin or the head of a snare drum). The reflections are generated when the waveforms bounce against the walls of an enclosure, such as a room, a stadium, or a theater. Depending on the size of the enclosure and the type of material that covers the walls, reverberation can change in length, clarity, and sonic characteristics. In a digital audio workstation, reverberation is usually added at the mix stage in order to place instruments accurately and realistically in a predetermined acoustic environment. The style of the project and the type of vocal ensemble are important factors in determining and properly choosing the best reverb type and in setting its parameters. Take a look again at some of the most common reverb parameters listed in Table 2.5.

Depending on the size of the vocal ensembles and the type of music you are producing, the ideal length of the reverb can vary considerably. In general, less reverb and shorter times transmit a sense of intimacy and directness, therefore making them suitable for smaller ensembles (lead only, or lead with one or two harmonies). Vice versa, for larger ensembles (such a large choir) we need to use more reverb and with longer time settings. In Table 6.4, you can see the most common options for adding reverb to vocal tracks based on the size of the ensemble.

Always use the reverb with parsimony. Don't over use it, since too much reverb (or too long) will add muddiness and will reduce the clarity and the definition of the tracks. Try to use only convolution reverbs if possible. They tend to sound more natural than synthesized ones, since convolution reverbs are based on impulse responses sampled from real spaces such as halls, churches, cathedrals, and studios.

Table 6.4 Reverb types and lengths for vocal tracks

Reverb length and type	Ensemble size	Comments
1 2 sec—room or small recording studio	Lead/solo vocal	A room reverb with a length of around 1.2 second works well for solo vocal track where it is important to reach an "in your face" effect. At low volume and dynamics, we can achieve a very intimate effect
1.5–1.6 sec—medium hall	Lead/solo vocal	A medium hall allows us to place a lead vocal track in a natural environment without removing it from the listener. The final result is still a direct vocal track that is a bit rounder than the previous reverb setting
2–2.2 sec—medium hall	Lead/solo. Lead with harmony	This solution allows us to place the lead vocal in a natural and soft environment. This setting works well for pop and modern productions that need to place the vocal track in a larger environment without sacrificing its presence
2.3–2.5 sec—large hall	Classical lead. Medium-sized choir. Background vocals	Large halls are ideal for solo classical voices. The space suddenly becomes part of the performance and helps to add character and depth to the vocal track. For medium-sized choirs, this setting works very well, helping to blend the different parts and voices
2.8–3 sec—large hall	Large choirs	A hall of these dimensions is ideal for large choirs that need the surrounding space to transmit a sense of realism and live performance. The extra "air" helps with the blending of the voices

6.16 Summary

Remember that, although there are general ranges for the standard SATB voices, each singer's instrument is unique. Furthermore, their instrument transforms with age. It is paramount that the music arranger engages with the singer in the present moment to determine the appropriate key. No instrumental writing should occur until this has been determined. Singers have a variety of experience and background; some are musically trained, others are strictly natural talents. Tone, style, and technique can vary widely so it is important to cast the proper vocal talent as a director would cast specific actors for a play or film.

Vocal music for a theater production will usually be sung by different singers throughout the history of the show. The arranger must write with this in mind by staying within a practical range as much as possible. Although most professional singers have a two-octave range, the common practical range is closer to one and a half octaves.

Remember that reading music is much more difficult for a singer than for an instrumentalist. The singer will be most successful if the writer can make each part as singable as possible. This means staying within the diatonic notes as much as possible. When chromaticism is necessary, those notes should resolve to the nearest diatonic neighbor tone. With chromatic regions (a phrase, or part of one, that remains outside the diatonic notes of the key), try to keep the chromatic notes in their own diatonic relationship (for example, A♭–E♭–C–A♭ resolving to A in the key of F major).

For vocal ensembles, each singer reads the full vocal score. He or she must see all of the parts to have a full reference of the notes being sung by the others. The piano reduction of the vocal parts is also helpful for the singers to see where their notes may rub with other singers in the ensemble. Remember that the tenor voices read treble clef but the notes are written an octave higher from where they sound (a small number 8 should appear directly below the clef).

Instrumental accompaniment must provide the entrance note for the singer(s). This is especially crucial in the introduction and following an instrumental interlude. The arrangement should underscore the dramatic aspect of the text (lyrics). When composing to a set of lyrics, they should be set rhythmically in a way that accommodates the natural flow of the language. Don't forget that singing multiple pitches on one syllable always requires a phrase mark (known as a melisma) over those notes.

One of the simplest and most direct ways to integrate vocal tracks in our sequences and productions is to use prerecorded sample libraries. Usually, these types of libraries work well for beat-based productions such as electronic dance music (EDM) or ethnic parts. When using these libraries, pay particular attention to have the tempo and key of the song matching the key and the tempo of the samples. For background parts where only ooh, ah, and mm are needed, sample libraries can produce fairly realistic results.

Adding realistic breathing pauses is essential in order to re-create renditions of vocal background parts. While sequencing each part, try to sing it and time the breathing pauses in a musical way. Layering a real vocalist on top of the MIDI parts can bring back to life a dull and artificial sampled (or synthesized) patch. Replacing the soprano MIDI part with a live recording can produce very good results, adding a more realistic touch to the production.

The combination of sample-based and physical modeling synthesis techniques allows us to create more flexible vocal phrases that can change and be adapted over time to re-create different phonetic sounds and syllables. This technique makes it fairly easy to have your "virtual choir" sing convincing parts with any lyric that you might need. For larger choirs, triple-tracking a live vocal (usually the soprano part) will guarantee a better blend with the MIDI parts.

To harmonize a prerecorded live vocal track, we can use a pitch shifter (or pitch tuner) and manually alter the pitches for the harmonized parts, or we can use a specialized plug-in that can automatically create two-, three- or four-part harmonies, such as Antares's Harmony Engine EVO.

When recording live vocal tracks, keep in mind some simple rules in order to achieve the best results: place the microphone stand on a rug or carpet; set the microphone about 12 inches from the singer; to further reduce plosives, you can use a pop filter; and place the singer about eight to 10 inches from the pop filter. To multiply a single vocal line, record between four and six takes. In order to minimize unpleasant phasing effects created by the fact that we are recording over and over the same vocal material, change the type of microphone between takes (switch between condenser, dynamic, and ribbon mics in order to create more variation). Try also to move the singer around the recording space. It will allow for the microphone to capture different reflections of the walls, injecting a series of differences in each take. Another way to enlarge the size of our vocal ensemble without actually recording a full choir is to use a "multiplier" plug-in.

When equalizing vocal tracks, always aim for clarity, intelligibility, and separation using a nice multiband equalizer that is transparent and that provides excellent flexibility. After applying the necessary equalization setting, I advise using a little bit of compression in order to achieve a more cohesive and solid vocal sound. I suggest starting with a low ratio and a fairly high threshold. This allows the compressor to "kick in" sporadically and to reduce the volume very gently. Reverberation will help you place your vocal ensemble in a natural space, adding character and realism to your vocal production. Less reverb and shorter times transmit a sense of intimacy and directness, while for larger ensembles (such a large choir) we need to use more reverb and with longer time settings. Always remember that too much reverb (or too long) will add muddiness and will reduce the clarity and the definition of the tracks.

6.17 Exercises

Exercise 6.1

Write a 30-second piece featuring a lead vocal. Record the lead vocal part and then create up to three background parts with ooh, ah, and mm. Sequence the background parts using a sample-based library.

Exercise 6.2

Write a one-minute piece featuring a lead vocal. Record the lead vocal part and then create a two-part harmony using any of the techniques you learned in this chapter.

Exercise 6.3

Using the previous exercise as a starting point, multiply the lead and the harmony parts for a medium or large choir.

7 Writing for Various Ensemble Sizes and Other Final Considerations

7.1 Atypical Instrumentation—Its Relation to Budget and Technology

The commercial music writer in today's production market must often conform to atypical instrumental combinations. This is due largely to a mix of multiple styles in one ensemble and, more importantly, budget restrictions. Even in high-profile venues such as a Broadway pit orchestra, quite often the size of the group has been substantially reduced from the full complement of musicians that might have been required for an original score of the 1960s or earlier. Electronics has had a significant effect on the musical environment too. Some acoustic instruments are miked while synthesizers replace other acoustic instruments. Thus the orchestrator must now work in two contexts: (1) the awareness of traditional orchestration; and (2) the awareness of how the traditional guidelines may be altered via electronics.

This scenario is also abundant in recording sessions, where overdubbing and layering is common.

In general, full orchestras are still used with large budget movies (as in Hollywood), video game soundtracks, and in live performances by major symphony orchestras. But most of the remaining productions require some deft juggling and altering of the instrumentation to conform to more humble budgets. The challenge for the writer is to provide the illusion of the large traditional ensemble (orchestra, big band, etc.). Although it is impossible to do this exactly, the writer strives to create a full ensemble sound that is gratifying in its own stature.

One of the authors recently attended a Harry Connick concert. His ensemble consisted of six strings (five violins and a cello), four horns (trumpet, trombone, tenor sax, and baritone sax with a "double" on flute), and a full rhythm section. Connick is a very skilled arranger who deftly created some "orchestral" pieces as well as some "big band" pieces in addition to his New Orleans-style blues genre (where the string and horn sections were also employed). So you can see that even big-name artists need to travel "lightly."

7.2   Recording Strings and Determining the Budget

The sound of a lush string orchestra is a most gratifying setting for a romantic or dramatic recording. But the cost of this wonderful sound can be staggering! Traditional orchestras can have as many as 60 string players! If you were fortunate enough to hire these players for $100 per individual per hour, the total would be $18,000 for a typical three-hour session—and that's a bargain price! Thirty players can still provide a lush sound but, with more realistic fees (at least $150), it would still cost at least $4,500 per hour or $13,500 for a 3-hour session. So you see that it is quite expensive to record in this format unless the budget is extremely large. With most recording musicians, but especially with string players, you must find the best players possible. Inferior players, even moderately adept ones, are usually not good enough to record. Maintaining good intonation on string instruments is always a challenge and only the best players can accomplish your musical goals in a gratifying way. Because you need the highest possible caliber of performer, you will also need to pay a reasonable fee for their service. A call to the local musicians union can inform you about the current hourly rate for a recording session so you can plan your budget realistically.

Here are some other financial considerations to think about:

For 30 string players, you will need quite a large recording room. This will also require assistants to the engineer to help with setting up chairs, stands, mics, headphones, etc. Of course, you will need a conductor, a copyist to prepare the manuscripts, and most likely a contractor. The contractor handles the union contracts for each player. Included are the salary, health benefits, and pension benefits. This information must eventually be submitted to the local musicians union on a special form that the players must sign. In addition, prior to the session, there are numerous e-mails and phone calls, etc. The work is very time-consuming and costly. Even if your session is "under the radar" (a non-union date), you will either need to do this work yourself or hire someone to do it for you. Be aware that many of the top-caliber string players may not accept your offer to do a non-union recording session.

7.2.1   Layering String Tracks

A decent compromise to the previous scenario is to take advantage of modern recording technology and layer the string tracks. This strategy reduces the number of string players considerably. Suppose you hired 10 players. With three layers, you can now create the illusion of having a 30-piece string section. Although this is not quite ideal, it will eventually create a big, lush sound. This approach, although effective and quite common today, has its challenges. The first layer must be pristine with regard to precision, intonation, and expression. Your players will be overdubbing to this first layer, so it is essential to get the best take possible. Once the first layer is recorded, the second layer is a bit easier but still a bit challenging for the players. They must listen carefully (through headphones) to their first layer and try to match it. Sometimes it may be possible for the players not to listen and simply play the second layer, but then you (or someone else) must make sure that both layers unite effectively. Upon completion you will notice the sound start to change favorably. But it is the third layer where the magic starts to happen. It is also a bit easier for the players because there are more recorded strings to blend with. I used this technique for a CD project titled *When Winter Comes*, which features compositions by guitarist and composer Fred Fried. You can find the recording on iTunes.

7.2.2 Layering during Rubato Tempo

Rubato passages can be quite tricky. You will usually need a warning count-off to find the first entrance and any fermata in the music will most likely need to be counted or timed specifically. Flexible click tracks (created via Tap Tempo in Digital Performer or similar methods in other digital audio programs) can be quite effective in this case. This technique was used for *When Winter Comes*, the title track of the Fred Fried's CD. If you find that an electronic click track is too confusing or too difficult to create, you can simply record yourself counting the beat numbers and use that for reference.

7.2.3 Determining the Appropriate Combination of Strings

The right combination of strings can be important. I have found that six violins, two violas, and two cellos work quite well. If you need the double bass, one will do with an overdub. This adds up to 10 players (11 with the double bass). Notice that the traditional proportion of each string section (found in the orchestra) is more or less preserved. As you layer, the proportion will remain intact. With six violins, you can record the first two layers using three players on violin 1 and three players on violin 2. This provides a good ensemble sound for each violin part during the critical first layer. Then add the second layer in the same fashion. For the third layer, have four violins play the violin 1 part and two violins play the violin 2 part. This will ultimately give you the sound of more violins on the violin 1 part (10 players on violin 1, eight players on violin 2).

There may be times where affording even 10 players is too costly. If you have only one player on each part, the outcome of the multiple layers will produce a bottom-heavy string section sound (too many basses, cellos, and violas, with a thin violin sound). In this context, it is important that there is enough isolation so the engineer can reduce the sound of the lower instruments and bring up the violins. As an extra safeguard, you might consider recording a separate track with only the violins. With a limited budget for only a string quartet/quintet, you can hear the overdubbing technique on some arrangements that the author did for Dominick Farinacci on his CD *Lovers, Tales, and Dances*, which is available on iTunes.

7.2.4 Overdub Strings with MIDI Tracks

When the budget is really small but the desire for a really big orchestral sound is great, another option is to use a handful of live players with your MIDI string orchestra tracks. A string quartet can add more realism and expression when layered over the MIDI tracks. For the most humble budget, even using one violinist and one cellist can help. You can hear this technique with the author's orchestrations for a theater production titled *Frankenstein—A New Musical*, which is available on iTunes.

7.3 Writing for a Handful of String Players in Live Performance

There are occasions where an arranger (or orchestrator) is required to write for only a few players within the context of a larger mixed ensemble during a live performance. As a precaution, the writer should not count on microphones to artificially establish "equity" with the brass or a rhythm section. They will help somewhat but it is better to assume that there will be no mics or that the sound amplification will be minimal. This will place you in the proper mindset to make musical choices that give the strings the most acoustic power possible. Another point to consider is the size of the ensemble and more importantly the other instruments that are included. Often, the style of music will inform the writer but it is good to get as much information as possible before proceeding.

7.3.1 Balancing a Few Strings with Horns and a Rhythm Section

Taking the most challenging scenario, let's assume that you have only a string quartet available and that it must compete (or try to) with a much stronger horn section and rhythm section. The quieter sections of music may pose little or no problem. But, in loud portions, the arranger must present the strings in the strongest way possible. The old adage "there is strength in numbers" definitely applies here. The idea is that one sure way to make your string quartet sound powerful is to write them in octaves. When four-part harmony is used, the quartet is reduced to "every man for himself." The result alone may be attractive but, in context with the ensemble, the sound may be thin and weak. It is much better to have both violins play in unison to create a violin "section" sound and have the viola double a line with the cello to create a "cello section" sound. The resulting two-part counterpoint presents a much stronger string section sound. It can become even more powerful when all four play only one line. Try it with an octave spread between the upper and lower string "sections." When the violins ascend into their highest register, try a two-octave spread.

7.3.2 Creating Contrast between the Horns and the Strings

There is no question that when the strings try to compete with the horns in forte passages they will lose the battle every time. It is best to create a contrast between both groups so the listener can make the distinction between the two. If the arranger has the strings playing the exact same thing as the horns, their sound will absorbed by the more powerful horns. Essentially the arrangement becomes monodimensional and ironically less gratifying. There are certainly places where a concerted approach can be dramatic but it is usually better to have the strings play something that is separate and distinct from what the horns are doing. Here are some ways to do that:

- have the strings move at a different speed (slower or faster) than the horns;
- keep the upper strings in a different register from the horns (violins above);
- keep the strings in monophonic or polyphonic texture while the horns are in homophonic texture.

Score 7.1 offers an example of a "rhythm and blues" rock style. Anticipating a live performance with the idea that the smallest string section may literally be one on each part, you will notice that the strings distinguish themselves by moving much faster than the horns. More importantly, the violins are scored in unison as are the viola and cello to create an upper and lower "section" of two. (There is a valid concern that having only two players in unison can expose problems with intonation but striving for a bigger sound is more important in this context.) In the first two bars, the upper and lower strings are separated by an octave. When the violins ascend to their highest register, the viola jumps to its highest register, an octave below the violins and an octave above the cello. The idea is to keep the lower strings in a robust register to support the much thinner and higher violins.

Score 7.2 shows a contrasting style but with the same concept. This example is similar to how the great arranger Nelson Riddle (and others like him) often scored strings with a big band behind Frank Sinatra. There are three strong, coexisting melodic ideas. The brass are placed in cup mutes for a warm sound and scored in typical big band swing style with close harmony in each section (tpts. and tbns.). The most rambunctious line is heard in the saxes; they are split in octaves. The strings start with an upward sweep that lands on the apex of their line and then moves slowly with a descending chromatic line. It is most ambient as the violins soar above the horns. The lower strings (15vb) add panorama to the string line and glide through the middle of the horn ensemble.

7.4 Small Horn Sections—Capturing Variety and a Big Sound

For those less familiar with jazz and pop music, "horns" in this context refers not to the French horns but most typically a varied combination of instruments that usually includes the saxophone, trumpet, and trombone. The two most famous jazz-rock bands to feature a small multihorn ensemble were Blood, Sweat and Tears and Chicago. There were also many jazz groups with a small multihorn instrumentation. Benny Golson and Oliver Nelson are two famous jazz composers and bandleaders. Art Blakey, though not a composer, led many outstanding multihorn groups. And Gil Evans, with Miles Davis, Gerry Mulligan, and John Lewis, made an innovative and critically acclaimed record called *The Birth of the Cool*. The instrumentation for this band was alto sax, baritone sax, French horn, trumpet, trombone, and tuba, and the sound along with the compositions was quite unique. The tuba was used in a very unconventional way—playing melodic lines in the baritone and tenor registers.

Certainly, there are musical contexts where a small horn section (two to three players) is all that is necessary or desired. The bandleader or composer may deliberately seek a smaller and more intimate sound. But, when a big band sound (or a similar large ensemble) is desired, the budget (similar to the problem with hiring a string orchestra) may be prohibitive. Once again, the modern arranger must come to the rescue and create the illusion of a big sound with only a few players. This is almost impossible to do with only three players but it is manageable with as few as four.

With reference to this context, extensive examples of Rich DeRosa's work can be found via many recorded arrangements for vocalist Susannah McCorkle. Her record company, Concord Records, could not afford to hire a big band so the instrumentation was set for four horns and four rhythm section instruments. The challenge quite often was to create the impression of a big band along with the many contrasts that larger horn sections can easily provide.

7.4.1   Choosing the Appropriate Horns

With only four horns, the trick is to choose ones that, in combination, cover the registers of the vocal choir (or middle registers of the piano). Specifically, those registers, from top to bottom, are soprano, alto, tenor, and baritone. There is no need for a bass instrument since the rhythm section instruments can cover this. In addition, the arranger has too few horns to waste them, so avoiding the bass register is a wise choice. It is more important to fatten the middle and upper registers, which also feature the melodic and harmonic aspects of the music. Directly below is an outline of the most typical instruments (within the popular and jazz styles) that fall within these registers:

- Soprano—flute, clarinet, soprano sax, trumpet
- Alto—clarinet, alto sax, flugelhorn
- Tenor—tenor sax, tenor trombone
- Baritone—baritone sax, bass trombone

Of course, with most of these selected instruments there is a significant overlap into other registers. When comparing them to a four-horn group with two trumpets and two alto saxes, the latter combination would certainly be less ideal because the tenor and baritone registers are not really available or, at best, they are weak.

7.4.2   An Ideal Horn Section Can Offer Many Combinations

Within the scope of work for vocalist Susannah McCorkle, the author's use of the trumpet, alto sax, tenor sax, and tenor trombone provided several ways to exhibit contrast and dimension. They are listed as follows:

- Create a distinctive brass section (tpt./tbn.) and a distinctive woodwind section (saxes). The brass can be in unison when in the alto register or in octaves when the trumpet ascends into the soprano register. The saxes can be in unison or octaves in similar fashion.
- Create a "soprano" (trpt. and alto sax) versus "tenor" (tenor sax and tbn.) sound.
- Darken the ensemble sound by having the trumpet player switch to flugelhorn.
- Feature one horn as a soloist and have the other three horns function as a unit of background horns.
- Muted brass can be very effective. Trumpet mutes in particular are handy because, unlike the trombone, the manipulation of the mute in and out of the horn is rather quick and easy.
- Ensemble tutti—where all the horns work in concert to create a "shout" (typical climactic moment in a jazz arrangement), lush pads, or punchy hits.

7.4.3 Specific Examples for Listening

Most of Susannah McCorkle's recordings are available through iTunes. The following arrangements offer specific references for the various ways that these four horns may be used:

They Can't Take That Away From Me—done as a sultry ballad, this features mostly low- to mid-register horns. The first phrase presents all four horns in harmony. They are kept close (within an octave). The second phrase keeps the same texture but the voicings are slightly wider as the trumpet melody moves upward. Shortly thereafter the "sax section" emerges, highlighted with Harmon-muted trumpet, and answered by the tenor voices (sax and tbn.) for a slightly darker and more robust sound. Continued listening will reveal unison counterlines in both the tenor/alto register and soprano register.

Let's Do It—the opening phrase begins with the horns in octaves and expands into four-part harmony. Here the horn ensemble is bright and full. All notes are in the alto and soprano registers. The voicing is spread (a bit beyond the range of an octave) to give a broad, panoramic sound, but not too wide as this would sound too threadbare. The lower horns provide more body while their darker hues nicely counterbalance the brighter tones of the trumpet and alto sax. The phrases that follow offer a playful commentary and interesting contrast—from high register to low, harmony to unison, low horns in unison and upper horns in unison. In the middle of the tune, the saxes and trombone create background pads for the trumpet player, who uses a plunger mute for some colorful solo commentary. Behind the tenor sax soloist, the brass players play a unison riff using plungers; the alto sax blends with the brass to help thicken the unison and strive to keep the illusion of a brass section. The trombone soloist uses the plunger for his improvisation; initially the saxes play a unison background riff in the tenor register to provide a darker hue for the high register trombone. Then, to increase the intensity, the saxes play another unison riff in the alto register. Near the end, there is a rousing "shout" section, where the trumpet goes up to a high C concert. Because of the extremely high register, the other three horns are placed considerably lower (in the alto range). This will retain the middle frequencies and the girth associated with bigger horn sections. (More specifics on this later in this chapter.)

7.5 Small Horn Section—Creating Stylistic Diversity and Greater Contrast

The utilitarian aspects of this horn section may be expanded further by broadening the instrumentation. Of course, this depends upon the versatility of the original four players. But most jazz saxophonists also play flute and/or clarinet. This is referred to as "doubling." With the brass, the use of various mutes can also provide a very distinctive and contrasting color from the mainstream sound of the four horns.

Directly below are four examples of arrangements for Susannah McCorkle:

Mixed styles—sometimes a certain arrangement may feature two distinctive styles. What is usually required is the vacillation of one style to the other with continued exchanges back and forth. With large ensembles, there are enough players and instruments to easily create this contrast. With only four players, the challenge is greater. These players may need to have other instruments available and brass players at the very least may need mutes. This scenario occurs in the arrangement of *Cheek to Cheek*, where there is a vacillation from a rather cute and light Latin (Cha-cha) flavor to a more energetic and soulful swinging big band sound. The introduction features the four mainstream horns (tpt., a.sx., t.sx., tbn.) voiced for a strong opening statement. During the resting period that follows, the alto saxophonist switches to the flute while the trumpet player grabs the Harmon mute. The trombonist uses a cup mute for a darker, more intimate timbre and the tenor saxophonist uses a technique called "subtone," which in effect gives a more intimate sound as it blends beautifully with the muted trombone when they play in unison. The contrast is most striking and shows the possibilities of implementing color changes—provided that the arranger leaves enough time for the players to manage the physical changes.

Broadway—the traditional sound of a Broadway score usually calls for flutes and clarinets instead of saxophones. This arrangement of *There's No Business Like Show Business* depicts a more reflective and contemplative mood that once again uses the cup-muted trombone and Harmon-muted trumpet sound. A clarinet in unison with the trombone creates a warm and rich tone, while a flute is paired with the Harmon-muted trumpet for a bright and glitzy timbre. A bit later, the flute is heard with flugelhorn for yet another timbral color. The final section of the arrangement becomes somewhat grandiose, which requires the woodwind players to switch to saxophones.

Dixieland (or New Orleans style)—this style of music was most popular in America in the 1920s. The typical instrumentation calls for clarinet, trumpet, and trombone. The extra saxophone is not really used but it could be used in imaginative ways. For instance, this music also featured the sound of a tuba playing bass lines. To a certain degree, it would be possible for the trombone to emulate the tuba. In this case, the saxophone (particularly a tenor) could assume the role that the trombone usually has in this style.

The song *A Friend Like Me*, from the Disney movie *Aladdin*, possesses some of the same qualities found in the Dixieland style as well as the slightly more contemporary swing style. The arrangement for Susannah McCorkle needed to capture the same energy and zaniness as the original soundtrack—but with only four horns and four rhythm!

Brazilian Bossa—this style is most readily defined by the great composer Antonio Carlos Jobim. His primary arranger was Claus Ogerman, who used a string section to accompany the vocalist. The rhythm section included nylon string guitar and delicate and sparse piano commentary, with subtle and quiet support from an acoustic bass and brushes on drums. The music is extremely legato, sensual, and intimate. But Ogerman's most obvious signature sound really came from several low- to mid-register flutes, which provide a cool "whispering" texture. Today, you can also hear his great arrangements for singer Diana Krall on her CD *The Look of Love*. The author's arrangement of *Let's Face the Music and Dance* for Susannah McCorkle strives for this sound. The saxophonists play a C flute and an alto (G) flute respectively. This creates the illusion of a flute "section" when they are played in unison. A bucket mute is used on the trombone, which darkens the tone considerably and makes it closer to the sound of a French horn. The flugelhorn is blended with the trombone at the unison to create the illusion of a "section" of French horns. Similar to the small string section, sometimes it is better to use a unison texture for an ironically fuller sound. By the way, the strings that are heard on this track emanate solely from a synthesizer! There was no budget for real strings.

7.6 Small Horn Section Writing—Effective Voicings and Other Techniques

With only a few horns to work with, styles of music that require more complex harmony present a greater challenge. In these cases, the arranger should select chord tones that offer the fullest sound. Doubling the bass will result in a thin sound. Essentially, the arranger should rely on the harmonic foundation provided by the rhythm section and embellish with the horns beyond that.

7.6.1 The Overtone Series—a Guide to Harmonic Voicings

With reference once again to the overtone series (at the piano), let this be your guide for creating robust four-horn voicings. The first thing to cross off your list of choices is any consideration of placing the horns with the bass register. So the lowest register to be used should be the tenor register (the third octave of the piano). In jazz, when the horns are primarily in the tenor and alto registers, it is usually necessary for two of the horns to have the third and seventh of the chord. The next choices would be extended chord tones (ninth, eleventh, thirteenth) or alterations of the fifth where necessary. Unless the melody note happens to be a root, and especially when using a dominant chord, do not use the root for harmonic support. The exception is for a major seventh chord, where the root creates a favorable tension with the seventh. Figure 7.1 illustrates a variety of voicings (on different chord qualities) that lie within the alto and tenor registers.

Figure 7.1 Voicings within the tenor and alt registers

Notice that all of the horns are placed above the bass register. The bottom two are in the tenor register and the top two are in the alto register. The opening Cma7 chord contains a root to more clearly establish the sound and to also instill a bit of tension between the B and the C. There are no roots in any of the other voicings. All voicings contain the third and seventh of each chord, with the exception of the Dmi11 chord in the fourth bar. The omission of the note C (the seventh) is acceptable for several reasons: (1) C is the tonic note of the key; (2) the previous chord voicings contain C, so the sound remains in the environment; (3) it is nice to offer different pitches in each horn part so here is a chance for the line to move away from C; (4) the voicing is actually more balanced with A instead of C. Try recording these voicings into your sequencer and assign each voice to its respective instrument. Then add an acoustic bass playing the root of each chord in half notes. After you've listened to the ensemble with the bass, mute the track and listen to just the horns. I think you'll discover that the sound of the horns still implies what the roots are. This is how good voicings work when they occur in the tenor and alto registers.

When the lead horn line needs to ascend to the upper range (particularly the trumpet), it is important not to write too high. In general, most of the notes in the lead voice should be placed a bit lower than those that a typical lead trumpet part might have within the context of a big band. With many horns, it is easy to support very high notes in the lead trumpet. With only four horns, there is a greater chance that the ensemble will lose its homogenized sound. Voicing the other horns too closely will result in a very shrill sound.

Figure 7.2 shows a typical loud and climactic moment. This occurs naturally whenever the horns are placed in the alto and soprano registers. The trick is to leave most of the horns in the alto range. The trumpet part sits at the top of the staff and climaxes on the high C concert. Notice that the two lowest horns remain in the alto register (high enough in their range to provide a bright and exciting tone but low enough in relation to the horn

Figure 7.2 Voicings within the alto and soprano registers

ensemble sound to retain body). The alto sax remains in the alto register for most of the time and only moves into the soprano register near the end of the phrase. Collectively, all of the horns work together to create a bright and powerful sound that is characteristic of the "shout" within the big band style.

Notice that the only roots present are the ones used in the melody (the last two pitches). Next locate the thirds and sevenths of each chord voicing. After that, look for the extended chord tones that are suggested by each chord symbol (and please allow for enharmonic spellings in some cases). Notice the wide gaps that appear between the trumpet and alto sax. This is acceptable because (1) it is more important to preserve the mid-register sound and (2) the sonic principles of the overtone series help to fill the gap (if you had another horn available you could easily place it inside these large gaps—the overtone series suggests this missing part). The extra horn would of course thicken the sound but the idea is that you can create the impression of a larger horn section if you voice it right!

Another common scenario is when the horns are capturing the upper impression of the harmony while letting the rhythm section preserve its fundamental aspects (more about this in the next topic). The horn voicing is actually abstract in the sense that the notes do not necessarily suggest the fundamental chord. Figure 7.3 illustrates this concept. The music example is set in an aggressive funk-jazz style that is reminiscent of the music by the Brecker Brothers. (For further study, they captured this concept with only three horns on their famous recording *Some Skunk Funk*.) Notice that the guitar, piano, and bass are working together to capture the fundamental harmonic sound of C7+9. Altogether, they provide the root, third, fifth, seventh, and raised ninth. This harmony contains quite a few chord tones but the complete sound within this example is actually based on the notes of the C Super Locrian scale (known more commonly as the "altered scale"). The specific scale tones are C–Db–D#–E–Gb–Ab–Bb. Look for the remaining tones in the horn parts. You'll hear how their addition to the rhythm section is essential to portray this

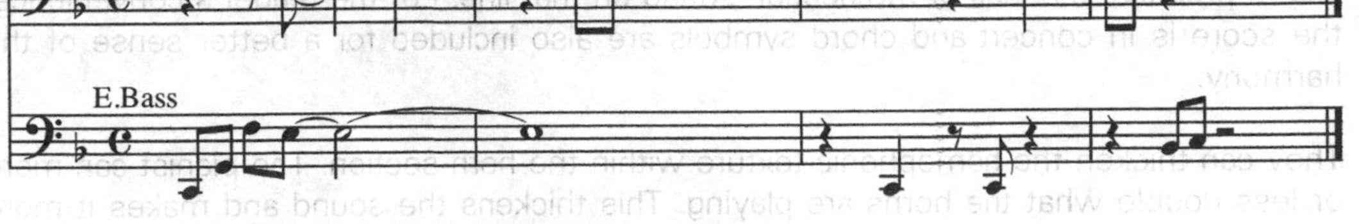

Concert sketch

Figure 7.3 Abstract voicings

complex harmonic sound. Also play the horn parts alone and notice how they can sound abstractly apart from the fundamental harmonic sound. Try playing them over an E♭ bass and you'll notice how the sound changes completely by portraying the sound of E♭ Dorian minor! This is how a small horn section can really extend the harmonic color and capture a big sound.

7.7 Utilizing the Rhythm Section Most Effectively

If you've had a chance to listen to the arrangements for Susannah McCorkle, you may have noticed that the horns do not play constantly. This puts the spotlight on the rhythm section, which allows the horns to rest or change equipment; it also creates a refreshing contrast for the listener. Personally, I think of a rhythm section as having the dual function of also being a string section. Although the strings are plucked or hammered, they nonetheless offer a way of thinking beyond the primary aspect of playing chords and grooves. The two "rhythm" instruments within the McCorkle ensemble that have the most versatility are the guitar and piano. Examples are as follows:

They can play a line in unison or octaves—in effect creating a string section soli (refer to Score 7.3). This is an extreme example of how you can really feature the rhythm section in a melodic way. The score is transposed. A careful analysis will reveal that all three instruments are heard in unison. The drummer supports this monophonic texture with a delicate, swinging brush pattern on the snare.

They can highlight horn lines—(refer to Score 7.4). This score is a continuation of Score 7.3 where the four horns now join the rhythm (string) section with lines of their own. Along with the preexisting line, there are now three additional lines. Each line emerges separately for a layered effect. In a big band arrangement, these lines could easily be covered by several horns on each line. But, with so few horns, it is now necessary to double these lines with the supporting instruments from the rhythm section. A careful analysis will show that the original string line is maintained throughout by the bassist and reinforced with the pianist's left hand. The guitarist continues in similar fashion with the bassist and pianist as the trumpet and alto sax parts emerge in unison with one new counterline. When the tenor sax part emerges with another new line, a section sound is still preferred, so the guitarist abandons the string line and couples at the unison with the tenor sax. Soon after, the guitarist couples with the newly emerging trombonist and finishes the phrase. In the meantime, the pianist's right hand couples with the tenor sax to preserve a sense of the section sound on that line. For the reader's convenience, the score is in concert and chord symbols are also included for a better sense of the harmony.

They can thicken the homophonic texture within the horn section. The pianist can more or less double what the horns are playing. This thickens the sound and makes it more percussive. It can also add the lower parts of a voicing when the horns are placed in a higher register (refer to Score 7.5). This is not always necessary but it is an option that the arranger should consider. In this example, the rhythm (string) instruments are designed for maximum flexibility. The chord symbols in the parts enable the players to create their own part should the arranger request that. Otherwise, the specific notes (in the guitar and piano parts especially) allow those players to unify with the horn section. The guitar plays the melody in octaves. This not only reinforces the high trumpet (one and two octaves lower) but also makes the sound of the guitar itself much bigger. The pianist's right hand has a similar function but its top octave is in unison with the lead trumpet while the lower octave helps to strengthen the piano sound. Down in the bass clef, the pianist's left hand acts as a lower section of horns that expands the sound of the small horn ensemble. With an overall view of the score, you will see the textural contrast between the first two bars and the last two. In the first two, most of the band is in uniform activity. Then the saxophones answer with a unison melodic comment while the guitar resumes its more common role as a "rhythm" instrument. But also notice that the piano part is designed to create an answer that has a strong musical character. It's almost like an echo of the horn ensemble sound.

They can create interesting counterpoint to the horn section (refer to Score 7.6). In this example, you can see how the guitar and piano create interesting melodic answers in bars 2 and 4 to the main melody in the horn section. In particular, notice that the trumpet has been replaced by the more mellow flugelhorn. This sound is more suited to the mood of the Brazilian bossa. Where the melody ascends to its zenith in bar 4 is about the highest you would want to write for the flugelhorn. This example represents a grandiose moment in the arrangement. The guitar abandons its main function as a "comping" instrument and instead joins with the piano to create a strong counterline against the horns. Notice how the notes (G and A) in the first bar highlight the main melody. The piano part also has chord symbols and, because the horns are in octaves initially, the pianist will need to create voicings to supply the harmonic texture at this point. Also notice that the horns expand into four-part harmony as they approach bar 2. A careful inspection of the chord tones will reveal that the seventh of the chord is absent. But this is acceptable because the horns are mostly in the alto register. In addition, the pianist will supply the seventh (F). Plus, that note is also in the counterline!

7.8 Alternate Instrumentation

Unorthodox musical settings can evolve because of limitations by the performers or unavailable instruments. To accommodate an unusual situation, the arranger may have to choose a secondary instrument to represent the desired one. This happens quite often in theatrical productions but the arranger should anticipate this in general. When there is a need to create a section sound and there is only one of the desired instruments, try using the alternative instrument with the desired one. For example, if you need two flutes but only one is available, try doubling the flute with a cup mute trumpet. Table 7.1 shows a graph with the desired instrument choice and its possible alternative.

Table 7.1 The desired instrument choice and its possible alternative

Desired Instrument	Alternate Instrument
Flute	Cup mute trumpet—must be played softly; works best in the flute's mid-register
Oboe	Soprano sax—for the oboe's mid to high register
	Straight mute—trumpet for oboe's low to mid-register
Bassoon	Baritone sax—played softly; best in upper register
French horn	Bass or tenor trombone with a bucket mute
Flugelhorn	Trumpet with a bucket mute

7.9 Final Considerations on Sequencing and Producing for Acoustic Ensembles

If you have followed methodically the book up to this point, you should have a detailed picture of the process and techniques required to write, sequence, and produce for acoustic ensembles. At this point, I believe it is important to wrap things up with a "check sheet" that includes the major point to keep in mind when sequencing for acoustic ensembles. Sometimes it is easy to be lost in the details and forget the main points, goals, and purposes we originally set. You can think of it as a "cheat sheet" to always keep next to your keyboard or your computer. Or think of it as an overall summary that can help you achieve quickly excellent results. Let's start!

7.9.1 Relationship between Writing and Sequencing

Always write for MIDI settings as you would write for real ensembles. I mentioned this point several times in the book but I can't stress enough how important it is to make sure that your parts are playable and realistic, even if you plan to eventually use MIDI and libraries to produce them. At least at the beginning of your career it is important to have the self-discipline of following this rule. Your productions will be more realistic and more balanced. If you are more experienced, though, I recommend (when appropriate) stretching this rule and, eventually, being flexible and open to experimentation. There are situations in which it is fine (and sometimes even encouraged) to create ensembles and parts that, though not realistically playable by real instruments, can produce very effective results in a MIDI/sound library production environment. A typical example could be the creation of a movie soundtrack where a larger orchestra with a deeper and more dramatic bass section is necessary. In this case, we can for example double the bass part an octave lower, using notes that might be outside the range of the acoustic bass. Also, always be open to take advantage of specific sounds and effects that your libraries provide. Sometimes interesting combination of instruments and articulations can be created in the MIDI domain that could not be executable in an acoustic setting.

7.9.2 The Best Tools for the Job

As much as we need to concentrate on the musical aspect of the writing and the production, the tools you use are important. They can have a huge impact on the final result. My recommendation is to always use the best libraries, DAW, effects, speakers, and gears you can afford. While it is not always necessary to own the most expensive ones, you should try to use the ones that best fit the task and production you are working on. In other words "the best tools for the job" should be your mantra! Do your research before starting. Make sure you have enough variety of sounds, articulations, and colors so that you won't have to scramble later. You will be able to concentrate more on the music aspect of the project.

7.9.3 Be in Control!

Always try to use the best alterative MIDI controllers for the task/part you are working on. Breath and wind controllers are ideal for wind instrument parts. They will guarantee you realistic phrasing and execution, while guitar/bass MIDI controllers are perfect for sequencing realistic voicings and grooves that would be hard to create with a regular MIDI keyboard controller. In particular, I recommend the use of MIDI pads or MIDI drums for drum and percussion parts. Not only will they improve the realism of your productions but also they will be more fun and musical to use!

In addition to dedicated hardware controllers, it is crucial that you take full advantage of the power and flexibility of the most common MIDI CC messages. In particular, use CC#7 and 11 to control, respectively, the overall volume and dynamics of a part. Use CC#72 and CC#73 for fine-tuning, respectively, release and attack parts that require flexible phrasing. Remember also to combine dynamics (most often CC#11) with sample switching (often assigned to CC#1). Constant micro and macro changes in dynamics are essential in order to sequence realistic parts. For string parts that are particularly exposed, use the "up/down-bow technique"; this will guarantee a smoother transition between notes.

7.9.4 Adding Some Life into It

Remember that, even though you have followed all the tips and techniques involved in sequencing your parts, adding some real instruments to your productions is essential in order to obtain the best final product. If possible, always record the top voice of a section with a real instrument. For example, in the string section of an orchestra, overdub (triple-track) violin 1. For small sections, instead of overdubbing, replace the top instrument. When the budget allows, add a second real instrument and assign it to the third voice, creating a "sandwich" of real–MIDI–real–MIDI. To add more realism, record and accurately place performance noises such as page turns and breathing. Pay particular attention to the placement of the noises. Use quantization only if necessary (usually for more rhythmic and active parts or passages). Too much quantization can make the part too stiff and unnatural. Make sure to use the advanced quantization parameters (strength, sensitivity, swing, etc.) to create smoother renditions, while maintaining rhythmic accuracy.

7.9.5   Mix Is the Key

Always spend as much time at the mix stage as you spent writing and sequencing the piece. A bad mix can easily sink hours and hours of good writing and sequencing. Remember that, when mixing, you are wearing a "different hat." Focus on reaching clarity, intelligibility, and balance. Carefully use equalization, pan, and reverb to place instruments and sections in the correct location on the virtual stage. A flat placement of your virtual ensemble can surely make it unnatural-sounding and artificial. Try to use a good convolution reverb with the right space and size. For small and medium ensembles, you can use a two- and four-zone reverb, as learned in the previous chapters, but for larger ensembles (such as a full orchestra) I recommend using a six-zone reverb, as shown in Figure 7.4.

As you can see, using a six-zone reverb makes it simpler to program reverb parameters for each of the sections: strings, WW, and brass/percussion. In this case, we need to make sure that each zone is slightly different from the others in order to re-create the feeling of a natural hall. Notice how in Figure 7.5 I have differentiated the three main areas that span from front to back of the virtual stage (areas A, B, and C). Use the length of the reverb and the EQ./damping parameter to differentiate between these three areas, as shown in Table 7.2.

7.9.6   Most Important of All . . .

Well, you did it! You have carefully read this book and worked on all the exercises. You have learned a great deal of information, tips, and tricks, and your writing and production skills have improved considerably. Congratulations! Before letting you go back to write some amazing music, I would like you to remember that, no matter how many techniques, tools, or equipment you have, what really matters, at the end, is the music and how much fun you have while writing it, producing it, and having it performed. Never let tools and techniques get in the way of your art. Let the inspiration run free and always follow the music!

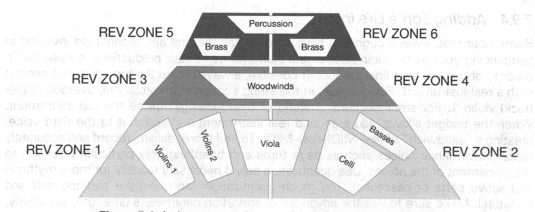

Figure 7.4 A six-zone reverb example applied to a large orchestra

Table 7.2 Reverb settings for a six-zone virtual stage for large ensembles

Area	Zone #	Reverberation time	Filter setting	Comments
A	1	2.6 seconds	Flat	Zones 1 and 2 are for the string section
A	2	2.3 seconds	Damp -2db around 100Hz	The slightly shorter reverb time and the filter on the low end of the spectrum gives a bit more clarity to basses and cellos
B	3	2.9 seconds	Damp -3db around 150Hz	Zones 3 and 4 are normally occupied by the WW
B	4	2.8 seconds	Damp -3db around 200Hz	
C	5	3.2 seconds	Make the reverb slightly darker for sections 5 and 6. You can use a low shelving eq. with a gain of -3db with a frequency of 8.5kHz	Zones 5 and 6 are reserved for the brass and percussion
C	6	3.1 seconds		

7.10 Summary and Final Considerations

Today, it is important for the arranger to think about orchestration in two distinct environments. The first aspect follows the tradition of the great symphonic writers who skillfully balanced instruments using only natural acoustics. This awareness helps the modern writer in live performances of music to get the best sound possible without the aid of any microphones for amplification purposes. The second aspect is to take advantage of modern technology via microphones, recording isolation, overdubbing, and MIDI soundtracks mixed with acoustic recording. Practicality (production budget, instrument availability, performance ability, rehearsal time, etc.) must usually be considered in conjunction with each artistic endeavor. This is particularly true with most recording sessions. But it is becoming more and more of a priority with live performance as well. It is simply too costly for most artists to tour with a huge ensemble. The skillful arranger who can deliver artistic gratification blended with practicality and efficiency has the best chance of developing a productive career.

When downsizing a string section, remember to preserve the proportion that exists in the full orchestra. More violins are needed than the number of lower strings. In some cases, particularly in loud pop music, the violas are omitted. Although they are wonderfully versatile in a traditional orchestral setting, they are often too subtle in other styles. When layering strings in a recording session, make sure that the first take is as good as it can possibly be. Usually, three layers are required to get a lush string section sound. When recording only with a string quartet or quintet, more layers may be needed. But remember that the number of lower strings will eventually outweigh the violins. So it is a good idea to record one or two extra layers of violins by themselves for greater control in the mix. Mixing real players with prerecorded MIDI tracks is also another alternative. One violin and one cello playing along with the soundtrack can add more realism and greater expression.

When smaller horn sections are required, remember that the most versatile combinations of horns are ones that provide a span of the following registers: soprano, alto, tenor, and baritone. Avoid the bass register and rely on your rhythm section to cover the fundamental aspects of harmony. Try to use your horn section in various combinations. Don't have them play together all the time. Create different section sounds (WW versus brass, alto line versus tenor line). Other contrasts include monophonic texture and polyphonic texture, as well as homophonic texture. When the lead horn ascends into the soprano range, remember to keep the lower horns mostly in the alto range. This will offer a more robust sound and balance better with the high instrument.

Think of your rhythm section in a more utilitarian way; the piano, guitar, and bass are not just for chords and rhythm but are also string instruments. Use them in that way to create melodic as well as harmonic information. Pair them with horns to highlight a melodic line. There are many possibilities. With small and diverse ensembles, the challenge is on the arranger to use more imagination and skill to come up with solutions that will create gratifying music to the point where the listener does not miss the large ensemble.

7.11 Exercises

Exercise 7.1

Write a short passage with an aggressive rhythmic feel for four horns, string quartet, and a four-piece rhythm section and incorporate the following characteristics:

- have the strings move at a different speed (slower or faster) than the horns;
- keep the upper strings in a different register from the horns (violins above);
- keep the strings in monophonic or polyphonic texture while the horns are in homophonic texture.

Exercise 7.2

Sequence and produce the piece written in Exercise 7.1 using the following techniques:

- Have two real horns and two MIDI horns (use the techniques you learned in Table 5.6 in Chapter 5 to choose the correct assignments).

- Have the string quartet produced using one real violin playing both violin 1 and 2, and MIDI viola and cello.

Exercise 7.3

In the staff below (Figure 7.5), add three more voices under the melodic line. These notes will represent a four-horn ensemble within a jazz arrangement. Do not use any roots and avoid placing notes in the bass register. Unless the third or seventh of the chord is in the melody, make sure that those chord tones are included. Then add extended chord color to provide the fullest sound possible. When you're working at the keyboard, keep two notes in each hand. This will help to keep a naturally good spread for the horn sound.

Figure 7.5 Exercise 7.3

Exercise 7.4

(a) Write a bass line using the notes of a minor pentatonic scale set in a funk groove.
(b) Add guitar to the bass line to create a rhythm section soli sound. The piano can also play the line or add another dimension. Superimpose a section of three horns playing another simple melody in counterpoint to the rhythm section line. Keep the horns more abstract by voicing them in various triads with each melody note. If you wish to try this with four horns, simply have the fourth horn play the lead horn part down an octave. Eventually, you can try adding a different note on the fourth part to expand the voicing.

Appendix A: Music Scores and Examples

Here you can find the scores mentioned in the previous chapters. For your convenience, you will find a digital version of the scores on the website too.

Chapter 2

Score 2.1

Score 2.1 Continued

Score 2.1 Continued

Score 2.2

Score 2.2 Continued

Score 2.3

Score 2.4

Score 2.4 Continued

Score 2.4 Continued

Chapter 3

Score 3.1

Score 3.2

Score 3.3

Score 3.4

Score 3.5

Score 3.6

Score 3.7

Score 3.8

Score 3.9

Score 3.10

Chapter 4

Score 4.1

c = closed voicing
d = drop 2 voicing
i = independent voicing

Score 4.2

Score 4.3

Score 4.3 Continued

Score 4.4

(non-transposed score)

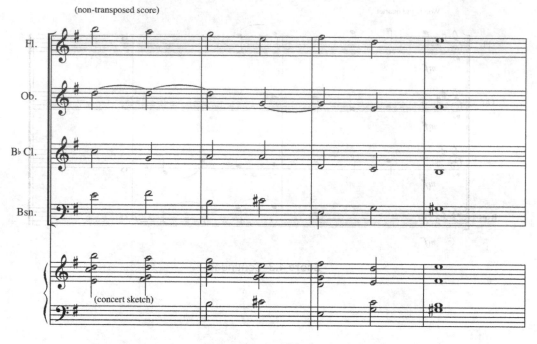

Score 4.5

(non-transposed score)

Score 4.6

Score 4.7

Chapter 5

Score 5.1

Score 5.2

Score 5.3

Score 5.3 Continued

Score 5.4

Score 5.5

Chapter 6

Lady Mine

H.E. Clarke

music by
Richard DeRosa

Score 6.1

Score 6.1 Continued

Score 6.1 Continued

Score 6.2

The Light of Other Days

Thomas Moore

music by
Richard DeRosa

Score 6.3

Score 6.3 Continued

Chapter 7

Score 7.1

Score 7.2

Score 7.3

Score 7.4

Score 7.4 Continued

Score 7.5

Score 7.6

Index

Please note: figures are displayed in *italics*, tables displayed in **bold**